International Review of

Cytology

A Survey of

Cell Biology

VOLUME 119

International Review of Cytology

A Survey of Cell Biology

Edited by

K.W. Jeon

Department of Zoology
The University of Tennessee
Knoxville, Tennessee

M. Friedlander

Jules Stein Eye Institute
UCLA School of Medicine
Los Angeles, California

VOLUME 119

Academic Press, Inc.
Harcourt Brace Jovanovich, Publishers
San Diego   New York   Berkeley   Boston   London   Sydney   Tokyo   Toronto

ACADEMIC PRESS, INC.
San Diego, California 92101

United Kingdom Edition published by
ACADEMIC PRESS LIMITED
24-28 Oval Road, London NW1 7DX

LIBRARY OF CONGRESS CATALOG CARD NUMBER: 52-5203

ISBN   0-12-364519-0   (alk. paper)

PRINTED IN THE UNITED STATES OF AMERICA
90 91 92 93   9 8 7 6 5 4 3 2 1

Contents

CONTENTS

Contributors

Numbers in parentheses indicate the pages on which the authors' contributions begin.

B. B. AMATI (57), *Swiss Institute for Experimental Cancer Research (ISREC), CH-1066 Epalinges, s/Lausanne, Switzerland*

M. E. CARDENAS (57), *Swiss Institute for Experimental Cancer Research (ISREC), CH-1066 Epalinges, s/ Lausanne, Switzerland*

S. M. GASSER (57), *Swiss Institute for Experimental Cancer Research (ISREC), CH-1066 Epalinges, s/Lausanne, Switzerland*

J. F.-X. HOFMANN (57), *Swiss Institute for Experimental Cancer Research (ISREC), CH-1066 Epalinges, s/ Lausanne, Switzerland*

WILLIAM R. JEFFERY (151), *Center for Developmental Biology, Department of Zoology, University of Texas at Austin, Austin, Texas 78712*

HELMUT PLATTNER (197), *Faculty of Biology, University of Konstanz, D-7750 Konstanz, Federal Republic of Germany*

A. WAYNE VOGL (1), *Department of Anatomy, Faculty of Medicine, University of British Columbia, Vancouver, British Columbia, Canada*

KATSUTOSHI YOSHIZATO (97), *Developmental Biology Laboratory, Department of Biology, Faculty of Science, Tokyo Metropolitan University, Fukazawa 2-1-1, Setagaya-ku, Tokyo 158, Japan*

INTERNATIONAL REVIEW OF CYTOLOGY, VOL. 119

Distribution and Function of Organized Concentrations of Actin Filaments in Mammalian Spermatogenic Cells and Sertoli Cells

A. Wayne Vogl

Department of Anatomy, Faculty of Medicine, University of British Columbia, Vancouver, British Columbia, Canada

I. Introduction

Cells of the seminiferous epithelium possess elaborate cytoskeletal systems that change their patterns of organization during spermatogenesis. One of the most interesting components of the cytoskeletal system in these cells is filamentous actin (F-actin). In this review, I will describe the arrangement of F-actin in the seminiferous epithelium and discuss current concepts of how it may contribute to some of the processes that occur during spermatogenesis.

The review begins with a brief description of the mammalian seminiferous epithelium and a short discussion of the structural and functional properties of F-actin in cells in general. These introductory sections are followed by a more comprehensive discussion of organized concentrations of F-actin in major cell types of the seminiferous epithelium.

A. Organization of the Seminiferous Epithelium in Mammals

In mammals, spermatozoa are produced by an epithelium that lines numerous small tubules in the testis. These sperm-producing or seminiferous tubules are connected at both ends to a collecting chamber, called the rete testis, that ultimately communicates via a series of other duct elements with the urethra.

The seminiferous epithelium that lines the seminiferous tubules is one of the most complex epithelia in the body and is composed of two populations of cells, spermatogenic cells and Sertoli cells (Fig. 1) (Fawcett, 1975a). Sertoli cells are nondividing cells that form the major structural elements of the epithelium. They are irregular columnar cells and extend from the base to the apex of the epithelium. Unlike in other epithelia, the junctional complex between Sertoli cells occurs near the base of the seminiferous epithelium. Extensive tight junctions in this network are the major ele-

1

1. Tubulobulbar complex   3. Subacrosomal space
2. Intercellular bridge   4. Ectoplasmic specialization

FIG. 1. Schematic diagram of the mammalian seminiferous epithelium. No attempt has been made either to depict accurately a single stage of spermatogenesis or to illustrate the cell types of a single species. Rather, the diagram depicts the general architecture of the mammalian epithelium and the sites of actin filament concentration (solid black areas). Two Sertoli cells are shown (shaded), together with their associated germ cells.

ments of the "blood–testis" barrier and divide the epithelium into a small basal compartment below the junctions and a large adluminal compartment above (Dym and Fawcett, 1970). These junctions enable the Sertoli cell to produce an adluminal environment that is essential for the normal differentiation of spermatogenic cells.

The spermatogenic cells lie between and are attached to the Sertoli cells. The proliferate as relatively undifferentiated cells (spermatogonia) in the basal compartment of the epithelium. As the cells become committed to meiosis (spermatocytes), they pass through the blood–testis barrier into the adluminal compartment. Here, meiosis is completed and the resulting haploid spermatids differentiate into cells that are ultimately released as spermatozoa into the duct system of the male reproductive tract. During this process of differentiation, known as spermiogenesis, an acrosome and flagellum develop, the nucleus condenses, and all excess or residual cytoplasm is packaged into a structure called the residual lobe that is phagocytosed by Sertoli cell when the spermatid is released from the epithelium.

Spermatids begin differentiation as round cells that are embraced by lateral processes or lamellae of adjacent Sertoli cells. As the cells elongate and a distinct head (acrosome and nucleus) and tail (flagellum) develop, they become situated in invaginations of the apical surfaces of Sertoli cells. These apical crypts, together with their attached spermatids, project deep within the epithelium at one point during spermatogenesis and then become situated at progressively more luminal positions as associated spermatids mature and acquire a species-specific form. Prior to sperm release, the crypts and attached spermatids occur in Sertoli cell processes attached by stalks to the apex of the epithelium (Russell, 1980, 1984).

Many of the morphological changes that occur in the seminiferous epithelium during spermatogenesis, such as the translocation of spermatids and the movement of spermatogenic cells through the blood–testis barrier, are thought to involve elements of the cytoskeleton. One of the major components of this cytoskeletal system is F-actin.

B. ACTIN IN GENERAL

Actin exists in two basic forms within cells, globular (G) and filamentous (F). G-Actin consists of single polypeptides that form the subunit pool from which F-actin is polymerized. The polymerization state of actin, the size of actin filaments, and the relationships of one filament to another are controlled by a plethora of actin-binding proteins (Craig and Pollard, 1982; Weeds, 1982; Stossel et al., 1985; Pollard, 1986).

Actin filaments relate to each other and to associated proteins to form two basic structural units, networks and bundles. Together with other elements of the cytoskeleton, actin filaments form extensive three-dimensional networks that fill the cytoplasm and in which actin filaments are one of the major components. An example of this arrangement occurs in the spread areas of cultured cells (Schliwa and Van Blerkom, 1981). They can also participate in forming networks of a more two-dimensional nature in which actin filaments are short and serve to link other components of the system together. A classic example of this type of network is the membrane skeleton of red blood cells (Goodman and Shiffer, 1983).

Bundles of actin filaments occur in two general forms. In one form the filaments have a mixed or random polarity, and usually have contractile properties; in the other, the filaments are unipolar, are tightly crosslinked into paracrystalline arrays, and are not contractile. Examples of the former type of bundle include contractile rings associated with the cleav-

age furrow in dividing cells and with the zonulae adherens of most epithelial cells. Stress fibers are also bundles of this type. Examples of tightly
crosslinked and noncontractile bundles are the cores of intestinal microvilli (Mooseker, 1985).

Functionally, actin filaments are implicated in numerous motility-
related events such as endocytosis, secretion, phagocytosis, cell translocation, and intracellular vesicular movement. They also appear to perform
more structural roles as in the cores of microvilli and stereocilia. Many of
these functions, both motility-related and structural, appear to involve an
interaction either with the plasma membrane or with intracellular membranes. The presence of an indirect or direct attachment of actin to specific
molecules in a membrane, together with the presence of numerous classes
of actin-binding proteins that function to change either the polymerization
state of the actin or the way in which one actin filament interacts with
another filament or with the membrane itself, may provide the cell with
a mechanism of (1) establishing specific domains within the membrane,
(2) internalizing specific regions of a membrane, (3) moving a membranous
organelle from one region of a cell to another, (4) generating changes in the
contour or shape of the cell, and (5) moving the cell from one point to
another.

Actin filaments are concentrated at specific sites in the seminiferous
epithelium (Fig. 1). As we shall see, these actin filaments illustrate many of
the organizational features of similar filaments described in numerous
other systems and perhaps have many of the same functional attributes.
Though similar to other systems, there are some unique organizational
features of F-actin in Sertoli cells and spermatogenic cells that are most
likely linked to functional demands placed on the system during the elaborate, cyclic, and highly controlled process called spermatogenesis.

II. Spermatogenic Cells

Actin filaments occur at numerous sites in spermatogenic cells. They
occur among mitochondria in the forming sperm tail (Baccetti et al., 1980)
and may occur in association with the chromatid body (Walt and Armbruster, 1984). In general, actin filaments also occur in the spindles of dividing
cells (Barak et al., 1981), where their function is controversial.

Actin filaments are concentrated in two major locations in spermatogenic cells: in intercellular bridges, connecting adjacent cells, and in
the subacrosomal space, situated between the nucleus and developing
acrosome.

A. Intercellular Bridges

Unlike in most somatic cell types in animals, the male germ-cell line is characterized by incomplete cytokinesis; in other words, after mitotic and meiotic divisions, daughter cells tend to remain connected by cylindrical channels of cytoplasm (Fig. 2). These channels form when the cleavage furrow constricts toward the midbody, or remnant of the spindle, found

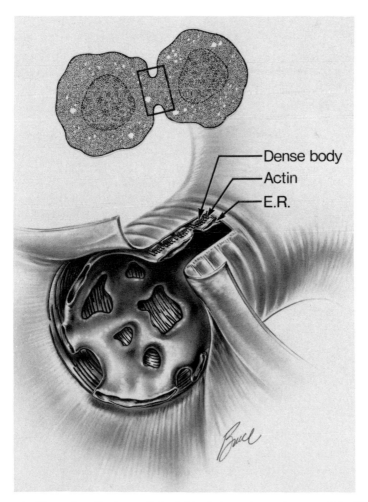

FIG. 2. Illustration of an intercellular bridge between two early spermatids. At this stage, the complex associated with the bridge wall consists of bridge densities, actin filaments, and endoplasmic reticulum (ER).

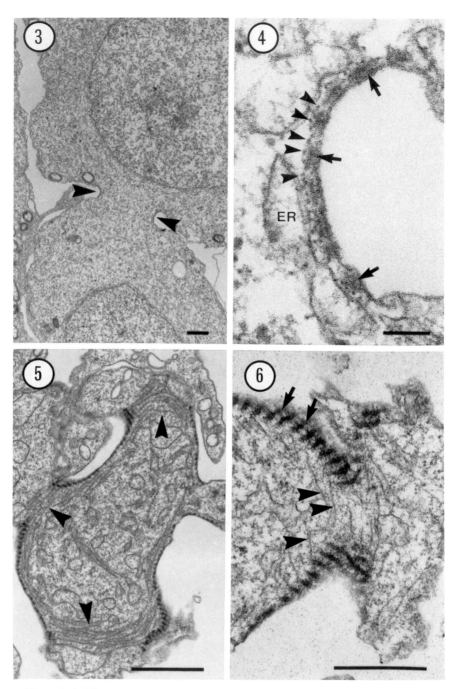

FIGS. 3–6. Electron micrographs of intercellular bridges. Fig. 3. An intercellular bridge (arrowheads) between two early round spermatids of the rat. Notice that, at this magnification, the membrane of the bridge appears thickened. ×5800, bar = 1 μm. Fig. 4. A higher

between the two daughter cells. As the midbody disassembles, the cleavage furrow apparently becomes stabilized by material that accumulates immediately beneath the plasma membrane (Fawcett, 1961; Weber and Russell, 1987). The resulting cylindrical channel of cytoplasm, together with its associated plasma membrane and underlying substructure, is termed an intercellular bridge. Detailed information on intercellular bridges in male germ cells and in other cell types is contained in papers by Fawcett *et al.* (1959), Fawcett (1961), Dym and Fawcett (1971), Gondos (1973, 1984), and Weber and Russell (1987).

Although intercellular bridges occur early in gonadal development (Gondos and Conner, 1973), they become particularly obvious during spermatogenesis. At this time, they persist and accumulate during the proliferative mitotic divisions of spermatogonia and the subsequent meiotic divisions of spermatocytes, ultimately to link hundreds (Dym and Fawcett, 1971) and potentially thousands (Huckins, 1978) of spermatids together. The function of intercellular bridges in these clones or cohorts of cells is not entirely clear; however, the most popular hypothesis is that they serve to maintain the remarkable degree of synchrony that occurs among the large numbers of differentiating spermatogenic cells (Fawcett *et al.*, 1959).

Intercellular bridges in the mammalian testis tend to have a characteristic morphology. Their most distinguishing feature is the presence of an electron-dense layer immediately beneath the plasma membrane. This "bridge density" tends to result in an apparently "thickened" membrane in regions of the bridge, particularly when viewed at low magnification (Fig. 3). Associated with this dense material are filaments of 5–7 nm in diameter that are oriented in a beltlike fashion around the bridge (Fig. 4) (Weber and Russell, 1987). In addition, and at certain stages of spermiogenesis, elements of the endoplasmic reticulum (ER) lie juxtaposed to these filaments. At these times the substructure underlying the plasma membrane of the bridge consists of dense material, filaments, and a cistern of ER.

---

magnification of a bridge similar to that shown in Fig. 3. Notice that clumps of dense material (arrows), occur adjacent to the plasma membrane of the bridge. Adjacent to this dense material is a layer of actin filaments (arrowheads) and a cistern of endoplasmic reticulum (ER). ×130,000, bar = 0.1 μm. Fig. 5. Shown here is a section through a bridge complex joining two or more residual lobes of elongate spermatids of the ground squirrel. Obvious in this micrograph are a series of filaments (arrowheads) associated with the bridge densities. The nature of these filaments is not known; however, they are too large to be actin filaments and appear to be of the intermediate type. ×19,000, bar = 1 μm. Fig. 6. A section through an intercellular bridge in the rat similar to that of the ground squirrel shown in Fig. 5. Bridge densities are indicated by the arrows. The arrowheads indicate filaments that appear to be of the intermediate type. ×40,000, bar = 0.5 μm. Figure 6 courtesy of B.D. Grove.

The dimensions and morphology of intercellular bridges change during spermatogenesis. The most profound morphological change occurs during cell division and involves preexisting bridges, that is, bridges that were formed during previous cells divisions. As cells in the cohort divide again, a parallel series of "disklike" cisternae of ER form within any preexisting bridge channels (Dym and Fawcett, 1971; Weber and Russell, 1987). In any single bridge, these disks lie perpendicular to the long axis of the bridge and their rims extend to the layer of dense material underlying the plasma membrane. The disks are separated from each other by a thin layer of fibrillar material that appears to contain, at least in cells undergoing meiosis, filaments of 5–7 nm diameter (Weber and Russell, 1987). This network of reticulum and associated material, termed a "bridge-partitioning complex" by Weber and Russell (1987), resembles a "stack of coins" that morphologically occludes the cylindrical intercellular bridge channel. The function of these complexes is not known, although they are thought to stabilize preexisting bridges in some way (Weber and Russell, 1987).

Other changes that occur during spermatogenesis in the rat include a general increase in diameter from 1 to 3 μm, a decrease in length from ~1 to 0.4 μm, a change in the nature and distribution of the plasma membrane-associated dense material, and the appearance of 10- to 12-nm filaments in bridge channels of step 10–17 spermatids (Weber and Russell, 1987). Similar filaments occur in bridges of ground squirrel spermatids. Representative examples of these filaments in the squirrel and rat are shown in Figs. 5 and 6, respectively. During spermiation in the rat, bridges either attenuate and break, or enlarge. The latter allows the cytoplasmic lobes of different spermatids within a cohort to coalesce (Weber and Russell, 1987).

That actin filaments occur in intercellular bridges has been documented by Russel et al. (1987). Intercellular bridges stain with NBD-phallacidin when viewed with fluorescence microscopy, and thin filaments or "periodic densities" associated with the bridge density and encircling the bridge channel bind subfragment 1 (S1) of the myosin molecule.

In general, it is well accepted that cytokinesis is effected by the action of a contractile ring situated at the cleavage furrow. The presence of actin and myosin at this location have been well established (Shroeder, 1973; Fujiwara and Pollard, 1976), and the observation that microinjection with antimyosin antibodies prevents cytokinesis (Mabuchi and Okuno, 1977; Kiehart et al., 1982) is consistent with the argument that an actin–myosin interaction is the motor for cytokinesis. Though not yet specifically studied in mammalian spermatogenic cells, a typical contractile ring occurs in crane fly spermatocytes (Forer and Behnke, 1972), and it is likely that a similar ring occurs during the initial stages of cytokinesis in mammalian spermatogenic cells. Detailed changes in constituents that occur in the contractile ring and cleavage furrow that result in the formation of a stable intercellular bridge during spermatogenesis are not known. What is known is that actin filaments appear to persist around the circumference of the bridge channel and that they are associated with the dense material lining the plasma membrane. Myosin has not yet been demonstrated to be a component of the bridge substructure, nor have any other actin-binding proteins been identified.

Actin filaments may participate in stabilizing intercellular bridges (Russell *et al.*, 1987). Results consistent with this hypothesis come from experiments involving the use of cytochalasin D. When this drug is injected intratesticularly, intercellular bridges progressively widen and eventually open, resulting in the formation of large multinucleate masses known as symplasts (Russell *et al.*, 1987). Experiments such as these must be interpreted with some degree of caution because numerous perturbants cause the appearance of symplasts in the testis; however, the presence of "actin foci" and "zeiotic blebs" in cytochalasin-induced symplasts does indicate a direct disruptive effect on actin.

The mechanism by which actin filaments could be involved with stabilizing intercellular bridges is not known. It has been suggested that as bridges form, the actin filaments become tightly crosslinked to each other and to adjacent structures (bridge density and ER) (Russell *et al.*, 1987). In this model, myosin need not be present for the actin filaments to function in structural reinforcement of the bridge. It has also been suggested that control of bridge substructure may involve elements of the ER that appear to become part of the plasma membrane-related bridge complex at certain stages of spermatogenesis, and that this control may be mediated by Ca^{2+} (Russell *et al.*, 1987). Although this hypothesis is reasonable, based on the concept that the ability to sequester and release Ca^{2+} may be a fundamental property of the ER (Somlyo, 1984) and the fact that the action of numerous actin-binding proteins are Ca^{2+}-dependent (Stossel *et al.*, 1985), it remains to be tested.

B. SUBACROSOMAL SPACE

Perhaps one of the most interesting locations, within spermatogenic cells, that contains actin is the subacrosomal space of differentiating spermatids (Fig. 7). In electron micrographs of tissue, prepared using standard techniques, the subacrosomal space appears filled with a flocculent filamentous material. That this material contains F-actin is indicated by its ability to bind NBD-phallacidin (Figs. 8, 9) (Welch and O'Rand, 1985; Vogl *et al.*, 1986; Halenda *et al.*, 1987; Masri *et al.*, 1987), immunological probes

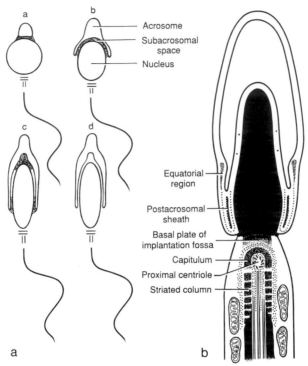

FIG. 7. (a) Progressive changes in subacrosomal actin filament distribution that occur during spermatogenesis (see text for description). (b) Diagram of a spermatozoon summarizing the sites at which actin (dots) has been detected mainly by immunofluorescence and immunogold techniques.

FIGS. 8–12. Fluorescence, phase, and electron micrographs demonstrating the presence of actin filaments in the subacrosomal space of early spermatids. Figs. 8 and 9. Fluorescence and corresponding phase micrographs of a ground squirrel spermatid that has been treated with NBD-phallacidin. The nucleus and acrosome of the cell are indicated by the N and A, respectively. The residual cytoplasm has been removed during the treatment protocol. Notice that a layer of diffuse fluorescence lies between the acrosome and nucleus. ×1080, bar = 10 μm. Fig. 10. A boar spermatid that has been labeled by post-embedding techniques with first a mouse monoclonal antiactin antibody and then a goat antimouse antibody conjugated to colloidal gold. Notice the well-defined labeling of material in the region between the nucleus (N) and acrosome (A). ×19,700, bar = 1 μm. Modified after Fig. 2a of Camatini *et al.* (1986). Fig. 11. Electron micrograph of an early elongate spermatid of the ground squirrel. The cell has been detergent-extracted and treated with the S1 fragment of the myosin molecule. The nucleus (N) and acrosome (A) are indicated, as is the field shown at higher magnification in Fig. 12. ×4600, bar = 1 μm. Fig. 12. Notice the presence of decorated filaments (arrowheads) in the subacrosomal space. ×50,100, bar = 0.5 μm.

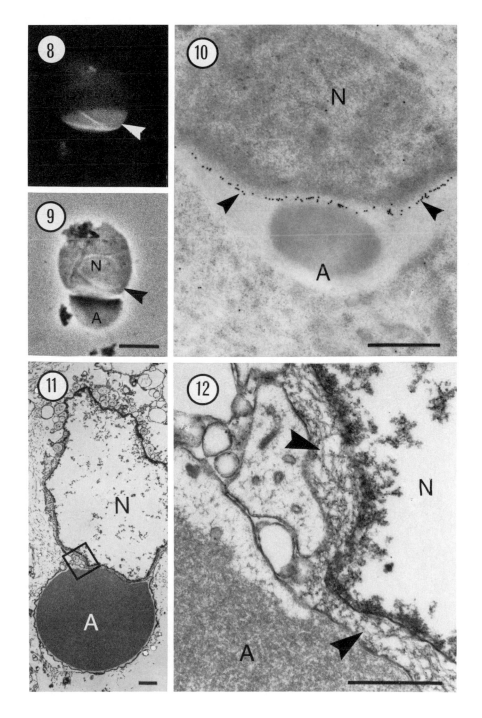

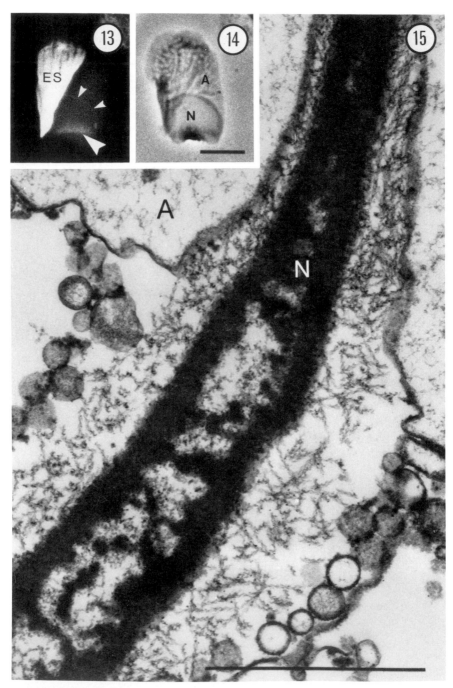

FIGS. 13–15. Fluorescence, phase, and electron micrographs illustrating the location of F-actin in elongate spermatids of the ground squirrel. Figs. 13 and 14. Fluorescence and phase micrographs, respectively, of a spermatid in which the ectoplasmic specialization (ES)

for actin (Fig. 10) (Camatini *et al.,* 1986, Fouquet *et al.,* 1987, 1989), and fragments of the myosin molecule (Figs. 11, 12) (heavy meromyosin: Campanella *et al.,* 1979; S1: Russell *et al.,* 1986; Vogl *et al.,* 1986; Masri *et al.,* 1987). The manner in which these actin filaments interact with each other and with adjacent membranes is not entirely clear; however, the staining patterns tend to indicate that the filaments are organized into a three-dimensional network.

The pattern of filament distribution in the subacrosomal space changes during spermatogenesis. These changes have now been documented in a number of species (ground squirrel: Vogl *et al.,* 1986; guinea pig: Halenda *et al.,* 1987; rat: Russell *et al.,* 1986; Masri *et al.,* 1987; pig: Camatini *et al.,* 1986; rabbit: Welch and O'Rand, 1985; Camatini *et al.,* 1987; rat, hamster, monkey, human: Fouquet *et al.,* 1989), and the following general story emerges from the data (Fig. 7a). Actin filaments become detectable in the region between the acrosome and nucleus as the acrosomic vesicle becomes closely juxtapositioned to the anterior pole of the nucleus. As the acrosome spreads over or "caps" the nucleus and the spermatid elongates, actin filaments become more prominent and their area of distribution expands posteriorly with the advancing edge of the acrosome. Later, as the spermatid acquires its species-specific shape, the filaments begin to disappear. In the ground squirrel, guinea pig, and rabbit, they persist for a short period in posterior regions of the subacrosomal space and as a rim at the most anterior extent of the space (Figs. 13–15) (Welch and O'Rand, 1985; Vogl *et al.,* 1986; Hallenda *et al.,* 1987), and may extend into postacrosomal regions (Welch and O'Rand, 1985; Vogl *et al.,* 1986). Eventually, all actin filaments disappear from the subacrosomal space as indicated by the lack of staining with NBD-phallacidin and absence of S1 binding.

An exception to the general pattern of actin filament loss during the latter stages of spermatogenesis occurs in the plains mouse. In this species, ventral processes or "hooks" that develop as extensions of the perinuclear space and postacrosomal region, extend from the spermatid head and contain well-formed bundles of actin filaments (Fig. 16–18)

---

of the associated Sertoli cell has been partially detached to reveal the location of F-actin in the underlying spermatid. The cell has been treated with NBD-phallacidin. Notice that fluorescence occurs in the subacrosomal space and is strongest in posterior regions (large arrowhead), where it occurs as a distinct band. Fluorescence also occurs in anterior regions, where it appears as a line rimming the disk-shaped nucleus (small arrowheads). The nucleus and acrosome of the spermatid are indicated by the N and A, respectively. ×1160, bar = 10 μm. Fig. 15. Electron micrograph of a detergent-extracted and S1-treated spermatid similar to that shown in Figs. 1 and 2. The region shown is approximately that indicated by the large arrowhead in Fig. 13. Notice the numerous decorated filaments situated between the acrosome (A) and nucleus (N). ×56,600, bar = 1 μm.

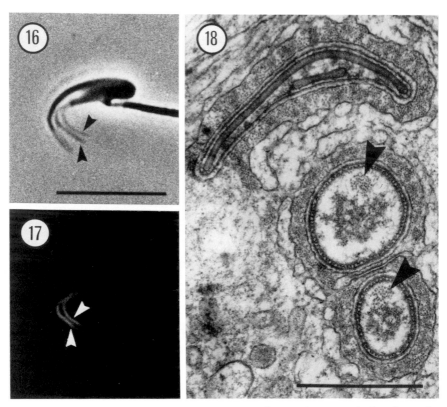

FIGS. 16–18. Fluorescence, phase, and electron micrographs, respectively, of maturing spermatids of the plains mouse. Figs. 16 and 17. The spermatid shown has been dissociated from the seminiferous epithelium and treated with NBD-phallacidin. Positive staining is present in the two ventral hooks (arrowheads) that develop as extensions of the perinuclear space and the postacrosomal region. ×2900, bar = 10 μm. Fig. 18. Electron micrograph of a spermatid at a similar stage of development; filament profiles (arrowheads) are obvious in the developing hooks. Unlike filamentous actin in the spermatozoa of most other mammals studied, these actin filaments persist in the spermatozoa of the plains mouse. ×66,160, bar = 0.5 μm. Modified after Figs. 21a, 21b, and 12 of Flaherty and Breed (1987).

(Flaherty and Breed, 1987). These filaments persist in spermatozoa (Flaherty *et al.*, 1983).

The presence of F-actin in the subacrosomal space during the period that the acrosome first contacts and caps the nucleus has led to suggestions that the actin may (1) play a role in the capping process (Welch and O'Rand, 1985), (2) stabilize protein domains in the inner acrosomal membrane (Welch and O'Rand, 1985), and (3) anchor the acrosome to the nucleus during the formative stages of the sperm head development and until superseded by other anchoring mechanisms (Russell *et al.*, 1986). None of these suggestions are mutually exclusive of each other.

If the anchoring hypothesis is true, then removal of the actin filaments should result in detachment of the developing acrosome from the nucleus. This prediction has been confirmed by Russell (1989). When filaments are pharmacologically disrupted with cytochalasin D administered intratesticularly, the acrosome separates to varying degrees from the nuclear surface (Figs. 19, 20). Also, segments of the acrosome in the advancing edge of the cap appear to be missing (Fig. 21)—an observation consistent with a filament role in the capping process.

There is now reasonably conclusive evidence that F-actin is present in the subacrosomal space of developing spermatids. The pattern of distribution and amount of this material changes during spermiogenesis. It first appears when the acrosomal vesicle contacts the nucleus, is most apparent during the capping process, and disappears during the maturation phase. Experimental evidence is consistent with a role in anchoring the acrosome to the nucleus during the capping process.

C. Postspermatogenic Cells (Spermatozoa)

Although the distribution of F-actin in spermatogenic cells is fairly well described and the pattern of distribution appears generally similar among the species that have been studied, the situation is somewhat different in postspermatogenic cells. The available evidence suggests that most mammalian spermatozoa contain at least some actin (Fig. 7b); however, the actin appears to be in a form that cannot normally be detected using NBD-phallacidin. Moreover, the amount and subcellular distribution of the protein appear to differ between species.

Results of biochemical studies are generally consistent with the conclusion that spermatozoa contain actin. When extracts of ejaculated or epididymal spermatozoa are electrophoresed on sodium dodecyl sulfate (SDS)–polyacrylamide gels, a protein band occurs at a similar molecular weight to that of muscle actin (Talbot and Kleve, 1978; Campanella *et al.*, 1979; Clarke *et al.*, 1982; Flaherty *et al.*, 1986; Lora-Lamia *et al.*, 1986). Furthermore, when transferred onto nitrocellulose, this protein band reacts with antiactin antibodies (Camatini *et al.*, 1986, 1987; Lora-Lamia *et al.*, 1986). Although it could be argued that the presumptive actin in these extracts may be due to contaminating cell types, similar results are obtained with most, but not all (Halenda *et al.*, 1987) species when precautions are taken to remove contaminants (Virtanen *et al.*, 1984; Flaherty *et al.*, 1988). Results of two-dimensional gel electrophoresis (Ochs and Wolf, 1985; Welch and O'Rand, 1985; Camatini *et al.*, 1987) and partial

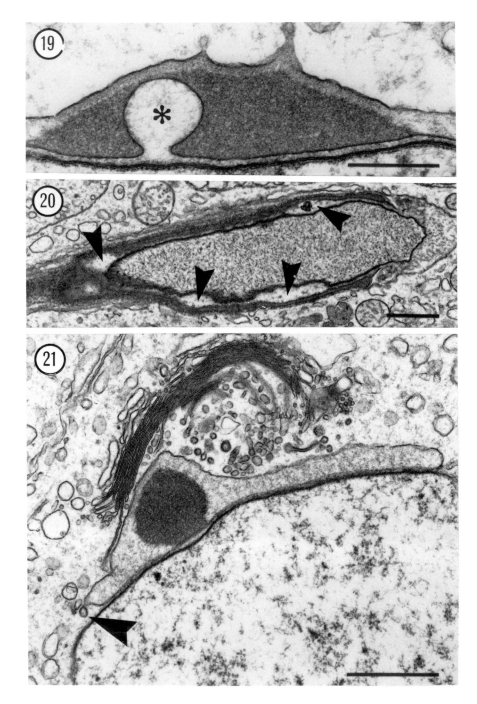

peptide mapping (Ochs and Wolf, 1985), and of myosin affinity studies (Clarke *et al.*, 1982), provide additional evidence that actin is present in most mammalian spermatozoa.

The actual location of actin in spermatozoa has been and continues to be somewhat controversial. Virtually all of our information on this subject comes from numerous studies in which immunological probes have been used at the light- (LM) and at the electron-microscopic (EM) levels. The variability in results appears to stem from (1) actual interspecies differences in actin distribution, and (2) the use by investigators of different antibodies, processing techniques, and levels of control. Although it is almost impossible to come to any general and conclusive statement regarding the distribution of actin in mammalian spermatozoa, the most commonly labeled sites include the posterior aspect of the head (equatorial segment and postacrosomal region) and the neck. Examples of labeling in the postacrosomal region and neck are shown in Figs. 22–24.

Although numerous immunofluorescence studies (Talbot and Kleve, 1978; Campanella *et al.*, 1979; Virtanen *et al.*, 1984; Camatini *et al.*, 1986; Camatini and Casale, 1987) tend to indicate that actin also occurs in the anterior aspects of sperm heads—presumably either between the acrosome and plasma membrane or in the subacrosomal space—these reports have generally proved difficult to confirm at the ultrastructural level (Flaherty *et al.*, 1988); however, there are a few notable exceptions. Hamster spermatozoa have a small region along the concave margin of the head between the plasma membrane and acrosome that reacts positively with immunological probes both at LM (Talbot and Kleve, 1978) and EM (Flaherty *et al.*, 1988) levels. Also, Camatini *et al.* (1986) report labeling of membranes over anterior aspects of acrosome-reacted boar sperm. These authors suggest that this result may indicate the presence of actin under the plasma membrane in intact sperm. They also report labeling associated with the subacrosomal space in these acrosome-reacted sperm. Material in blisterlike swellings of the subacrosomal space in intact rabbit sperm also reacts positively with antiactin probes (Figs. 25–27) (Camatini *et al.*, 1987).

In posterior head regions, labeling with antiactin probes generally occurs in the equatorial segment (Talbot and Kleve, 1978; Camatini *et al.*, 1986; Camatini and Casale, 1987) and postacrosomal region (Clarke and

---

FIGS. 19–21. Electron micrographs of cytochalasin D-treated spermatids of the rat. The actin-disrupting drug was administered intratesticularly. Figs. 19 and 20. Acrosome appears to be separating (asterisk and arrowheads) from the underlying nucleus. Fig. 21. Acrosome is not complete (arrowhead) on one side of the nucleus. Fig. 19, ×48,000, bar = 0.5 μm; Fig. 20, ×14,000, bar = 1 μm; Fig. 21, ×24,000, bar = 1 μm. Modified from Russel (1989).

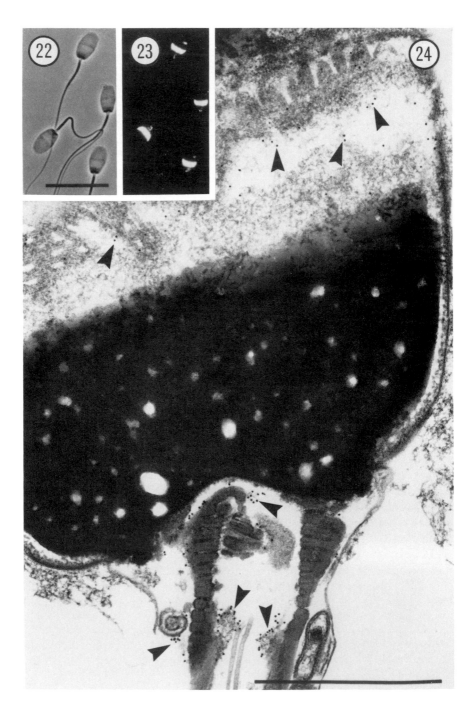

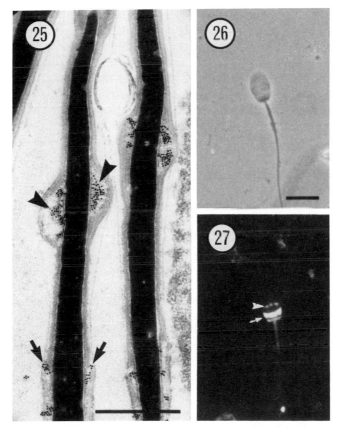

FIGS. 25–27. Immunogold, phase, and immunofluorescence micrographs, respectively, of rabbit spermatozoa labeled with antiactin antibodies. At both the EM and LM levels, labeling occurs in blisterlike swellings of the subacrosomal space (arrowheads) and in post-acrosomal regions (arrows). Fig. 25, ×22,800, bar = 1 μm; Figs. 26 and 27, ×850, bar = 10 μm. Modified after Figs. 5b, 5f, and 5g of Camatini *et al.* (1987).

FIGS. 22–24. Phase, immunofluorescence, and immunogold micrographs, respectively, of rabbit spermatozoa treated with antiactin antibodies. Figs. 22 and 23. Positive staining occurs mainly in the anterior postacrosomal region and in the neck. ×1670, bar = 10 μm. Fig. 24. This localization is confirmed at the ultrastructural level. The arrowheads indicate the position of colloidal-gold particles. ×49, 400, bar = 1 μm. Modified after Figs. 14a, 14b, and 15 of Flaherty *et al.* (1988).

Yanagimachi, 1978; Tamblyn, 1980; Welch and O'Rand, 1985; Flaherty *et al.*, 1986; Camatini *et al.*, 1987). In the equatorial region, labeling at the ultrastructural level is reported to occur mainly between the plasma membrane and outer acrosomal membrane (Camatini *et al.*, 1986; Lora-Lamia *et al.*, 1986; Camatini and Casale, 1987). In the postacrosomal region, positive labeling is reported between the postacrosomal sheath and nuclear membrane (Flaherty *et al.*, 1988), and between the sheath and plasma membrane (Lora-Lamia *et al.*, 1986).

In the sperm tail, the neck appears to label the most consistently with antiactin probes (Talbot and Kleve, 1978; Clarke *et al.*, 1982; Welch and O'Rand, 1985; Flaherty *et al.*, 1986, 1988, Camatini and Casale, 1987). It is in this region that the connecting piece at the root of the flagellum is attached to the implantation fossa of the head (Fawcett, 1975b). The connecting piece consists of a single caplike capitulum connected to nine striated or segmented columns that in turn are attached to the nine outer dense fibers of more distal regions of the tail. Embedded in the connecting piece are the proximal centriole, remnants of the distal centriole, and initial parts of the flagellar axoneme. Surrounding the connecting piece are variable amounts of cytoplasm. Numerous fine filamentous linkages have been observed between the various structures in the neck (Fawcett, 1975b; Escalier, 1984). At the ultrastructural level, and depending on the species, immunological probes for actin label granular material in and around the connecting piece, most notably in inner regions where the segmented columns merge with the outer dense fibers, in the gap between the capitulum and basal plant of the implantation fossa, on the external surfaces of the segmented columns, and in regions surrounding the proximal centriole (Flaherty *et al.*, 1988).

Actin is also reported to occur in regions of the tail other than the neck (Clarke *et al.*, 1982; Flaherty *et al.*, 1986; Talbot and Kleve, 1978; Campanella *et al.*, 1979; Virtanen *et al.*, 1984; Clarke *et al.*, 1982; Lora-Lamia *et al.*, 1986). At the ultrastructural level in human spermatozoa, the inner and outer surfaces of the fibrous sheath in the principal piece label with antiactin probes (Flaherty *et al.*, 1988).

The available morphological data just discussed can be summarized as follows. The distribution of actin in mammalian spermatozoa appears to have a high degree of species variability. The protein is most commonly reported to occur in posterior regions of the head and in the neck. In anterior regions of the head there is some evidence for its presence in the subacrosomal space and very little convincing morphological evidence for its presence between the acrosome and plasma membrane. In some species, such as the guinea pig, actin is either present in very low quantities and restricted to the tail in distribution (Flaherty *et al.*, 1986) or cannot be detected at all using immunological probes at the morphological or the biochemical level (Halenda *et al.*, 1987).

The polymerization state of actin is most sperm cells is not entirely clear. What is clear is that spermatozoa do not normally stain with NBD-phallacidin—an observation that has led most investigators to conclude that any actin present in the cells is probably in the G or monomeric form (Virtanen *et al.*, 1984), and that it may be sequestered by a binding protein similar to profilin (Greenburg and Tamlyn, 1981). Others have suggested that the actin may be associated with a spectrinlike protein (Virtanen *et al.*, 1984) and may exist in a form similar to that in red blood cells, that is, in short polymers. Several reports tend to indicate that actin is membrane-associated and is present in both monomeric and polymeric forms (Peterson and Hunt, 1987; Hunt *et al.*, 1988). These reports are generally consistent with detergent extraction experiments in which immunofluorescence is reduced, but not eliminated by extraction (Welch and O'Rand, 1985; Flaherty *et al.*, 1986).

The function of actin in spermatozoa is not known. Some authors argue that some of the actin present may simply be residual protein remaining after spermiogenesis (Halenda *et al.*, 1987; Flaherty *et al.*, 1988). It is also suggested that, because of the interspecies variability in content and distribution of the actin, any proposed function may not be universally applicable (Flaherty *et al.*, 1988). Nevertheless, the actin that has been reported in spermatozoa has been suggested to have numerous functions.

Many of the proposed functions for actin in spermatozoa are, at this point, speculative. Because (1) actin is involved with numerous membrane-related events in other systems, (2) elaborate membrane domains exist in sperm cells, (3) membrane domains in spermatozoa change during physiological maturation, capacitation, activation, and fertilization, and (4) actin is present in mammalian spermatozoa of most species, it is perhaps not surprising that actin is suggested to play a role in the numerous membrane-related events that occur during the life history of a sperm cell. Based on the appearance of NBD-phallacidin staining during capacitation and the fact that this change correlates with a change in position of a membrane protein, together with the observation that cytochalasin D prevents movement of the membrane protein, Saxena *et al.* (1986) suggest that actin may be involved in the capacitation process. There are also reports that "microfilaments" appear among membrane elements during the acrosome reaction in boar sperm (Peterson *et al.*, 1978), and that, in the same species, immunological probes for actin label disrupted membranes associated with acrosome-reacted sperm (Camatini *et al.*, 1986). In addition, cytochalasin D has been observed to inhibit fertilization by capacitated sperm (Rogers *et al.*, 1986).

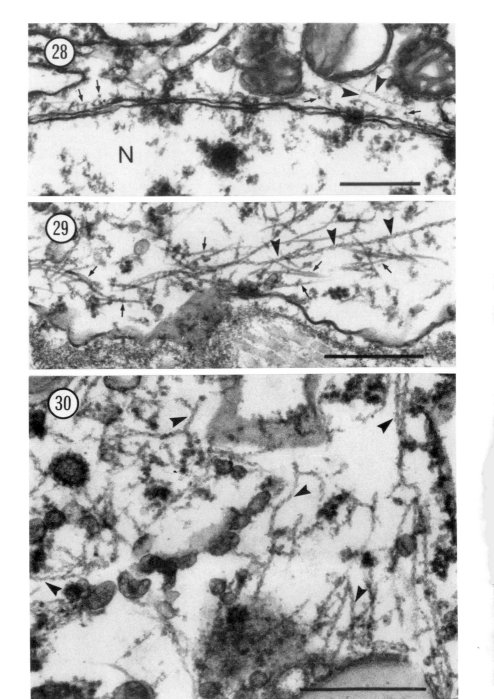

The possible function of actin in the tail has not been addressed, although "linking functions" appear reasonable to suggest.

Because of the lack of experimental data, it would be premature to come to any conclusion about the role of actin in spermatozoa. Suffice it to say that the most active areas of investigation are those dealing with the relationship between actin and associated membranes, particularly in the head.

III. Sertoli Cell

Actin filaments occur to some degree throughout the Sertoli cell (Vogl *et al.*, 1983a). Although intermediate filaments predominate around the nucleus (Fig. 28), at the base of the cell (Fig. 29), and in association with desmosome-like junctions, actin filaments can also be detected at these sites in detergent-extracted and S1-treated tissue (Vogl *et al.*, 1983a; Russell *et al.*, 1987). In the same material, actin filaments also occur among the remnants of membranous organelles in central regions of the cell (Fig. 30), and in more peripheral areas where they are found mainly in association with the plasma membrane (Vogl *et al.*, 1983a).

Although actin filaments appear diffusely distributed throughout the Sertoli cell, they are concentrated in two specific structures: tubulobulbar complexes and ectoplasmic specializations.

A. Tubulobulbar Complexes

Tubulobulbar complexes consist of elongate projections (tubulobulbar processes) of maturing spermatids together with the associated plasma membrane and substructure of the adjacent Sertoli cell invaginations into which the projections extend (Fig. 31). These complexes form in areas of intercellular attachment characterized by the presence of ectoplasmic specializations and may be part of the mechanism by which junctional

---

Figs. 28–30. A series of electron micrographs of detergent-extracted and S1-treated Sertoli cells of the ground squirrel. Fig. 28. Section through the nucleus (N) and adjacent region of the cell. Although intermediate filaments (small arrows) predominate in perinuclear regions, actin filaments (arrowheads) occasionally can be identified in these locations. ×43,200, bar = 0.5 μm. Fig. 29. Section through the base of a Sertoli cell and the adjacent extracellular matrix. Intermediate filaments (small arrows) are abundant in basal regions of the cell. In this micrograph, a decorated actin filament (arrowheads) is also evident. ×52,600, bar = 0.5 μm. Fig. 30. Section through the central region of a Sertoli cell. Decorated actin filaments (arrowheads) are obvious among remnants of the various membranous organelles. ×66,000, bar = 0.5 μm.

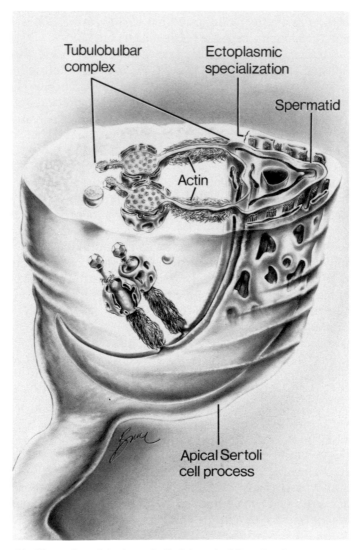

FIG. 31. Illustration of the lower half of the apical Sertoli cell process containing a late spermatid and showing the major organizational features of tubulobulbar complexes. Elongate processes develop from the spermatid head and protrude into corresponding invaginations of the adjacent Sertoli cell. Most of these processes have both tubular and bulbar regions as shown here. Networks of actin filaments occur in Sertoli cell regions surrounding tubular regions, while a cuff of endoplasmic reticulum surrounds the bulbar regions. What are apparently coated pits occur at the ends of the processes.

membrane domains are degraded prior both to spermiation and to movement of spermatocytes through the blood–testis barrier.

Tubulobulbar complexes have a characteristic morphology whether
formed in areas of adhesion to spermatids (Russell and Clermont, 1976;
Russell, 1979a; Russell and Malone, 1980) or in regions of the junctional
complex (blood–testis barrier) between neighboring Sertoli cells (Russell
and Clermont, 1976; Russell, 1979b). At the apex of the cell and at a stage
of spermatogenesis just preceding sperm release, spermatids develop
tubular projections of generally between 1 and 2 μm in length (Russell and
Malone, 1980) that invaginate the adjacent Sertoli cell (Figs. 32, 33). These
tubular projections eventually develop an expanded or swollen area along
their length (Fig. 34) that results in the projection having a tubular proximal
segment, a bulbar middle region, and a small tubular distal segment (Russell, 1979a). The plasma membrane of the adjacent Sertoli cell is separated
from the plasma membrane of the tubulobulbar process of the spermatid
by ~6–8 nm. Sertoli cell cytoplasm surrounding the tubular segments is
filled with a dense network of filaments, and adjacent to the bulbar segment
are elements of the ER. Adjacent to the tip of the process is what appears
to be a coated pit. In fact, this coated-pit region of the Sertoli cell is the first
element of the tubulobulbar complex that forms. Multiple generations of
tubulobulbar processes of the spermatid are formed during the period just
prior to spermiation, and each is phagocytosed and degraded by the adjacent Sertoli cell (Russell, 1979a).

Though shorter, tubulobulbar complexes formed basally at the blood–
testis barrier between adjacent Sertoli cells are generally similar to those
formed between spermatids and Sertoli cells at the apex of the seminiferous epithelium. In the rat, they are most abundant in early stages of the
cycle (stages II–V) and are degraded in midcycle (stages VI–VII) (Russell,
1979b). Apical movement of spermatocytes through the blood–testis barrier begins in late stage VII (Russell, 1977a).

Evidence that the filament network in Sertoli cell cytoplasm surrounding
tubulobulbar processes contains F-actin comes from experiments in which
NBD-phallacidin and S1 are used as probes. Adjacent to the concave
aspect of the late spermatid heads of the rat, two rows of columnar
structures are visible when fixed material is stained with NBD-phallacidin
(Figs. 35, 36). These structures occur in the same position that tubulobulbar complexes are observed in electron micrographs (Russell and Clermont, 1976). When similar tissue is detergent-extracted and decorated
with S1, filaments in Sertoli cell cytoplasm surrounding the tubulobulbar
processes of spermatids are heavily decorated (Figs. 37, 38).

The function of F-actin in tubulobulbar complexes remains to be
clarified. Their presence in regions of the cell that are initially associated

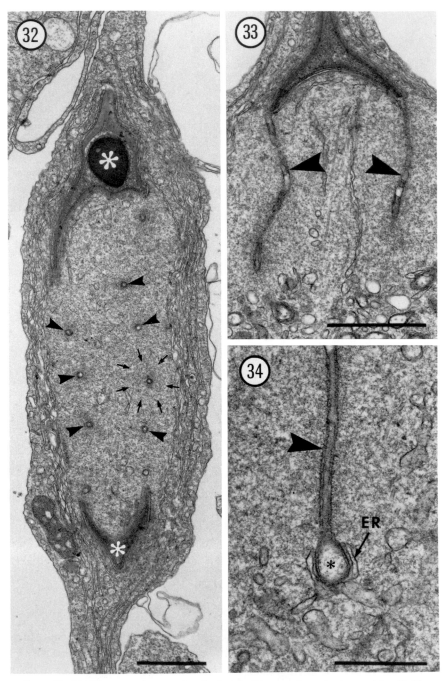

FIGS. 32–34. Electron micrographs of tubulobulbar complexes in the rat. Fig. 32. Cross section of a single Sertoli cell apical process containing a late spermatid. The plane of section cuts twice through the spermatid head (asterisks) and includes numerous cross-sectional

with ectoplasmic specializations indicates that the filament network may represent a restructuring of filaments that were originally part of, or derived from, actin bundles in ectoplasmic specializations, and that they may retain linkages to membrane elements with which they were originally associated. However, nothing is really known about how the filaments surrounding tubulobulbar processes are linked to each other and to adjacent membranes, or how the network is formed or degraded. Intratesticular injection of cytochalasin D prevents the formation of normal tubulobulbar complexes, indicating that actin may be involved in this process (Russell et al., 1989); however, the mechanism of their involvement remains to be clarified.

Tubulobulbar complexes have been suggested to have numerous functions. Because these structures occur just prior to sperm release and are often among the last elements of contact between Sertoli cells and germ cells, it has been suggested that they are intercellular attachment or "anchoring" devices, and that their degradation is part of the mechanism of spermiation (Russel and Clermont, 1976).

They have also been hypothesized to be a mechanism of eliminating the cytoplasm of spermatids during the maturation phase of spermiogenesis (Russell, 1979c). The time during which tubulobulbar complexes are formed and degraded correlates with the period during which the cytoplasmic volume of spermatids is reduced. Moreover, if tubulobulbar processes are prevented from forming, by the intratesticular injection of cytochalasin D, cytoplasmic volume is not reduced (Russell et al., 1989). Though consistent with the hypothesis, the latter observation must be interpreted with some degree of caution. When cytochalasin is administered intratesticularly, its effects are not confined to tubulobulbar complexes; rather, they occur throughout the organ. The observations that tubulobulbar complexes do not form and cytoplasmic volume of spermatids is not reduced may be the result of two independent effects.

One of the problems with all functions attributed to tubulobulbar complexes that occur between Sertoli cells and spermatogenic cells is that they do not easily account for the presence of morphologically similar structures at the blood–testis barrier. For example, it is difficult to reconcile the

---

profiles of tubulobulbar processes (arrowheads). Notice that a halo of filamentous material (small arrows) surrounds each tubulobulbar process. ×18,600, bar = 1 μm. Fig. 33. Longitudinal section of two tubulobulbar processes extending from a spermatid head and into the adjacent Sertoli cell. ×26,000, bar = 1 μm. Fig. 34. Longitudinal section of a tubulobulbar process in which both the tubular (arrowhead) and bulbar (asterisk) components are visible. Notice that elements of the endoplasmic reticulum (ER) are associated with the bulbar component. ×48,600, bar = 0.5 μm.

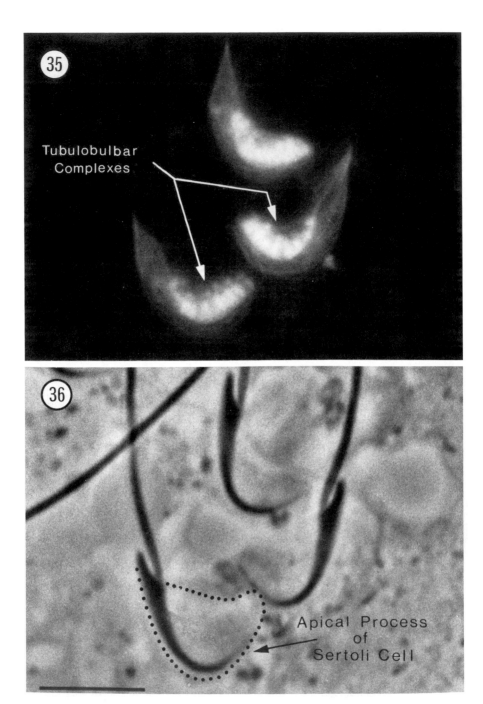

FIGS. 35 and 36. Fluorescence and corresponding phase micrographs of apical Sertoli cell processes and associated late spermatids of the rat. The tissue has been treated with the actin filament-specific probe NBD-phallacidin. Notice that Sertoli cell regions associated with the tubulobulbar complexes along the concave margins of the hook-shaped spermatid heads are strongly fluorescent. ×2800, bar = 10 μm. Modified after Figs 8e,e′ of Vogl et al. (1985).

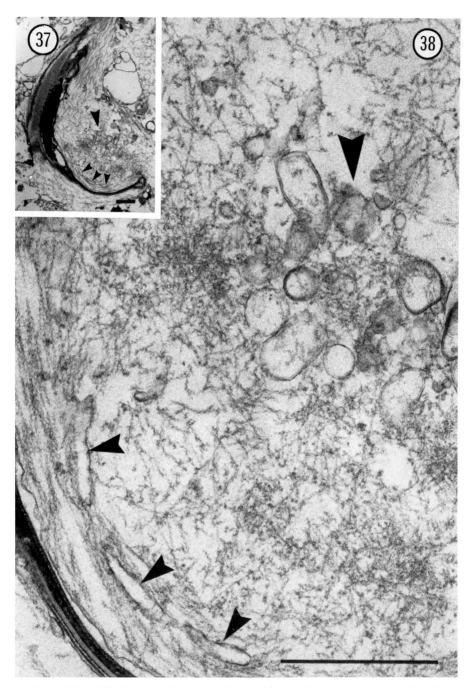

FIGS. 37 and 38. Low- and high-magnification images of a detergent-extracted and S1-treated apical Sertoli cell process similar to those shown in Fig. 36. Remnants of tubulobulbar processes (large arrowheads) are evident, as is a reasonably intact ectoplasmic specialization (small arrowheads). Notice that Sertoli cell regions around tubulobulbar processes are filled with decorated filaments. Fig. 37, ×4700, bar = 1 μm; Fig. 38, ×42,500, bar = 1 μm. Micrographs courtesy of L. D. Russell.

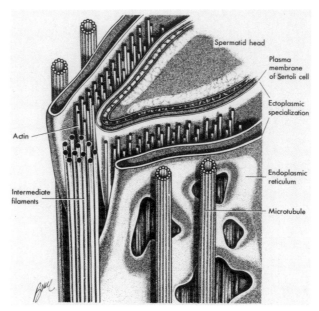

FIG. 39. Diagram showing the arrangement of cytoskeletal elements in and around a Sertoli cell ectoplasmic specialization that occurs adjacent to an apical crypt containing an elongate spermatid. The dorsal or convex aspect of a rat spermatid head and adjacent regions of the crypt are indicated in the figure. A defect or slit in the endoplasmic reticulum (ER) of the ectoplasmic specialization allows a bundle of intermediate filaments, which extends apically from perinuclear regions of the Sertoli cell, to associate closely with the underlying actin filaments and plasma membrane. Microtubules occur immediately adjacent to the cytoplasmic face of the ER of the ectoplasmic specialization and are oriented parallel to the long axis of the Sertoli cell.

presence of tubulobulbar complexes between two Sertoli cells with the hypothesis that their primary function is to eliminate excess cytoplasm.

Perhaps the most interesting hypothesis of function is that tubulobulbar complexes are primarily involved with the elimination, by Sertoli cells, of membrane domains involved with intercellular junctions (Russell, 1979b; Russel et al., 1988; Pelletier, 1988). Observations consistent with this hypothesis are the following. First, tubulobulbar processes occur only at

FIGS. 40 and 41. Electron micrographs of ectoplasmic specializations associated with the blood–testis barrier (Fig. 40) and in regions of attachment to spermatids (Fig. 41). At both sites, ectoplasmic specializations lie immediately adjacent to the plasma membrane and consist of a layer of hexagonally packed actin filaments and a cistern of endoplasmic reticulum (ER). At the blood–testis barrier, ectoplasmic specializations (arrowheads) occur in Sertoli cells on each side of the junction. In areas of attachment to spermatids, only the Sertoli cell contains an ectoplasmic specialization. The asterisk indicates the acrosome of a spermatid. Microtubules (Mt) are often closely related to the ER ectoplasmic specializations associated with apical crypts containing attached spermatids. Fig. 40, ×18,800, bar = 1 μm; Fig. 41, ×156,000, bar = 0.1 μm.

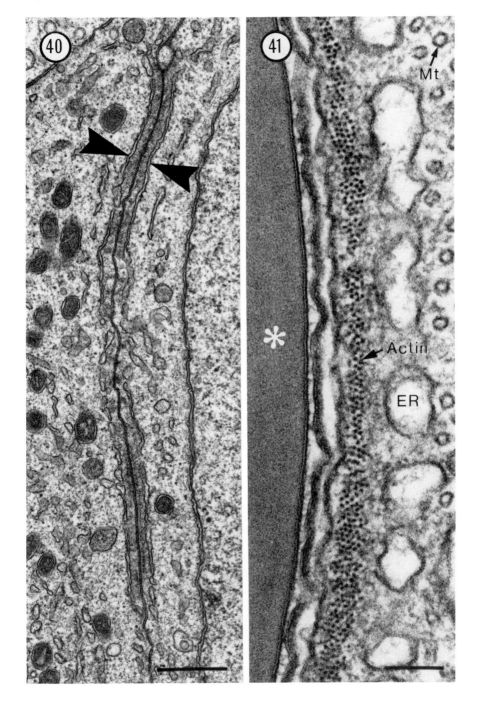

cell junctions. Specifically, they occur in regions initially associated with ectoplasmic specializations—that is, sites of adhesion between Sertoli cells and spermatids, and at the blood–testis barrier. Second, their appearance is associated with a local disruption of ectoplasmic specializations—structures implicated in stabilizing membrane domains involved with intercellular adhesion (Vogl *et al.,* 1986) (see Section III,B,3,a). Third, "filamentous" linkages similar to those observed by Hirokawa and Heuser (1981) between enterocytes at the zonula adherens occur between adjacent plasma membranes in the coated-pit region of tubulobulbar processes (Russell and Malone, 1980). Fourth, junctions (gap and tight) are visible in tubulobulbar processes at the blood–testis barrier (Russell, 1979b). Fifth, tubulobulbar processes formed at the apex of the epithelium appear as ectoplasmic specializations, begin to disappear, and are the last structures present prior to spermiation. This observation is consistent with the hypothesis that ectoplasmic specializations stabilize and tubulobulbar processes degrade junctional domains. Sixth, tubulobulbar processes are internalized and degraded by Sertoli cells. Finally, tubulobulbar processes appear prior to two events involving junction turnover: spermiation and movement of spermatocytes through the blood–testis barrier.

The hypothesis just outlined is generally consistent with all the information we currently have concerning tubulobulbar complexes and does not necessarily exclude the possibility that a secondary function, or perhaps a consequence of the phagocytotic process, is reduction of the cytoplasmic volume of spermatids. Among the roles played by actin filament network may be that of ensuring that specific areas of the plasma membrane are incorporated into the tubulobulbar complexes for eventual degradation.

B. Ectoplasmic Specializations

1. *Definition and Structure*

Ectoplasmic specializations are actin-containing structures found adjacent to specific sites of intercellular attachment (Fig. 39). Current evidence indicates that they may be involved primarily with stabilizing membrane domains involved with intercellular adhesion.

When visualized by EM, ectoplasmic specializations of eutherian mammals consist of a layer of filamentous material sandwiched between the plasma membrane and a cistern of ER (Figs. 40, 41) (Brokelmann, 1963; Flickinger and Fawcett, 1967; Nicander, 1967; Dym and Fawcett, 1970; Russell, 1977b). That the filaments are actin is indicated by their size and by their ability to bind fragments of the myosin molecule (Figs. 42, 43) (Tokyama, 1976; Vogl *et al.,* 1983a, 1986; Masri *et al.,* 1987), actin antibodies (Fig. 44) (Franke *et al.,* 1978; Camatini *et al.,* 1986, 1987), and

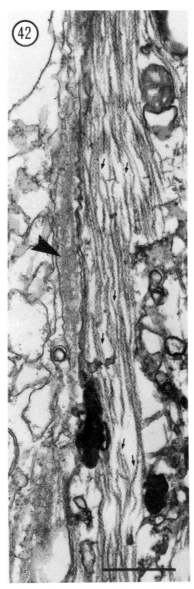

Fig. 42. Electron micrograph of detergent-extracted and S1-treated ectoplasmic specializations associated with the blood–testis barrier in the ground squirrel. The Sertoli cell on one
side of the junction has not been extracted sufficiently to allow penetration of S1 into the
ectoplasmic specialization (arrowhead). Filaments in the ectoplasmic specialization on the
other side of the junction are well decorated and appear arranged in a unipolar fashion as
indicated by the small arrows. ×38,675, bar = 0.5 μm. Modified after Fig. 6b of Vogl *et al.*
(1983a).

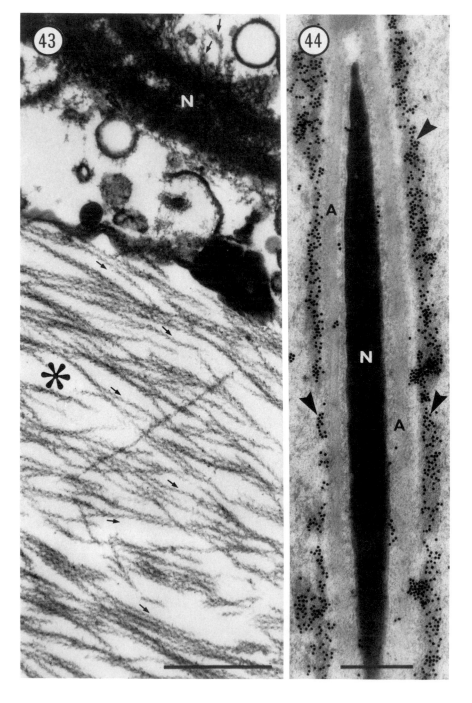

NBD-phallacidin (Suarez-Quian and Dym, 1984, 1988; Vogl and Soucy, 1985; Vogl *et al.*, 1985). The filaments are hexagonally packed, not unlike the situation in intestinal microvilli, and form a carpet that is adjacent to the plasma membrane and is approximately six filaments deep. Ectoplasmic specializations are found in two major locations in Sertoli cells: (1) adjacent to Sertoli–Sertoli junctions (blood–testis barrier) and (2) next to sites of adhesion to spermatogenic cells.

At the blood–testis barrier, ectoplasmic specializations ring the bases of Sertoli cells in a beltlike fashion (Weber *et al.*, 1983). In material that has been cross-sectioned, then stained with NBD-phallacidin, ectoplasmic specializations at the basal Sertoli–Sertoli junctions appear as linear streaks of fluorescence (Fig. 45). When similar regions are viewed *en face*, actin filaments in ectoplasmic specializations appear as polygonal rings of staining that together form a honeycomb pattern at the base of the seminiferous epithelium (Fig. 46). Ultrastructural evidence together with the fluorescence data indicate that filament bundles in ectoplasmic specializations adjacent to the blood–testis barrier run around the circumference of the Sertoli cells parallel to the base of the epithelium.

At sites of adhesion to spermatogenic cells, ectoplasmic specializations are most obvious in regions adherent to the heads of elongate spermatids. They first become evident as spermatids polarize and assume a position within Sertoli cell crypts (Vogl and Soucy, 1985; Suarez-Quian and Dym, 1988). They expand to surround the head of elongating spermatids and then regress during the late stages of spermiogenesis. Just prior to, or at the time of spermiation, they disappear from regions associated with the detaching spermatids.

The arrangement of actin filaments in ectoplasmic specializations associated with spermatids generally conforms to the contours of the crypt

FIGS. 43 and 44. Evidence that filaments within ectoplasmic specializations associated with spermatids are actin. Fig. 43. An elongate spermatid and attached ectoplasmic specialization of a ground squirrel were detergent-extracted and treated with the S1 fragment of the myosin molecule. The nucleus of the spermatid is indicated by the N. Notice that filaments in the ectoplasmic specialization (asterisk) of the Sertoli cell are all labeled with the S1 and that the filaments are arranged in a unipolar fashion; that is, the arrowhead patterns formed by the binding of S1 to the filaments point in the same direction, as indicated by the small arrows, on all the filaments. Also shown in this micrograph are decorated filaments (indicated by the two small arrows at the top of the micrograph) in the subacrosomal space of the spermatid. ×56,700, bar = 0.5 μm. Fig. 44. Illustration, at the ultrastructural level, of the binding of antiactin antibodies to ectoplasmic specializations of the boar. Areas of positive binding are indicated by the electron-dense colloidal-gold particles. Notice that ectoplasmic specializations are heavily labeled (arrowheads). The nucleus and acrosome of the spermatid are indicated by the N and A, respectively. ×38,000, bar = 0.5 μm. Figure 44 modified after Fig. 3c of Camatini *et al.* (1986).

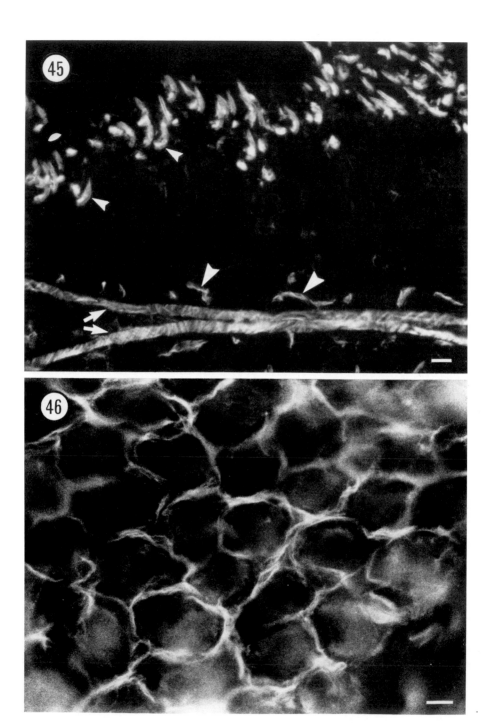

membrane and adjacent surfaces of the differentiating spermatid. As spermatid heads change shape during spermatogenesis, the adjacent bundles of actin filament in ectoplasmic specializations undergo remarkable changes in pattern (Figs. 47–54) (Vogl and Soucy, 1985; Vogl et al., 1985). This is particularly evident in species, such as the ground squirrel, that have large spermatogenic cells. In the ground squirrel, and during the early phase of elongation, the actin filaments encircle the area adjacent to the site of contact between the acrosome and nucleus. As spermiogenesis proceeds and the spermatid heads elongate and flatten, ectoplasmic specializations expand to line the entire crypt membrane adjacent to the heads. At this time, actin filament bundles in the ectoplasmic specializations become organized into a pattern resembling the ridges and bands on a scallop shell. In association with late spermatids, the filaments occur in bundles that run parallel to the rim of the saucer-shaped heads.

Although ectoplasmic specializations associated with the blood–testis barrier and with spermatids differ in shape and in the overall pattern of actin bundle distribution, the basic ultrastructure of the system is similar at the two sites; that is, the system consists of a layer of actin filaments and an associated cistern of ER that together lie against the plasma membrane in areas of intercellular attachment. The only noticeable difference in the system at the two sites is the presence of ribosomes on the cytoplasmic side of the ER associated with ectoplasmic specializations at the base of the cell.

Filaments in ectoplasmic specializations are linked to each other and to adjacent membranes. Evidence for the presence of linkages is both direct and indirect. In electron micrographs of the system, linkages can often be seen between filaments and adjacent membranes (Franke et al., 1978; Russell, 1977b), and between one filament and another. Moreover, when the seminiferous epithelium is mechanically disrupted or detergent-extracted, the plasma membrane of the Sertoli cell, the underlying carpet

---

FIGS. 45 and 46. Fluorescence and phase micrographs indicating the distribution of actin in ectoplasmic specializations of the seminiferous epithelium. Fig. 45. Fluorescence micrograph obtained by treating a section of fixed and frozen rat testis with NBD-phallacidin. Actin in ectoplasmic specializations associated with the basally situated blood–testis barrier appears as linear tracts of fluorescence (large arrowheads). Actin in ectoplasmic specializations associated with late spermatids occurs at the apex of the epithelium (small arrowheads). Also stained in this micrograph is actin in myoid cells (arrows) situated in the tubule wall. ×510, bar = 10 μm. Fig. 46. Phase micrograph of a sheet of ground squirrel seminiferous epithelium isolated from the tubule wall using EDTA. The epithelium was fixed, treated with NBD-phallacidin, then viewed en face. The focal plane passes through the blood–testis barrier. Actin in ectoplasmic specializations associated with the junctional sites outlines the borders of the Sertoli cells. ×700, bar = 10 μm.

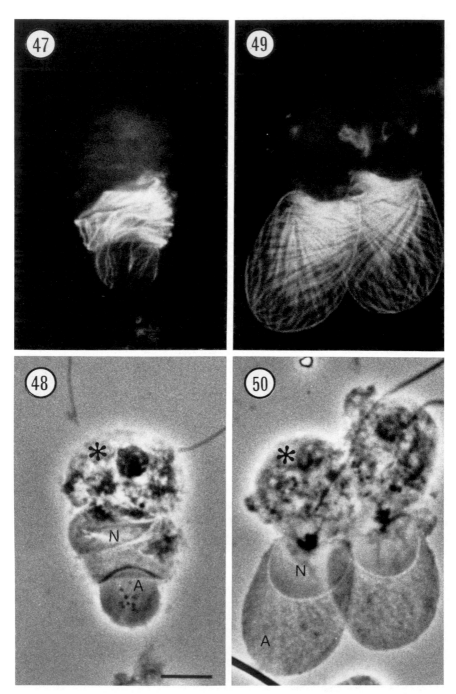

FIGS. 47–54. A series of fluorescence (Figs. 47, 49, 51, 53) and corresponding phase micrographs (Figs. 48, 50, 52, 54) illustrating changes that occur, during spermatogenesis, in the pattern of actin filament distribution in ectoplasmic specializations associated with spermatids of the ground squirrel. The spermatids and attached ectoplasmic specializations shown here were mechanically separated from the seminiferous epithelium, then treated with

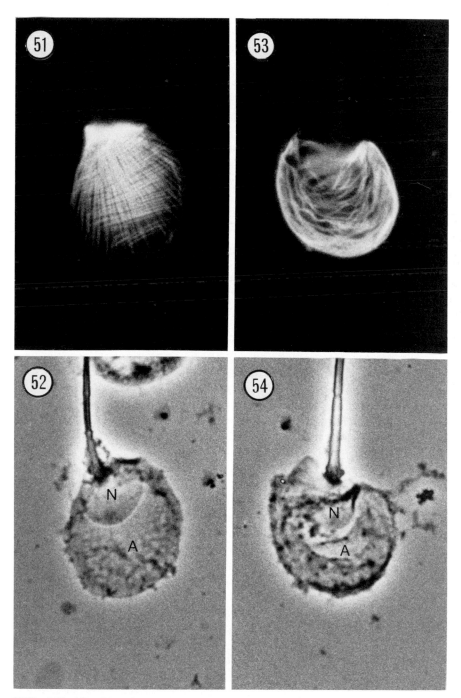

NBD-phallacidin. In each case, the acrosome (A) and nucleus (N) of the spermatid are indicated. The asterisks indicate residual cytoplasm. Notice that as the spermatids elongate, flatten, and mature, the patterns of actin filament distribution in the attached ectoplasmic specializations change. ×1370, bar = 10 μm of actin filaments, and the adjacent membrane of ER remain tightly attached to each other and isolate as structural units, indirectly indicating the presence of linkages between the three elements.

Little is known about the nature of these linkages. Immunofluorescence studies (Franke *et al.*, 1978; Russell and Goh, 1988) indicate that one linker that may be present is α-actinin, an actin-binding protein that links actin filaments into loose bundles (Jockusch and Isenberg, 1981) and that may also participate in linking actin filaments to the plasma membrane (Burn, 1988). The presence and exact location of α-actinin in ectoplasmic specializations have yet to be conclusively established.

Another actin-binding protein that may be present in ectoplasmic specializations is fimbrin. Fimbrin is a 68-kDa globular protein that crosslinks actin filaments into bundles in which all filaments have the same polarity (Stossel *et al.*, 1985). In addition to being found in the vicinity of some actin-associated adhesion sites (Bretscher and Weber, 1980), it is also a major element in the cores of intestinal microvilli (Mooseker, 1985). In immunoblots of testicular fractions enriched for ectoplasmic specializations, a band that corresponds in molecular weight to fimbrin reacts positively with an immunological probe for fimbrin (Grove and Vogl, 1986b). Because these fractions are crude and the results have not yet been verified at the morphological level, any conclusion regarding the presence or absence of fimbrin would be premature; however, the presence of this protein would be consistent with the unipolar arrangement and hexagonal packing of the actin filaments in ectoplasmic specializations.

One actin-binding protein that is not concentrated in ectoplasmic specializations is myosin. Two pieces of evidence support this conclusion. First, immunological probes for myosin do not label ectoplasmic specializations (Vogl and Soucy, 1985). Second, glycerinated models do not contract in the presence of exogenous ATP and Ca^{2+} (Vogl and Soucy, 1985), unlike the situation in junction-related actin networks (contractile rings) in other systems that do contain myosin (Hirokawa *et al.*, 1983; Owaribe *et al.*, 1981). The absence of myosin, together with the observations that the actin filaments are all unipolar and occur in hexagonal arrays, has led to the conclusion that ectoplasmic specializations are *structural* without being *contractile*.

There is one protein, ZO-1, that has been identified at basal junctional sites between Sertoli cells and at apical sites between Sertoli cells and some stages of spermatids (Byers *et al.*, 1988). Zonula occludens-1 is a 225-kDa peripheral-membrane protein that is specifically associated with tight junctions (Stevenson *et al.*, 1986). It is perhaps not surprising that it is present at the blood–testis barrier because of the extensive and elaborate tight junctions that occur at this site. Its presence in apical locations may reflect the presence of discontinuous tight junctions in these regions (Russell and Peterson, 1985). The relationship between ZO-1 and elements of ectoplasmic specializations present at similar locations currrently is not known.

2. Relationship to Cytoskeletal Elements Other than Actin

Ectoplasmic specializations situated adjacent to spermatids are related to both Sertoli cell microtubules (Mt) and intermediate filaments during spermiogenesis. The cytoplasmic face of the ER is associated with Mt throughout most phases of spermiogenesis. In fact, linkages have been observed between these two elements (Russell, 1977b). Sertoli cell Mt acting in concert with ectoplasmic specializations are thought to facilitate positioning of spermatids within the seminiferous epithelium (Russell, 1977b). A mechanistic model is presented in Section III,B,3,b.

In addition to being related to the cytoplasmic face of ectoplasmic specializations, Mt occasionally occur between the ER and plasma membrane among the actin filaments. Their function in this location is unknown.

Intermediate filaments are also intimately related to ectoplasmic specializations (Amlani and Vogl, 1988) at specific stages during the spermatogenic cycle. As is the case with Mt, intermediate filaments are thought to facilitate positioning of spermatids within the seminiferous epithelium.

3. Function

The numerous specific functions (reviewed in Sprando and Russell, 1987) attributed to ectoplasmic specializations generally fall into two basic categories: (1) those proposing that the structures in some way facilitate intercellular attachment, and hence are involved with spermiation and turnover of the blood–testis barrier, and (2) those postulating that, together with Mt and intermediate filaments, the structures are part of the mechanism by which germ cells are positioned within the seminiferous epithelium. Neither of these functions is mutually exclusive of the other. In fact, evidence indicates that both appear likely.

a. *Intercellular Attachment.*   Numerous pieces of evidence are consistent with the hypothesis that ectoplasmic specializations are associated with junctions and that this association may be primarily with adhesion junctions.

First, ectoplasmic specializations are primarily found at intercellular junctions (Russell and Peterson, 1985). At the base of the cell they are associated with the junctional complex between Sertoli cells, and at the apex of the cell they are found adjacent to sites of attachment to spermatids.

Second, the "disappearance" (either natural and pharmacologically induced) of ectoplasmic specializations is associated with a change in or a loss of intercellular junctions. During spermiation, the "dismantling" of ectoplasmic specializations is correlated with a loss of intercellular adhesion that eventually leads to sperm release (Vogl *et al.*, 1983a). Also, as spermatocytes move through the blood–testis barrier, junctions and ectoplasmic specializations appear to be dismantled above the cells while new ones are formed below (Russell, 1977a). Pharmacological disruption of ectoplasmic specializations with the actin perturbant cytochalasin D is correlated with an increase in permeability and a change in the morphology of the blood–testis barrier (Weber *et al.*, 1988), and a loss of "adhesiveness" between Sertoli cells and newly elongating spermatids (Russell *et al.*, 1988). These results are consistent with results in other epithelial systems in which changes in permeability and morphology of intercellular junctions have been correlated with the perturbation of junction-associated actin networks (Bentzel *et al.*, 1980; Elias, *et al.*, 1980; Cereijido *et al.*, 1981; Duffey *et al.*, 1981; Madara *et al.*, 1986; Madara, 1988; Meza *et al.*, 1980, 1982).

Third, when spermatogenic cells are mechanically dissociated from the seminiferous epithelium, the Sertoli cell plasma membrane and underlying ectoplasmic specializations remain tightly attached to the heads of the spermatogenic cells (Romrell and Ross, 1979; Franke *et al.*, 1978; Russell, 1977b; Vogl *et al.*, 1985, 1986; Masri *et al.*, 1987). This observation indicates that, at least in apical positions, ectoplasmic specializations are found in regions of intercellular adhesion junctions. The conclusion that these junctions are, in fact, a type of adhesion junction is reinforced by the observation of fine filamentous linkages between the plasma membrane of the Sertoli cells and the adjacent plasma membrane of the spermatids (Russell *et al.*, 1988). These linkages appear similar to those described at sites of the zonulae adherens between adjacent intestinal absorptive cells (Hirokawa and Heuser, 1981).

Finally, the fact that ectoplasmic specializations are found in association with spermatids, where the predominant intercellular junction appears to be of the "adhesive" type, indicates that cytoplasmic specializations may be most directly related to adhesion junctions. Although treatment with cytochalasin D results in altering the morphology and permeability of the blood–testis barrier, the relationship of filaments to tight junctions at the base of the cell may be secondary to their association with adhesive elements present at the same location. This argument is consistent with the view expressed by Gumbiner and Simons (1987), that in other epithelia it may be the zonula adherens that influences the assembly and/or position of the zonula occludens in apical regions of the cells, and with the observation that perturbation of actin, associated primarily (but not exclusively) with the zonula adherens in most systems, affects the structure and function of the zonula occludens. That adhesive junctional elements may be present at the blood–testis barrier is indicated by the occurrence at this site of intermembrane linkages (Pelletier, 1988) similar to those present between Sertoli cells and spermatids in regions of ectoplasmic specializations.

If ectoplasmic specializations are, in fact, a component of a specialized type of actin-associated adhesion site, then components that characterize actin-associated adhesion junctions in general should be present in ectoplasmic specializations.

In general, intercellular and cell–substratum adhesion sites that are associated with actin filaments tend to be characterized by the presence of at least one protein, vinculin (Drenckhahn and Franz, 1986). Vinculin is a 130-kDa (actually 116 kDa; Burridge, 1986) protein thought to be involved with binding actin to the cell membrane (Geiger, 1979, 1983; see also Burridge et al., 1988). The details of how this binding occurs are controversial. An additional protein of 82 kDa may also be specific to adhesion sites (Beckerle, 1986) and facilitate filament–membrane attachment.

Each of the two types of actin-associated adhesion sites (cell–cell and cell–substratum) is, in turn, associated with specific proteins. Talin, for example, a 215-kDa vinculin-binding protein, occurs in cell–substratum attachments and, together with the 82-kDa protein and vinculin, is postulated to be part of the mechanism by which actin filaments are linked to ECM receptors in the membrane (Beckerle, 1986). At cell–cell adhesion sites, actin filaments are most likely linked indirectly, via vinculin and other elements, to cell adhesion molecules (CAM) that facilitate intercellular binding (Gumbiner and Simons, 1987). At least two CAM have been identified in other systems at intercellular adhesion sites, and both are calcium-dependent [uvomorulin, or L-CAM, at zonulae adherens of intestinal enterocytes and most other polarized epithelial cells; A-CAM at adherens junctions between cultured lens epithelial cells, brain cells, cardiac muscle cells, and cultured kidney cells (Volk and Geiger, 1986)].

Accumulating biochemical and immunological evidence indicates that vinculin is present in ectoplasmic specializations. In testicular fractions enriched for ectoplasmic specializations associated with spermatids and examined by immunoblot analysis, a protein is present that reacts with antisera and affinity-purified antibodies raised against human platelet vinculin and that comigrates with vinculin isolated from human platelets (Grove and Vogl, 1989). When sections of fixed frozen sections of rat testis are treated with the same affinity-purified antibodies, specific fluorescence occurs in Sertoli cell regions both adjacent to spermatid heads at the blood–testis barrier (Figs. 55, 56) (Grove and Vogl, 1989). Presumably this

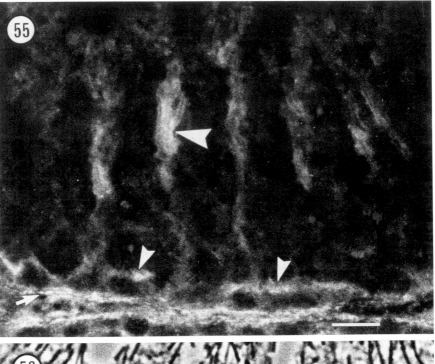

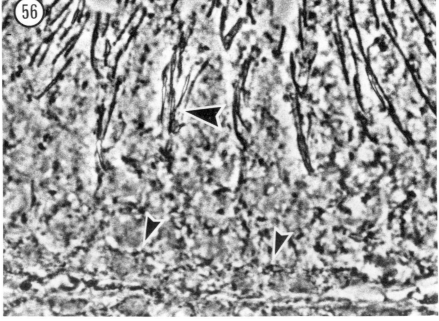

fluorescence is emitted by labeling of vinculin in ectoplasmic specializations, although this remains to be verified at the ultrastructural level. These data are consistent with the hypothesis that ectoplasmic specializations, both at the apex and at the base of the cell, are primarily a component of an actin-associated intercellular adhesion junction. The nature of the relationship between the membrane elements involved with intercellular adhesion and the adjacent actin filaments is not known, nor has the CAM present at the site been identified.

Although ectoplasmic specializations (in eutherian mammals) appear to have features in common with actin-associated junctional complexes in other systems, they do differ from these other systems in two important ways: (1) they are not contractile, and (2) the actin filaments are organized in paracrystalline arrays, not unlike the arrangement of actin filaments in microvilli.

If ectoplasmic specializations are indeed associated with intercellular adhesion and are therefore directly related to the process of spermiation and turnover of the blood–testis barrier, it becomes essential to understand how the system is regulated or controlled. At present, we know very little about this subject. Because many actin-binding proteins are Ca^{2+}-dependent and this ion is characteristically one of a number of second messengers in cells, Ca^{2+} is a likely candidate for being involved with controlling the structure and function of ectoplasmic specializations. Although the presence or absence of Ca^{2+} has been reported to have little direct effect on the structure of isolated ectoplasmic specializations (Grove and Vogl, 1986a), there is some preliminary evidence that Ca^{2+} may have an indirect effect on these structures via calmodulin. Treatment of seminiferous tubules with trifluoroperazine, a calmodulin inhibitor, results in disorganization of the actin filaments and displacement of ER (Franchi and Camatini, 1985a). Moreover, calmodulin is implicated in cytoskeletal control mechanisms in general (Dedman et al., 1979; Means and Dedman, 1980), and the concentration of this protein, in Sertoli cells, appears to vary during spermatogenesis (Mali et al., 1985). Calmodulin has not yet been localized in ectoplasmic specializations; however, the protein is codistributed with actin filament bundles in cultured cells (Dedman et

---

FIGS. 55 and 56. Immunofluorescence and corresponding phase micrographs documenting the presence of vinculin at the blood–testis barrier and at sites of attachment to spermatids. The section of rat testis shown here was fixed, frozen, then cut on a cryostat. The section was treated first with an affinity-purified polyclonal antibody raised in rabbits against human platelet vinculin, then with a goat anti-rabbit IgG conjugated to fluorescein isothiocyanate. Notice that fluorescence occurs in regions surrounding elongate spermatids (large arrowheads) and at the base of the epithelium in regions corresponding to sites of the blood–testis barrier (small arrowheads). ×765, bar = 10 μm.

al., 1978), where it is involved with regulation of myosin light-chain kinase activity.

By analogy with muscle and other cell systems, a strong candidate for regulating local concentrations of Ca^{2+} in ectoplasmic specializations is the cistern of ER that forms a component part of the structure (Means *et al.*, 1980). Franchi and Camatini (1985b), using pyroantimonate, have reported that Ca^{2+} is present in the ER of ectoplasmic specializations; however, the use of pyroantimonate is not a sensitive technique and their results will have to be verified using other methods.

The possible involvement of other second-messenger pathways, such as the inositol triphosphate pathway, in the control of ectoplasmic specializations is not known.

b. *Positioning of Spermatids in the Seminiferous Epithelium.* Because of their suspected association with adhesion sites and their close relationship with Mt and intermediate filaments in Sertoli cells, ectoplasmic specializations may participate in positioning spermatids within the seminiferous epithelium.

As spermatids elongate, they become situated in crypts or invaginations in the apices of Sertoli cells. As noted before, ectoplasmic specializations occur in regions of the crypt that are strongly adherent to spermatid heads. These apical crypts, together with their attached spermatids, assume a position deep within the epithelium at one point during spermatogenesis (stage V in the rat) and are translocated to the apex of the epithelium as spermatogenesis continues (Russell, 1980). "Ever since microtubules were first described in Sertoli cells, they have been suspected to play a role in adjusting the positions of spermatids during spermatogenesis" (Christensen, 1965). Fawcett (1975a) suggested that the Mt generate movement of cytoplasm, in a fashion analogous to that which occurs in neurons, and that this movement of cytoplasm is ultimately responsible for the translocation of developing spermatids. Russell (1977b) later refined this hypothesis and suggested that ectoplasmic specializations move along Mt tracts and carry with them the attached spermatids. According to the proposed hypothesis, the cytoplasmic side of the ER is free to move along adjacent Mt tracts. The other side, or ectoplasmic membrane, is anchored by the layer of actin filaments to the plasma membrane that, in turn, is adherent to the spermatid. As the ER moves along the Mt, it carries with it the adjacent spermatid" (text modified slightly from Vogl, 1988). The exciting concept that emerges here is that an intracellular membranous organelle (the cistern of ER) is anchored to an area of the cell involved with intercellular attachment. Work done to translocate the cistern of ER intracellularly along Mt would ultimately result in the translocation of the attached spermatogenic cells within the seminiferous epithelium. Hence a basic mechanism of intracellular motility (Mt-based organelle movement) may be directly coupled, via a junction network, to an extracellular event—that is, movement of spermatogenic cells within the seminiferous epithelium. An analogous situation occurs in skeletal muscle, where an actin-based intracellular motor is coupled, via an association with cell–substratum attachment sites, to an extracellular event (movement of joints).

The hypothesis that the translocation of ectoplasmic specializations together with their attached spermatids is Mt-based is consistent with current concepts of Mt-based organelle movement in general (Vale *et al.*, 1985a,b; Amos and Amos, 1987; Ogawa *et al.*, 1987; Paschal and Vallee, 1987; Paschal *et al.*, 1987; Vale, 1987; Pratt, 1980, 1986). In other systems, most notably in neurons, vesicular movement appears to involve mechanoenzymes, related to myosin and flagellar dynein, that convert chemical energy to mechanical force and that move vesicles in either an anterograde or retrograde direction along Mt (Vale *et al.*, 1985a,b; Vale, 1987; Paschal and Vallee, 1987). At least one of these motors, kinesin, has been well studied (Vale, 1987) and has been identified in cells other than neurons (Neighbors *et al.*, 1988).

Membrane-linked mechanoenzymes, such as kinesin and cytoplasmic dyneins, that interact with Mt to generate movement in cells appear to be a general phenomenon (Vale, 1987; Neighbors *et al.*, 1988; Ogawa *et al.*, 1987; Paschal and Vallee, 1987). Terasaki *et al.* (1986) have presented evidence to support the hypothesis that the ER is positioned in cells via a Mt-based mechanism.

Evidence that a Mt-based mechanism, involving the movement of ectoplasmic specializations along Mt tracts and the use of mechanoenzymes similar to those identified in other systems, may be responsible for translocating spermatids within the seminiferous epithelium is very preliminary and indirect. Microtubules are numerous adjacent to apical crypts, and their orientation is consistent with the observed direction (apical–basal) of spermatid movement (Amlani and Vogl, 1988; Christensen, 1965; Fawcett, 1975a; Russell, 1977b; Vogl *et al.*, 1983a; Vogl, 1988). Also, linkages have been observed between the ER of ectoplasmic specializations and Mt surrounding the crypts (Russell, 1977b). In addition, MAP-1c, a cytoplasmic dynein and a retrograde motor (Paschal and Vallee, 1987), occurs at high levels in testicular homogenates (Collins and Vallee, 1988) and appears concentrated in Sertoli cells (Neely and Boekelheide, 1988). The hypothesis of Mt-based movement of ectoplasmic specializations has yet to be tested directly.

If Mt interact, in some way, with ectoplasmic specializations to position spermatids within the epithelium, disruption of either Mt or ectoplasmic specializations should result in malorientation of the spermatids. Such effects have been observed after perturbation of actin with cytochalasin D (Russell *et al.*, 1988) and of Mt with taxol (Russell *et al.*, 1989 and colchicine (Vogl *et al.*, 1983b). Although the results of such experiments are consistent with the view that ectoplasmic specializations and Mt are involved with positioning of spermatid in the seminiferous epithelium, they indicate very little about the mechanism of this involvement.

Intermediate filaments may also participate with ectoplasmic specializations and Mt in the positioning of spermatids within the seminiferous epithelium. When spermatids are at their deepest position within Sertoli cell crypts (stage V), bundles of 8–12 intermediate filaments extend apically from perinuclear regions and into Sertoli cell regions adjacent to the dorsal aspect of the spermatid heads (Figs. 57–60). In this position, the filaments pass through a slitlike defect in the ER of ectoplasmic specializations and acquire a close association with the underlying actin filaments and plasma membrane. It has been suggested that the intermediate filaments may facilitate anchoring the crypts and attached spermatids in a basal position within the epithelium (Amlani and Vogl, 1988). The observation that these spermatids, together with their attached ectoplasmic specializations and associated bundles of intermediate filaments, remain firmly attached to the epithelium and in a supranuclear position in detergent-extracted material consistent with this argument (Fig. 59) (Amlani and Vogl, 1988).

c. *Unifying Hypothesis.*   Based on our current knowledge of ectoplasmic specializations and by analogy with other cell systems, the following working hypothesis can be generated. Ectoplasmic specializations may primarily function to establish and/or stabilize intercellular adhesion domains in the plasma membranes of Sertoli cells. Fundamental to this function is some form of "linkage" between adhesion molecules in the plasma membrane and the adjacent actin filaments. Crosslinking of the actin filaments into paracrystalline arrays adjacent to the plasma membrane reinforces the adhesion domains. Control of ectoplasmic specializations, and therefore of adhesion domains, may involve the cistern of ER that forms an intricate component of the system. Ectoplasmic specializations may be structural and functional intercellular adhesion units that can interact with (1) other cytoskeletal elements in the cell (Mt and intermediate filaments) to orient and position spermatogenic cells in the seminiferous epithelium, and (2) other junctional components ("occludens units") in the membrane to facilitate the formation of the blood–testis barrier.

Many elements of this hypothesis have yet to be critically tested. Although the presence of vinculin in ectoplasmic specializations is consistent

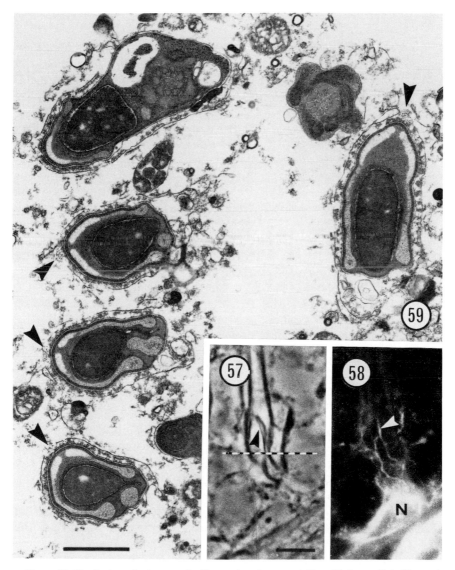

FIGS. 57–59. Series of micrographs illustrating the association of intermediate filaments with Sertoli cell crypts containing elongate spermatids in the rat. Figs. 57 and 58. Phase and immunofluorescence images of a Sertoli cell and associated spermatids at a stage of spermatogenesis when spermatids are situated deep within the epithelium. The tissue has been treated first with a mouse monoclonal antivimentin antibody then with goat anti-mouse IgG conjugated to fluorescein isothiocyanate. Notice that perinuclear filaments extend into apical Sertoli cell cytoplasm adjacent to dorsal regions of the spermatids (arrowheads). The Sertoli cell nucleus is indicated by the N. ×1100, bar = 10 μm. Modified after Figs. 6a,a′ of Amlani and Vogl (1988). Fig. 59. An electron micrograph of a detergent-extracted Sertoli cell and attached spermatids. The plane of section, indicated by the segmented line in Fig. 57, cuts through a number of apical crypts containing elongate spermatids. Notice that a group of intermediate filaments (arrowheads) occurs in Sertoli cell regions immediately adjacent to the dorsal aspect of each spermatid. ×17,800, bar = 1 μm.

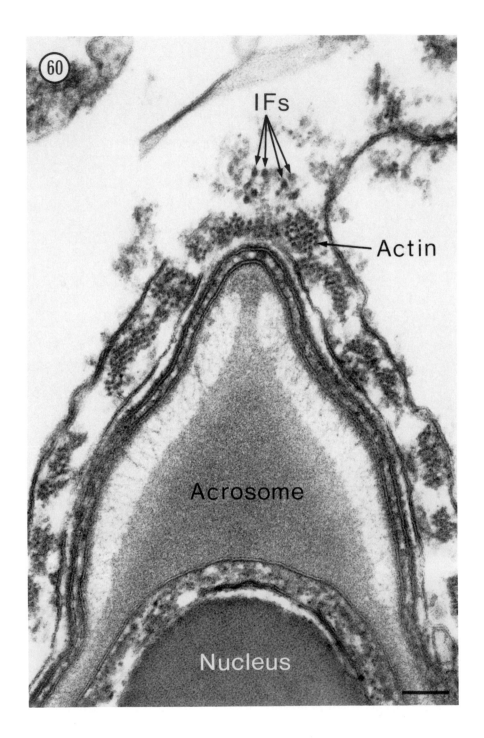

with the argument that these structures are a form of actin-associated intercellular adhesion site, adhesion molecules in the membrane have not been identified; however, it is reported that N-CAM (neural cell adhesion molecule) is present testis (Edelman, 1985). Also, we do not have much conclusive information about components, other than vinculin, that are involved in linking the actin filaments to each other or to the adjacent plasma membrane and cistern of ER. We know nothing about how the different junctional components (tight and adhesion) are related to or influence each other, except perhaps that work in other systems indicates that they probably do (Gumbiner *et al.,* 1988). We know little about the mechanisms by which cytoskeletal elements such as Mt and intermediate filaments actually influence ectoplasmic specializations, and we know even less about how ectoplasmic specializations are controlled.

4. *Ectoplasmic Specializations in Nonmammalian Vertebrates*

Ectoplasmic specializations do occur in nonmammalian vertebrates, particularly in those species in which developing spermatids become situated in crypts or invaginations within support cells (Sprando and Russell, 1987). They differ from the classically described mammalian system in two major ways. First, elements of the ER are not always associated with the system. Second, the actin filaments are not generally organized into paracrystalline arrays; rather, they appear to form loosely arranged bundles. The latter observation would tend to indicate that the actin-binding proteins present in these bundles may differ from those present in the mammalian system.

One actin-binding protein that may be present in the ectoplasmic specializations of some nonmammalian vertebrates is myosin. Antimyosin antibodies label the actin filament bundles associated with spermatids of the ratfish (Stanley and Lambert, 1985). Moreover, these filament bundles appear to associate at one end with a specific area of the Sertoli cell plasma membrane that is attached by dense periodic linkages to the adjacent spermatid, and at the other end with the base of the Sertoli cell. It is suggested that, in this species, the "ectoplasmic specializations" function directly, via an actomyosin mechanism, to orient and pull spermatids to the base of the epithelium (Stanley and Lambert, 1985). Although the loose organization of actin filaments in ectoplasmic specializations of some other

---

FIG. 60. Electron micrograph of a detergent-extracted rat Sertoli cell and adjacent dorsal region of an elongate spermatid. Intermediate filaments (IFs) are intimately related to actin-containing ectoplasmic specializations surrounding the crypt. ×128,000, bar = 0.1 μm. Modified after Fig. 10g of Amlani and Vogl (1988).

nonmammalian vertebrates is consistent with there being myosin present, this has not been confirmed.

From the limited amount of comparative data available, it is tempting to speculate that mammalian ectoplasmic specializations evolved from a junction-related actin network in which myosin was present. Hence, the system initially had contractile properties, not unlike the contractile rings associated with the zonulae adherens in most other epithelia in general. In mammals, myosin has been lost and the actin filaments have been cross-linked into noncontractile arrays, perhaps to add greater reinforcement to the intercellular adhesion sites. With the loss of contractile properties and the formation of a discrete structure or mantel around apical crypts, the ectoplasmic specializations now interact with other components of the cytoskeletal system, such as Mt and intermediate filaments, to position spermatogenic cells in epithelium.

IV. Summary

Actin filaments are concentrated in specific regions of spermatogenic cells and Sertoli cells. In spermatogenic cells they occur in intercellular bridges and in the subacrosomal space. In Sertoli cells they are abundant in ectoplasmic specializations and in regions adjacent to tubulobulbar processes of spermatogenic cells. At all of these sites, the filaments are morphologically related to the plasma membrane and/or intercellular membranes, and, as in many other cell types, are arranged in either bundles or networks.

In at least two of the locations just indicated (ectoplasmic specializations and intercellular bridges), elements of the ER are closely related to the actin filaments. In tubulobulbar complexes, ER is present but is more distantly related to the filaments. Elements of the ER, when present, may serve a regulatory function.

The filaments in ectoplasmic specializations and in regions adjacent to tubulobulbar processes of spermatogenic cells are suspected to be in-volved with the mechanism by which intercellular junctions are estab-lished, maintained, and degraded. In intercellular bridges, actin filaments may serve to reinforce and perhaps regulate the size of the cytoplasmic connections between differentiating germ cells. Filaments in the subacro-somal space may serve as a linking network between the acrosome and nucleus and may also be involved in the capping process.

Because of the possibility that the actin filaments discussed before may be related to specific membrane domains involved with intercellular or interorganelle attachment, and that changes in these membrane domains

are prerequisite to processes such as sperm release, turnover of the blood–
testis barrier, formation of the acrosome, and coordination of spermatogenic cell differentiation, an understanding of exactly how these actin
filaments are related to elements in the membrane and how this interaction
is controlled is fundamental to our understanding, and perhaps our manip-
ulating, of male fertility. I suspect that working out the molecular organization of these actin filament-containing sites and determining how their
organization is controlled will be the major focus of research in this field
over the next few years.

References

Amlani, S., and Vogl, A. W. (1988). *Anat. Rec.* **220**, 143–160.
Amos, L. A., and Amos, W. B. (1987). *J. Cell Sci.* **87**, 1–2.
Baccetti, B., Bigliardi, E., Burrini, A.G., and Pallini, V. (1980). *Gamete Res.* **3**, 203–209.
Barak, L. S., Nothnagel, E. A., De Marco, E. F., and Webb, W. W. (1981). *Proc. Natl. Acad. Sci. USA* **78**, 3034–3038.
Beckerle, M. C. (1986). *J. Cell Biol.* **103**, 1679–1687.
Bentzel, C. J., Hainau, B., Ho, S., Hui, S. W., Edelman, A., Anagnostopoulos, T., and Benedetti, E. L. (1980). *Am. J. Physiol.* **239**, C75–C89.
Bretscher A., and Weber, K. (1980). *J. Cell Biol.* **86**, 335–340.
Brokelmann, J. (1963). *Z. Zellforsch. Microsk, Anat.* **59**, 820–850.
Burn, P. (1988). *TIBS* **13**, 79–83.
Burridge, K. (1986). *Cancer Rev.* **4**, 18–78.
Burridge, K., Fath, K., Kelly, T., Nuckolls, G., and Turner, C. (1988). *Annu. Rev. Cell Biol.* **4**, 487–525.
Byers, S. W., Graham, R., and Stevenson, B. (1988). *J. Cell Biol.* **107**, 170a.
Camatini, M., and Casale, A. (1987). *Gamete Res.* **17**, 97–105.
Camatini, M., Anelli, G., and Casale, A. (1986). *Eur. J. Cell Biol.* **42**, 311–318.
Camatini, M., Casale, A., and Cifarelli, M. (1987). *Eur. J. Cell Biol.* **45**, 274–281.
Campanella, C., Gabbiani, G., Baccetti, B., Burrini, A. G., and Pallini, V. (1979). *J. Submicrosc. Cytol.* **11**, 53–71.
Cereijido, M., Meza I., and Martinez-Palomo, A. (1981). *Am. J. Physiol.* **240**, C96–C102.
Christensen, A. K. (1965). *Anat. Rec.* **151**, 335a.
Clarke, G. N., and Yanagimachi, R. (1978). *J. Exp. Zool.* **205**, 125–132.
Clarke, G. N., and Clarke, F. M., and Wilson, S. (1982). *Biol. Reprod.* **26**, 319–327.
Collins, C. A., and Vallee, R. B. (1988). *Int. Congr. Cell Biol. 4th* p. 2.9.2.
Craig, S. W., and Pollard, T. D. (1982). *TIBS* **7**, 88–92.
Dedman, J. R., Welsh, M. J., and Means, A. R. (1978). *J. Biol. Chem.* **253**, 7515–7521.
Dedman, J. R., Brinkley, B. R., and Means, A. R. (1979). *Adv. Cyclic Nucleotide Res.* **11**, 131–174.
Drenckhahn, D., and Franz, H. (1986). *J. Cell Biol.* **102**, 1843–1852.
Duffey, M. E., Hainau, B., Ho, S., and Bentzel, C. J. (1981). *Nature (London)* **294**, 451–453.
Dym, M., and Fawcett, D. W. (1970). *Biol. Reprod.* **3**, 308–326.
Dym, M., and Fawcett, D. W. (1971). *Biol. Reprod.* **4**, 195–215.
Edelman, G. M. (1985). *In* "The Cell in Contact" (G. M. Edelman and I.-P. Thiery, eds.), pp. 139–168. Wiley, New York.
Elias, E., Hruban, Z., Wade, J. B., and Boyer, J. L. (1980). *Proc. Natl. Acad. Sci. USA* **77**, 2229–2233.

Escalier, D. (1984). *Biol. Cell.* **51,** 347–364.

Fawcett, D. W. (1961). *Exp. Cell Res., Suppl.* No. 8, 174–187.

Fawcett, D. W. (1975a). *In* "Handbook of Physiology" (R. O. Greep, ed.), Sect. 7, Vol. 5, pp. 21–55. Williams & Wilkins, Baltimore, Maryland.

Fawcett, D. W. (1975b). *Dev. Biol.* **44,** 394–436.

Fawcett, D. W., Ito, S., and Slautterback, D. (1959). *J. Biophys. Biochem. Cytol.* **5,** 453–460.

Flaherty, S. P., and Breed, W. G. (1987). *Gamete Res.* **17,** 115–129.

Flaherty, S. P., Breed, W. G., and Sarafis, V. (1983). *J. Exp. Zool.* **225,** 497–500.

Flaherty, S. P., Winfrey, V. P., and Olson, G. E. (1986). *Anat. Rec.* **216,** 504–515.

Flaherty, S. P., Winfrey, V. P., and Olson, G. E. (1988). *Anat. Rec.* **221,** 599–610.

Forer, A., and Behnke, O. (1972). *Chromosome,* **39,** 175–190.

Flickinger, C., and Fawcett, D. W. (1967). *Anat. Rec.* **158,** 207–222.

Fouquet, J. P., Kann, M.-L., and Dadoune, J.-P. (1989). *Anat. Rec.* **223,** 35–42.

Franchi, E., and Camatini, M. (1985a). *Tissue Cell.* **17,** 13–25.

Franchi, E., and Camatini, M. (1985b). *Cell Biol. Int. Rep.* **9,** 441–446.

Franke, W. W., Grund, C., Fink, A., Weber, K., Jockusch, B. M., Zentgraf, H., and Osborn, M. (1978). *Biol. Cell.* **31,** 7–14.

Fujiwara, K., and Pollard, T. D. (1976). *J. Cell Biol.* **71,** 848–875.

Geiger, B. (1979). *Cell* **18,** 193–205.

Geiger, B. (1983). *Biochim. Biophys. Acta* **742,** 129–134.

Gondos, B. (1973). *Differentiation* **1,** 177–182.

Gondos, B. (1984). *In* "Ultrastructure of Reproduction" (J. Van Blerkon and P. M. Motta, eds.), pp. 31–45. Nijhoff, Boston, Massachusetts.

Gondos, B., and Conner, L. A. (1973). *Am. J. Anat.* **136,** 23–42.

Goodman, S. R., and Shiffer, K. (1983). *Am. J. Physiol.* **244,** C121–C141.

Greenberg, B. J., and Tamblyn, T. M. (1981). *J. Cell Biol.* **92,** 191a.

Grove, B. D., and Vogl, A. W. (1986a). *Anat. Rec.* **214,** 45a.

Grove, B. D., and Vogl, A. W. (1986b). *J. Cell Biol.* **103,** 538a.

Grove, B. D., and Vogl, A. W. (1989). *J. Cell Sci.* **93,** 309–323.

Gumbiner, B., and Simons, K. (1987). *CIBA Found. Symp.* No. 125, 168–181.

Gumbiner, B., Stevenson, B., and Grimaldi, A. (1988). *J. Cell Biol.* **107,** 1575–1587.

Halenda, R. M., Primakoff, P., and Myles, D. G. (1987). *Biol. Reprod.* **36,** 491–499.

Hirokawa, N., and Heuser, J. E. (1981). *J. Cell Biol.* **91,** 399–409.

Hirokawa, N., Keller, T. C. S., III, Chasan, R., and Mooseker, M. S. (1983). *J. Cell Biol.* **96,** 1325–1336.

Huckins, C. (1978). *Am. J. Anat.* **153,** 97–122.

Hunt, W. P., Peterson, R. N., and Bozzola, J. J. (1988). *FASEB J.* **2,** 318A.

Joskusch, B. M., and Isenberg, G. (1981). *Proc. Natl. Acad. Sci. USA* **78,** 3005–3009.

Kiehart, D. P., Mabuchi, I., and Inoue, S. (1982). *J. Cell Biol.* **94,** 165–178.

Lora-Lamia, C., Castellani-Ceresa, L., Andreetta, F., Cotelli, F., and Brivio, M. (1986). *J. Ultrastruct. Mol. Struct. Res.* **96,** 12–21.

Mabuchi, I., and Okuno, M. (1977). *J. Cell Biol.* **74,** 251–263.

Madara, J. L. (1988). *Cell* **53,** 497–498.

Madara, J. L., Barenbery, D., and Carlson, S. (1986). *J. Cell Biol.* **102,** 2125–2136.

Mali, P., Welsh, M. J., Toppari, J., Vihko, K. K., and Parvinen, M. (1985). *Med. Biol.* **63,** 237–244.

Masri, B. A., Russell, L. D., and Vogl, A. W. (1987). *Anat. Rec.* **218,** 20–26.

Means, A. R., and Dedman, J. R. (1980). *Nature (London)* **285,** 73–77.

Means, A. R., Dedman, J. R., Tash, J. S., Tindall, D. J., van Sickle, M., and Welsh, M. J. (1980). *Annu. Rev. Physiol.* **42,** 59–70.

Meza, I., Ibarra, G., Sabanero, M., Martinez-Palomo, A., and Cereijido, M. (1980). *J. Cell Biol.* **87,** 746–754.

Meza, I., Sabanero, M., Stefani, E., and Cereijido, M. (1982). *J. Cell Biochem.* **18,** 407–421.

Mooseker, M. S. (1985). *Annu. Rev. Cell Biol.* **1,** 209–241.

Neely, M. D., and Boekelheide, K. (1988). *J. Cell Biol.* **107,** 1767–1776.

Neighbors, B. W., Williams, R. C., Jr., and McIntosh, J. R. (1988). *J. Cell Biol.* **106,** 1193–1204.

Nicander, L. (1967). *Z. Zellforsch. Microsk. Anat.* **83,** 375–397.

Ochs, D., and Wolf, D. P. (1985). *Biol. Reprod.* **33,** 1223–1226.

Ogawa, K., Hosoya, H., Yokota, E., Kobayashi, T., Wakamatsu, Y., Ozato, K., Negishi, S., and Obika, M. (1987). *Eur. J. Cell Biol.* **43,** 3–9.

Owaribe, K., Kodama, R., and Eguchi, G. (1981). *J. Cell Biol.* **90,** 507–514.

Paschal, B. M., and Vallee, R. B. (1987). *Nature (London)* **330,** 181–183.

Paschal, B. M., Shpetner, H. S., and Vallee, R. B. (1987). *J. Cell Biol.* **105,** 1273–1282.

Pelletier, R.-M. (1988). *Am. J. Anat.* **183,** 68–102.

Peterson, R. N., and Hunt, W. P. (1987). *Biol. Reprod., Suppl.* No. 1, p. 142a.

Peterson, R. N., Russell, L. D., Bundman, D., and Freund, M. (1978). *Biol. Reprod.* **19,** 459–465.

Pollard, T. D. (1986). *J. Cell. Biochem.* **31,** 87–95.

Pratt, M. M. (1980). *Dev. Biol.* **74,** 364–378.

Pratt, M. M. (1986). *J. Cell Biol.* **103,** 957–968.

Rogers, B. J., Bastias, C., and Russell, L. D. (1986). *Biol. Reprod. Suppl.* No. 1, p. 54a.

Romrell, L. J., and Ross, M. M. (1979). *Anat. Rec.* **193,** 23–42.

Russell, L. D. (1977a). *Am. J. Anat.* **148,** 313–328.

Russell, L. D. (1977b). *Tissue Cell* **9,** 475–498.

Russell, L. D. (1979a). *Anat. Rec.* **194,** 213–232.

Russell, L. D. (1979b). *Anat. Rec.* **155,** 259–280.

Russell, L. D. (1979c). *Anat. Rec.* **194,** 233–246.

Russell, L. D. (1980). *Gamete Res.* **3,** 179–202.

Russell, L. D. (1984). *In* "Ultrastructure of Reproduction" (J. Van Blerkon and P. M. Motta, eds.), pp. 46–66. Nijhoff, Boson, Massachusetts.

Russell, L. D. (1989). *Anat. Rec.* **223,** 99A.

Russell, L. D., and Clermont, Y. (1976). *Anat. Rec.* **185,** 259–278.

Russell, L. D., and Goh, J. C. (1988). *In* "Development and Function of the Reproductive Organs" (M. Parvinen, I. Huhtaniemi, and L. J. Pelliniemi, eds.), Serono Symp. Rev., Vol. 2, pp. 237–344. Ares–Serono Symp. Via Ravenna, 8-Rome.

Russell, L.D., and Malone, J. P. (1980). *Tissue Cell* **12,** 263–285.

Russell, L. D., and Peterson, R. N. (1985). *Int. Rev. Cytol.* **94,** 177–211.

Russell, L. D., Weber, J. E., and Vogl, A. W. (1986). *Tissue Cell* **18,** 887–898.

Russell, L. D., Vogl, A. W., and Weber, J. E. (1987). *Am. J. Anat.* **180,** 25–40.

Russell, L. D., Goh, J. C., Rashed, R. M. A., and Vogl, A. W. (1989). *Biol. Reprod.* **39,** 105–118.

Russell, L. D., Saxena, N. K., and Turner, T. T. (1989). Submitted.

Saxena, N., Peterson, R. N., Saxena, N. K., and Russell, L. D. (1986). *J. Exp. Zool.* **239,** 423–427.

Schliwa, M., and Van Blerkom, J. (1981). *J. Cell Biol.* **90,** 222–235.

Shroeder, T. E. (1973). *Proc. Natl. Acad. Sci. USA* **70,** 1688–1692.

Somlyo, A. P. (1984). *Nature (London)* **309,** 516–517.

Sprando, R. L., and Russell, L. D. (1987). *Tissue Cell* **19,** 479–493.

Stanley, H. P., and Lambert, C. C. (1985). *J. Morphol.* **186,** 223–236.

Stevenson, B. R., Siliciano, J. D., Mooseker, M. S., and Goodenough, D. A. (1986). *J. Cell Biol.* **103,** 755–766.

Stossel, T. P., Chaponnier, C., Ezzell, R. M., Hartwig, J. H., Janmey, P. A., Kwiatkowski, D. K., Lind, S. E., Smith, D. B., Southwick, F. S., Yin, H. L., and Zaner, K. S. (1985). *Annu. Rev. Cell Biol.* **1,** 353–402.

Suarez-Quian, C. A., and Dym, M. (1984). *Ann. N.Y. Acad. Sci.* **438,** 476–480.

Suarez-Quian, C. A., and Dym, M. (1988). *Int. J. Androl.* **11,** 301–312.

Talbot, P., and Kleve, M. G. (1978). *J. Exp. Zool.* **204,** 131–136.

Tamblyn, T. M. (1980). *Biol. Reprod.* **22,** 727–734.

Terasaki, M., Chen, L. B., and Fujiwara, K. (1986). *J. Cell Biol.* **103,** 1557–1568.

Toyama, Y. (1976). *Anat. Rec.* **186,** 477–492.

Vale, R. D. (1987). *Annu. Rev. Cell Biol.* **3,** 347–378.

Vale, R. D., Reese, T. S., and Scheetz, M. P. (1985a). *Cell* **42,** 39–51.

Vale, R. D., Schnapp, B. J., Michison, T., Stever, E., Reese, T. S., and Sheetz, M. P. (1985b). *Cell* **43,** 623–632.

Virtanen, I., Badley, R. A., Paasivuo, R., and Lehto, V.-P. (1984). *J. Cell Biol.* **99,** 1083–1091.

Vogl, A. W. (1988). *Anat. Rec.* **222,** 34–41.

Vogl, A. W., and Soucy, L. J. (1985). *J. Cell Biol.* **100,** 814–825.

Vogl, A. W., Lin, Y. C., Dym, M., and Fawcett, D. W. (1983a). *Am. J. Anat.* **168,** 83–98.

Vogl, A. W., Linck, R. W., and Dym, M. (1983b). *Am. J. Anat.* **168,** 99–108.

Vogl, A. W., Soucy, L. J., and Lew, G. J. (1985). *Anat. Rec.* **213,** 63–71.

Vogl, A. W., Grove, B. D., and Lew, G. J. (1986). *Anat. Rec.* **215,** 331–341.

Volk, T., and Geiger, B. (1986). *J. Cell Biol.* **103,** 1441–1450.

Walt, H., and Armbruster, L. (1984). *Cell Tissue Res.* **236,** 487–490.

Weber, J. E., and Russell, L. D. (1987). *Am. J. Anat.* **180,** 1–24.

Weber, J. E., Russell, L. D., Wong, V., and Peterson, R. N. (1983). *Am. J. Anat.* **167,** 163–179.

Weber, J. E., Turner, T. T., Tung, K. S. K., and Russell, L. D. (1988). *Am. J. Anat.* **182,** 130–147.

Weeds, A. (1982). *Nature (London)* **296,** 811–816.

Welch, J. E., and O'Rand, M. G. (1985). *Dev. Biol.* **109,** 411–417.

Studies on Scaffold Attachment Sites and Their Relation to Genome Function

S. M. Gasser, B. B. Amati, M. E. Cardenas, and
J. F.-X. Hofmann

*Swiss Institute for Experimental Cancer Research (ISREC), CH-1066
Epalinges s/Lausanne, Switzerland*

I. Introduction

The DNA of a eukaryotic cell must be compacted roughly 250-fold in length to fit into the interphase nucleus, and roughly 8000- to 10,000-fold to accommodate the condensed structure of mitotic chromosomes (Paulson, 1988). The core histones alone are able to organize DNA into the 11-nm nucleosomal fiber, and in association with histone H1 this fiber coils into a 30-nm fiber, allowing for an overall compaction of ~40-fold (for solenoid model see Finch and Klug, 1976). The next level of organization is thought to come about through the binding of nonhistone proteins that bind genomic DNA at dispersed sites, and fold the chromatin fiber into individual looped domains, which could then further coil, or condense upon themselves (Paulson and Laemmli, 1977; Marsden and Laemmli, 1979; Rattner and Lin, 1985). This review will focus on the evidence for loops in the genome, and on the potential function of this higher-order folding.

The loop hypothesis for the organization of DNA arose largely from observation, with the electron microscope, of loops emanating from histone-depleted chromosomes and nuclei (Cook and Brazell, 1975, 1976; Benyajati and Worcel, 1976; Paulson and Laemmli, 1977; Vogelstein *et al.*, 1980). It was clear from sedimentation studies that the DNA was constrained in histone-depleted nuclei and metaphase chromosomes, and that that constraint required nonhistone proteins (Adolph *et al.*, 1977; Hancock and Hughes, 1982). The observation that halo size changed upon addition of ethidium bromide demonstrated that the individual loops could become supercoiled, and thus were topologically independent domains.

The proteinaceous structures that remained after membrane and histone extraction were variously called scaffolds (Paulson and Laemmli, 1977; Adolph, 1980), nuclear matrices (Berezney and Coffey, 1974), cages (Cook and Brazell, 1980), nucleo- or karyoskeletons (Miller *et al.*, 1978; Hancock, 1982), or the nuclear matrix–pore complex–lamina fraction (NMPCL; Fisher *et al.*, 1982). As we will describe, the composition of these various substructures depends largely on the protocol

57

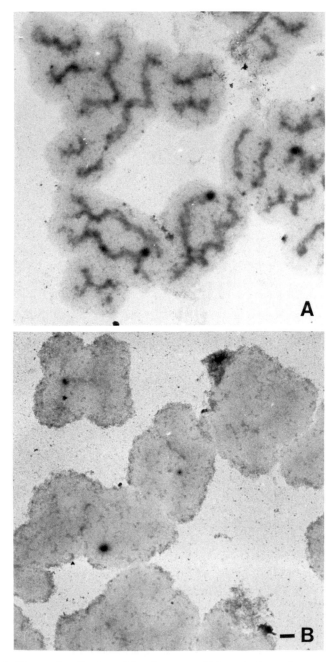

FIG. 1. Immunolocalization of topoisomerase II in metaphase chromosomes. (A) The antibody directed against human topoisomerase II (Sc-1 protein, M_r 170,000) identifies an axial element that extends the length of the chromatid of histone-depleted metaphase chro-

used for nuclear isolation and for histone depletion. Only in the case of the nuclear lamina is it clear that the proteins that are isolated as a nuclear substructure actually form an analogous structure in the cell (reviewed in Franke, 1987; Gerace and Burke, 1988; Nigg, 1989). Since the various nuclear subfractions are operationally defined, we will first describe what we mean by each of these terms. The influence of different parameters of isolation will be discussed in more detail in Section II.

A. THE METAPHASE CHROMOSOMAL SCAFFOLD

The metaphase chromosomal scaffold is the least complex structure in terms of protein composition, largely because the metaphase chromosome itself contains far fewer nonhistone proteins than the interphase nucleus (Lewis and Laemmli, 1982). The scaffold results from a two-step operation: (1) the stabilization of the isolated chromosomes by exposure to Cu^{2+} or Ag^+ at 4°C, and (2) extraction of histones and many nonhistone proteins by 2 M NaCl, dextran sulfate–heparin, or low concentrations of lithium 3′,5′-diiodosalicylate (LIS) in a low- or medium-salt buffer. The metal stabilization can be eliminated if the chromosomes are isolated in the presence of Mg^{2+}, rather than in buffers containing polyamines and ethylenediamine tetraacetate (EDTA).

In the case of metaphase scaffolds the means of histone extraction (whether salt, polyanion, or LIS) appears to have little or no influence on the complexity of the remnant structure. In all cases, a protein of 170,000 MW is the major component of the metaphase scaffold from HeLa chromosomes (Lewis and Laemmli, 1982; Gasser et al., 1986). This protein, originally called Sc-1, has now been unequivocally identified as the enzyme topoisomerase II (Earnshaw et al., 1985; Gasser et al., 1986). Immunolocalization of topoisomerase II to an axial core of the unextracted metaphase chromosome, exending from tip to tip of the chromatid, supports the idea that the scaffold reflects an organizing structure that exists in untreated chromosomes (Fig. 1) (Earnshaw and Heck, 1985; Gasser et al., 1986). Besides topoisomerase II, metaphase scaffolds contain a number of less abundant high molecular weight proteins in the range from 90,000 to 135,000 (Lewis and Laemmli, 1982).

---

mosomes from HeLa cells (Gasser et al., 1986). A peroxidase-labeled secondary antibody was used in the staining reaction, with both immune (A) and preimmune (B) sera. (B) Control panel showing slight staining of kinetochores but not the axial element. The immunostaining was done by T. Laroche in the laboratory of U. K. Laemmli.

B. Nuclear Matrices

The nuclear matrix generally refers to an interphase nuclear structure that is resistant to extraction by nonionic detergent and high concentrations of NaCl. One major point of variation in different preparations of nuclear matrices is the introduction of a nuclease digestion step of the isolated nuclei prior to histone extraction. The digestion is generally done in the presence of Mg^{2+} or Ca^{2+} at 37°C; this temperature and certain divalent cations induce many otherwise soluble proteins to become insoluble (Section II). For the purposes of this review we will use the terminology introduced by Lebkowski and Laemmli (1982a,b) of matrix types I and II, which allows differentiation between stabilized and nonstabilized structures.

Type I nuclear matrices result from extraction of histones with 1 or 2 *M* NaCl after a stabilization step that involves a brief exposure to Cu^{2+}, Ca^{2+}, or 37°C, or combinations of divalent cations and heat. Type I matrices have a complex protein composition, and constrain DNA in a fast-sedimenting structure (15,000S: Lebkowski and Laemmli, 1982a). Electron microscopy of these structures reveals a well-defined nucleolar structure immersed in a fibrous internal network of proteins, which is surrounded by the nuclear lamina (Lewis *et al.*, 1984; Jackson and Cook, 1988). The term "nuclear matrix" alone usually refers to type I matrices.

Type II matrices result from histone extraction of isolated nuclei that have never been exposed to oxidizing conditions, heavy metals, or temperatures >30°C. Morphologically these appear as empty spheres, whose shape is defined by the nuclear lamins and pore complex proteins that form a lattice just inside the nuclear membrane. Type II matrices have a relatively simple protein pattern, consisting largely of the nuclear lamins and pore complex proteins (Lebkowski and Laemmli, 1982b; Hancock and Hughes, 1982). The presence of a residual nucleolar structure depends largely on whether or not the nuclei were treated with RNase prior to extraction (Hancock, 1982; Bouvier *et al.*, 1982; Lewis *et al.*, 1984). Genomic DNA is constrained, but as a slow-sedimenting structure (8500S), with a larger halo of DNA. The difference in sedimentation velocity and hence apparent loop size between type I and type II matrices suggests that there are two levels of protein–DNA interaction: one with components of the nuclear interior (the matrix or scaffold), and another with the nuclear lamina or lamin-associated proteins (Lebkowski and Laemmli, 1982b; Hancock, 1982).

C. The Nuclear Scaffold

In 1984, a protocol was developed for the isolation of a nuclear scaffold, which entailed stabilization of nuclei by a brief exposure either to Cu^{2+} or to 37°C, followed by extraction of histones by LIS (Mirkovitch *et al.,* 1984). The latter is a detergentlike salt that can be used at low concentrations in either low-salt or physiological salt buffers to solubilize efficiently histones and other nonhistone proteins. The resulting scaffold has a complex protein pattern, like the nuclear matrix type I, but maintains specific interactions with dispersed sequences in the genome; these were called scaffold-attached regions (SAR). With this method consistent results have been obtained for mapping precisely the sites that presumably form the bases of the DNA loops observed in extracted chromosomes and nuclei (Sections IV and V).

II. Parameters That Affect Scaffolds and Matrices

As mentioned before, the complexity and character of the nuclear matrix or scaffold depends largely on the conditions of isolation. Critical examination of several steps in these protocols very often led to the opinion that artifactual results had been produced. However, some of these residual structures bind a specific class of related DNA fragments and retain proteins important for nuclear functions. Hence, it seemed likely that a critical study of such extracted nuclei would nonetheless shed light on the role of DNA compaction and organization in the regulation of transcription and DNA replication. This section discusses the influence of some physicochemical parameters on the isolation of nuclei and nuclear substructures.

A. Temperature

A number of reports in the literature have shown that the composition and structural stability of karyoskeletal protein-enriched fractions isolated from eukaryotic cells can be profoundly affected either by brief incubation of nuclei at moderately elevated temperatures *in vitro* (Lebkowski and Laemmli, 1982a,b; Evan and Hancock, 1985; McConnell *et al.,* 1987) or in response to heat shock conditions *in vivo* (Littlewood *et al.,* 1987; McConnell *et al.,* 1987).

Izaurralde *et al.* (1988) tested *in vitro* the effect of temperature on the isolation of *Drosophila* and rat liver scaffolds. In the scaffold isolation procedure nuclei are usually extracted with LIS after the nuclei have been stabilized for 20 minutes at 37°C. If the stabilization step is omitted, scaffold structures are obtained that have a very simple protein pattern, almost exclusively composed of the three lamin proteins A, B, and C (equivalent to matrix type II). Incubation of nuclei at 37°C (*Drosophila*) or 42°C (rat liver) before extraction increases the protein complexity of the scaffolds, and only these complex structures bind specific *Drosophila* DNA fragments (SAR). This shows that different species seem to have slightly different temperature optima (corresponding to heat shock conditions) for stabilizing the proteins involved in specific DNA binding.

Similarly, studies on human embryonic kidney cells (Littlewood *et al.*, 1987) showed that the complex type I matrices could be isolated from kidney cell nuclei after incubation at 37°C. These authors went on to show that identical results were obtained by incubation of intact cells for 1 hour at 40.5°C prior to nuclear isolation, as long as cells did not recover for >2.5 hours at 37°C after the heat shock treatment. Cofractionation of c-myc and v-myc proteins with the matrix was observed after either *in vivo* or *in vitro* heat shock (Evan and Hancock, 1985; Littlewood *et al.*, 1987). The induction of this insoluble state occurs in a temperature- and time-dependent fashion, and was not inhibited by the presence of dithiothreitol (DTT) and EDTA, which argues against the involvement of disulfide bonds or protein–metal interactions. These results also imply that the *in vitro* complex is related to a phenomenon that occurs in intact cells.

Despite this correlation, the mechanism by which the heat shock complex is formed is unclear. One possibility is that hydrophobic regions of proteins become exposed at elevated temperatures, causing them to aggregate. Treatment of cells with amino acid analogs, arsenite, or hypoxia are also capable of inducing the formation of nuclear complexes, albeit of somewhat different composition (Littlewood *et al.*, 1987). Notably all conditions that result in stabilized nuclear substructures are conditions inducing stress response in the cell.

McConnell *et al.* (1987) have also studied the thermal stabilization of nuclear proteins in embryos of *Drosophila melanogaster*. Their results indicated that the association of two lamin proteins (MW 74,000 and 76,000, respectively), of the nuclear pore glycoprotein gp188, and of topoisomerase II with the NMPCL fraction are temperature-dependent phenomena. Either the exposure of nuclei to 37°C for increasing periods of time or an extended incubation on ice stabilized these proteins with the matrix. This is in contrast to $p62^{c-myc}$, which remains soluble after incubation on ice (Evan and Hancock, 1985).

Similar heat-stabilized structures can be isolated from nuclei of budding and fission yeasts, by incubation at 37°C in the presence of 1 mM Cu^{2+} (Amati and Gasser, 1988; S. M. Gasser, unpublished observations). Scaffolds could also be isolated from *in vivo* heat-shocked yeast cells that showed similar DNA-binding specificity.

B. Metals and Reducing Agents

One of the most convincing studies on metalloprotein interactions in chromosomes and their implications in the structure of metaphase scaffolds was done by Lewis and Laemmli (1982). The authors show that histone-depleted chromosomes sediment more slowly in a sucrose gradient after addition of 2-mercaptoethanol or hydrophobic chelators. This appears not to be due to a reduction of chromosomal proteins, but to result from metal depletion, since $NaBH_4$, a strong reducing agent, had no effect on the sedimentation behavior of chromosomes.

Metal-depleted, slowly sedimenting chromosomes could be restored into the compact, fast-sedimenting structures by addition of low concentrations of copper (10^{-8} M). Other metals did not show this effect, except for Ca^{2+} (10^{-4} M), whose stabilization was less specific. This effect was reversible by chelators under nitrogen atmosphere and very likely due to an interaction of Cu^{2+} with sulfhydryl (SH) groups, because metal-depleted chromosomes could not be "rescued" by Cu^{2+} when previously treated with Hg^{2+} or iodoacetamide.

Van Straaten and Rabbitts (1987) have similarly demonstrated that c- and v-myc can be stabilized in the interphase scaffold by Cu^{2+} interaction. These proteins were not recovered with nuclear matrices that were incubated with the chelator 8-hydroxyquinoline prior to Cu^{2+} addition, suggesting that the link of c-myc with the nuclear matrix is Cu^{2+}-mediated. Although c-myc was also matrix-associated after a mild oxidation of isolated nuclei with dithiopyrimidine, the mechanism of association of c-myc with the nuclear matrix appears to be different depending on the stabilization step. In the Cu^{2+}-mediated stabilization the nuclear matrix-associated c-myc is recovered in nonreducing gels as a single polypeptide of 63 kDa, whereas in the oxidant-stabilized matrix it is recovered as a large complex that fails to migrate into the gel. The observation that chelators can prevent but not reverse the copper stabilization suggests that certain nuclear proteins bind Cu^{2+} very tightly, and that the presence of the ligand alters their solubility. In this case, however, Cu^{2+} does not induce disulfide crosslinking.

With respect to the interphase nucleus, Dijkwel and Wenink (1986) have proposed that thiol reagents have both a reducing effect on the nuclear lamina and a chelating effect on the internal matrix. These authors, like

Lebkowski and Laemmli (1982a), propose that chromosomal DNA is attached to the nuclear matrix both internally and peripherally; they propose that disulfide bonds might stabilize the nuclear lamina and divalent cations the internal matrix. Since the effects of thiol alkylating reagents on the composition of the nuclear matrix has been reviewed extensively by Kaufmann *et al.* (1986a), it will not be discussed further here.

Little is known about the exact mechanism of Cu^{2+} stabilization, although the observed results suggest that some nuclear proteins have a very high affinity for this metal. Conservation of some features of metal-binding domains in nucleic acid-binding proteins have been proposed (Berg, 1986), and a copper-binding transcription factor that meets these proposed requirements has been described (Fürst *et al.*, 1988). This factor activates transcription only in the presence of copper and becomes more resistant to proteolysis after addition of the metal. Examination of solubilized scaffold proteins and their binding affinities to copper in relation to their DNA or protein interactions might reveal the mechanism of copper stabilization.

Regarding the results mentioned earlier, it is remarkable that heat stabilization, applied to nuclei prior to scaffold isolation, can be replaced by a stabilization with Cu^{2+} to produce the same results in experiments that map scaffold-attached DNA (Mirkovitch *et al.*, 1984). Consistently, the distribution of certain proteins into scaffold or matrix and soluble fractions seems not to vary with different stabilization procedures, as exemplified by topoisomerase II (Lewis and Laemmli, 1982, Gasser *et al.*, 1986) and c-myc (Littlewood *et al.*, 1987; Van Straaten and Rabbitts, 1987). These results suggest that some of the interactions induced by metals can be mimicked by heat shock conditions.

C. CONDITIONS OF HISTONE EXTRACTION

Polyanions were used by Adolph *et al.* (1977) and Lewis and Laemmli (1982) to remove histones from HeLa metaphase chromosomes. Histone-depleted chromosomes isolated on a sucrose gradient contained a highly folded DNA chain and <0.1% of the original histone content. These structures were stable in gradients containing 2 *M* NaCl, which had previously been shown to dissociate histones from DNA (Spelsberg and Hnilica, 1971), thereby showing that histones did not contribute structurally to the complex.

Identical results were obtained for the sedimentation of histone-depleted metaphase chromosomes and for the scaffolding-protein pattern when either low-salt buffers containing dextran sulfate and heparin, or high-salt extraction buffers were used (Paulson and Laemmli, 1977; Lewis and Laemmli, 1982). Extraction occurs at low ionic strength by competition of the negatively charged polymers for proteins associated with DNA, while in 2 M NaCl extraction occurs primarily by dissociation of electrostatic interactions.

In the case of interphase nuclei, however, there is evidence that the method of histone extraction may generate artifacts. In particular, high salt concentrations (2 M NaCl) appear to result in the precipitation of transcription complexes onto the nuclear matrix (Kirov *et al.*, 1984; Mirkovitch *et al.*, 1984; Roberge *et al.*, 1988). In salt-extracted matrices it was not possible to map closely specific sites in the genomic DNA that bind to the resulting structure; rather, transcribed regions were found associated with the matrix in a transcription-dependent manner (reviewed in Jackson, 1986).

In an attempt to avoid artifactual precipitation, low-salt buffers were used in the nuclear extraction step during scaffold isolation (Mirkovitch *et al.*, 1984). These buffers contain the detergentlike reagent LIS, which had been used previously in the isolation of glycoproteins from cell membranes (Marchesi and Andrews, 1971). This reagent is extremely efficient at "salting in" or solubilizing proteins (Robinson and Jencks, 1965), and at concentrations <25 mM it efficiently dissociates nucleosomes and other protein complexes. In scaffold preparations LIS is usually used in conjunction with a low level of a nonionic detergent such as digitonin, which enhances both the solubility of LIS itself and the efficiency of extraction of membrane proteins and histones. While these conditions result in the release of polymerases (see later), they allow a highly specific scaffold–DNA interaction.

III. Specific Protein Components of Scaffolds and Matrices

Over the last few years an increasing number of nuclear proteins have been identified as forming part of the nuclear matrix or scaffold. Verheijen *et al.* (1988) have reviewed in detail the matrix association of the major components of the nuclear pore complex, the lamina and nucleolus, as well as nuclear actin and proteins involved in RNA processing and transport. Functional nuclear proteins that have been linked to the stabilized nuclear matrix include DNA polymerases α and β (Nishizawa *et al.*, 1984; Smith *et al.*, 1984; Foster and Collins, 1985; Tubo and Berezney, 1987a,b), DNA methylase (Burdon *et al.*, 1985; Tubo and Berezney, 1987a,b), DNA primase (Wood and Collins, 1986; Tubo and Berezney, 1987c), ribonuclease H (Tubo and Berezney, 1987a), topoisomerase II (Berrios *et al.*, 1985; Berrios and Fisher, 1988), RNA polymerases (Lewis *et al.*, 1984; for an alternative view see Roberge *et al.*, 1988), and hormone receptors (Kaufmann et al., 1986b; Kaufmann and Shaper, 1986, reviewed in Nelson et al., 1986).

In the following paragraphs we will focus on a few recent studies of nuclear proteins that we think may be particularly pertinent to the question of chromatin organization. Among these are the proteins that are common to both interphase and metaphase scaffolds (topoisomerase II and p62). We also discuss RNA polymerases I and II because transcription has been proposed to be matrix-associated, c-myc because it is thought to function in DNA replication, and the abundant yeast nuclear protein repressor–activator binding protein (RAP-1) because it is the only protein to date purified by its affinity to scaffold-attached DNA sequences.

A. PROTEINS COMMON TO METAPHASE AND INTERPHASE SCAFFOLDS

1. *Topoisomerase II*

Topoisomerase II has been unambiguously identified as the major component of the metaphase scaffold from chick (Earnshaw et al., 1985) and HeLa cells (Gasser et al., 1986). Its abundance (roughly three copies per 70-kb loop of DNA), and the immunolocalization of the enzyme to an axial core of unextracted chromosomes are consistent with a role in stabilizing the bases of DNA loops in metaphase chromosomes (Fig. 1) (Gasser et al., 1986; Earnshaw and Heck, 1985). Evidence that the enzyme is important for proper chromosome condensation as well as chromatid segregation was provided by an elegant study in *Schizosaccharomyces pombe* making use of conditional mutants in topoisomerase II and β-tubulin (Uemura et al., 1987). By combining cold-sensitive mutants in topoisomerase II and β-tubulin, they were able to show clearly that topoisomerase II was required for chromosome condensation, as well as for the proper decatenation of replicated chromosomes prior to segregation (DiNardo et al., 1984; Uemura and Yanagida, 1984, 1986; Holm et al., 1985).

Topoisomerase II is also a component of the heat-stabilized interphase scaffolds or matrices from *Drosophila* embryos (Berrios et al., 1985), *Drosophila* tissue culture cells (Mirkovitch et al., 1988), and yeast (Berrios and Fisher, 1988; J. F. X. Hofmann and S. M. Gasser, unpublished observations). If interphase nuclei from *Drosophila* cells are heat-stabilized, 60–70% of topoisomerase II is resistant to extraction with LIS. In unstabilized nuclei, between 10 and 15% remains associated with the resulting type II matrices or scaffolds (Mirkovitch et al., 1988). Since topoisomerase II is known to be a target for protein kinases and poly(ADPribose) polymerase *in vitro*, this distribution may reflect differently modified subpopulations of the enzyme, which are perhaps found in different complexes in the nucleus (reviewed in Wang, 1985).

Additional evidence suggesting that topoisomerase II has a structural role in both interphase and metaphase scaffolds, comes from the mapping of DNA regions that specifically bind the scaffold. These SAR were found to contain a significant enrichment of the putative consensus sequence for topoisomerase II cleavage published by Sander and Hsieh (1985). Moreover, Udvardy *et al.* (1985) have shown that topoisomerase II cleaves preferentially in two of these scaffold-bound fragments (those at the histone and *hsp70* loci of *Drosophila*) *in vitro,* and in some cases also in intact cells treated with drugs that specifically inhibit the religation action of topoisomerase II (Udvardy *et al.,* 1986; Rowe *et al.,* 1986).

While the evidence implicating topoisomerase II in the maintenance of metaphase chromosomal structure is quite strong, it is important to note that the level of topoisomerase II appears to vary greatly with the proliferative capacity of the cell. Differentiated tissues, such as chick myotubes, appear to have <300 copies of topoisomerase II per cell (Heck and Earnshaw, 1986). If topoisomerase II does have a role in the organization of interphase chromatin, it is clearly not the only element involved. One could imagine that as cells stop dividing and heterochromatin levels increase, the internal scaffolding might lose importance and be replaced by a less dynamic interaction of condensed heterochromatin with the nuclear periphery (see model, Fig. 2).

2. *p62*

Kaufmann and Shaper have developed a method for the preparation of nuclear matrices from rat liver nuclei that are stabilized by a SH-crosslinking reagent, sodium tetrathionate (Kaufmann and Shaper, 1984). By using this new approach they have identified a 62-kDa nonlamin, nonhistone component of the nuclear matrix. A polyclonal antibody raised against the 62-kDa protein was used in cellular immunolocalization studies (Fields and Shaper, 1988). This protein shows an exclusively nuclear distribution. In interphase hepatocytes it was concentrated at the nuclear periphery and distributed less densely along the chromatin in the nuclear interior. During mitosis the 62-kDa protein remains tightly associated with the condensed chromatin, and fractionates with a crude metaphase scaffold preparation. Its tight association with metaphase chromosomes and scaffolds led to the proposal that it is involved in maintenance of chromosomal structure (Fields and Shaper, 1988), and clearly distinguishes it from a 62-kDa pore complex protein identified by Blobel and co-workers (Davis and Blobel, 1986).

B. RNA POLYMERASES

In 1984, Lewis *et al.* observed that RNA polymerase II was quantitatively recovered in nuclear matrices from polyamine-isolated nuclei that

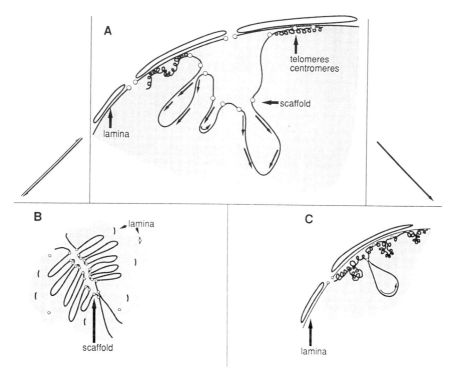

FIG. 2. Model of chromosome organization with respect to the scaffold and nuclear lamina, showing hypothesized chromatin organization in (A) interphase cells, (B) metaphase cells, and (C) terminally differentiated, nondividing cells. DNA is indicated by a solid black line that is folded into loops. Coding sequences of transcribed genes are indicated by arrows along the DNA fiber, and sites of association with nonhistone proteins of the nuclear scaffold are indicated by open circles. Open circles also indicate proteins of the nuclear pore structure which, in some instances, copurify with the scaffold. Condensed inactive chromatin is represented by a coiled DNA fiber, indicating that this DNA may be tightly compacted into a 30-nm solenoid fiber. Telomeres and centromeres of higher eukaryotic cells are also maintained as compact structures, probably associated with the nuclear periphery. This may be achieved by binding the nuclear lamina directly or through an intermediary protein. Active regions are held in a more expanded form by proteins of the nuclear interior, some of which may be scaffold components (open circles). (B) In metaphase the nuclear envelope breaks down into small membrane vesicles and the lamina depolymerizes. We propose that the metaphase chromosome organization is in part maintained by the same scaffold proteins that organize interphase chromatin, for example, topoisomerase II. The loops are shown here decondensed, although in mitosis they would be coiled into 30-nm fibers that may fold again upon themselves. (C) Chromatin in nondividing or terminally differentiated cells. Since topoisomerase II and an internal matrix structure appear to be largely absent from these nuclei, we propose that the majority of the DNA is condensed in inactive domains, associated with the nuclear periphery (see coils). The genes that are expressed in such cells may have an open conformation maintained by scaffold proteins associated with the nuclear lamina. Actively cycling cells such as tissue culture cells may only vary between the types of organization seen in A and B.

had been stabilized with Cu^{2+} and extracted with 2 M NaCl (type I). In contrast, this polymerase was absent from type II matrices (see earlier) and metaphase scaffolds. In a more extensive study, Roberge *et al.* (1988) have examined the distribution of active and inactive polymerases I and II identified by a photoaffinity-labeling technique, between nuclear scaffold, matrices, and the corresponding soluble fractions. The studies were done on matrices prepared by high-salt and medium-salt extractions and on scaffolds isolated with 25 mM LIS. This analysis showed that <10% of the RNA polymerases I and II were associated with the LIS-extracted nuclear scaffolds. Type I matrices prepared by extraction with 2 M NaCl retained a significant proportion of the polymerases, yet this association could be avoided by washing the nuclei at intermediate salt concentrations before the high-salt extraction. Because 0.4 M KCl resulted in the release of the RNA polymerases, the authors concluded that they are not integral components of either nuclear matrices or scaffolds. These results conflict with a model for RNA transcription that envisons the nuclear matrix as an anchoring framework for RNA polymerases, with the transcribed DNA being reeled through the complex (for reviews see Nelson *et al.*, 1986; Jackson, 1986).

Fisher *et al.* (1989) came to a similar conclusion. They found that RNA polymerase II was recovered with the nuclear matrix after heat shock of intact *Drosophila* embryos, as well as after incubation of isolated nuclei at 37°C. Association with the high-salt matrix thus correlates with inactivation of the enzyme, arguing against a model in which transcription occurs at the nuclear matrix. These results are consistent with the finding of SAR in nontranscribed sequences (see Section IV).

C. c- AND v-myc

The mechanisms involved in the association of c-myc and v-myc to the nuclear matrix have been studied in detail (Evan and Hancock, 1985; Littlewood *et al.*, 1987; Van Straaten and Rabbitts, 1987). Evan and Hancock (1985) described a discrete subset of nuclear proteins that includes c- and v-myc, and that becomes insoluble and irreversibly linked to the nuclear scaffolds upon incubation of isolated nuclei at temperatures of ≥35°C. Van Straaten and Rabbitts (1987) observed that c-myc cofractionates with the nuclear matrix that has been stabilized with Cu^{2+}. While it is not clear how either heat or copper induces insolubilization, in the case of c-myc heat apparently mimics the effect of Cu^{2+}, as mentioned earlier. The correlation of matrix-bound c-myc and DNA replication will be discussed later (Section VI).

D. Nuc2 Protein

Using a genetic approach Hirano et al. (1988) have characterized a temperature-sensitive mutant, nuc2-663, in S. pombe, which is defective in anaphase spindle movement. At the restrictive temperature these mutants arrest in metaphase—that is, their chromosomes remain condensed—yet do not undergo the characteristic chromosome separation of early anaphase. The gene nuc2 was cloned and antibodies against a lacZ–nuc2$^+$ fusion protein have identified a protein of 67 kDa in wild-type S. pombe extracts. Based on its resistance to extraction by 2 M NaCl, 25 mM LIS, or 1% Triton X-100, this polypeptide has been identified as a component of the nuclear scaffold or matrix fraction. Interestingly, it is part of the LIS- and salt-insoluble pellet even in the absence of heat or copper stabilization. At nonpermissive temperature a soluble polypeptide of 76 kDa, which appears to be a precursor to the 67-kDa polypeptide, accumulates in the nuc2 mutant.

These observations suggest that the 67-kDa nuc2 protein is essential for anaphase spindle movement, and that it may be analogous to kinetochore proteins (CEN-B, CEN-C). These proteins may promote the association of the kinetochore with centromeres, and were identified as components of the metaphase scaffold of human chromosomes (Earnshaw et al., 1984).

E. Repressor–Activator Binding Protein-1 (RAP-1)

In order to purify proteins from the yeast nuclear scaffold, Hofmann et al. (1989) have developed a procedure to solubilize the scaffold fraction in urea, remove DNA by centrifugation, and then renature the proteins to form a soluble extract. This scaffold extract has been used for gel retardation assays with several SAR DNA probes, and for biochemical purification of the proteins involved in SAR binding. By monitoring the gel retardation of the HMR-E silencer and of deletion derivatives (Brand et al., 1987), it could be shown that the abundant nuclear factor RAP-1 (Shore and Nasmyth, 1987) is a component of the heat-stabilized interphase scaffold. Moreover, it appears to play a key role in the silencer–scaffold interaction. Immunoblotting of scaffold-bound and scaffold-released proteins confirmed that >90% of RAP-1 is recovered in the stabilized nuclear scaffold. The factor RAP-1 (also termed TUF: Huet et al., 1985; or GRF-1: Buchman et al., 1988a) not only binds to the cis-acting silencers that flank the silent mating-type genes (Shore and Nasmyth, 1987), but also to telomeres, and numerous upstream activating sequences involved in transcriptional regulation (Buchman et al., 1988a,b). The occurrence of RAP-1-binding sites in both regulatory and structural DNA elements suggests that it may indeed be important in the proper organization of chromatin.

We have chosen these proteins to illustrate a few of the approaches that may yield useful results for identifying proteins involved in chromatin folding. Certainly further studies are needed to establish their roles in the intact cell. The rest of this review will focus on the DNA fragments that bind the scaffold or matrix, which presumably define the bases of DNA loops.

IV. Specific Scaffold-Associated DNA Regions

As a test for the loop model one might hope to find specific regions spaced along the DNA at which interaction with the scaffold occurs. Such specific SAR have been identified with the help of the LIS extraction procedure (Mirkovitch *et al.*, 1984). As described before, LIS extracts histones and other nuclear proteins. After repeated washing, the extracted nuclei are digested to completion with restriction enzymes. A restriction fragment that contains a scaffold-binding site will sediment with the scaffold pellet, while an unbound "loop" fragment will be recovered in the supernatant (Fig. 3). This procedure was originally developed for *Drosophila* cells (Mirkovitch *et al.*, 1984, 1986, 1988; Gasser and Laemmli, 1986a,b) but has since been applied to nuclei from mouse cell lines (Cockerill and Garrard, 1986a; Cockerill *et al.*, 1987; Bode and Maass, 1988). Chinese hamster ovary (CHO) cells (Käs and Chasin, 1987; Dijkwel and Hamlin, 1988), chicken cells (Phi-van and Stratling, 1988), human cells (Jarman and Higgs, 1988; Sykes *et al.*, 1988), and yeast (Amati and Gasser, 1988; Conrad and Zakian, 1988). The most extensive characterization of SAR has been made in *Drosophila,* with nearly 20 SAR mapped and over 450 kb of the genome screened for scaffold-binding sites (reviewed in Gasser and Laemmli, 1987). Many of the characteristics conserved among *Drosophila* SAR fragments have been observed in other systems as well. In the following discussion we will stress these conserved features and use examples other than *Drosophila* whenever possible. The results obtained with yeast will be presented separately.

In brief, SAR appear to have the following characteristics:

1. Scaffold-binding sites can be mapped to fragments ranging from 300 to 1000 bp, and these all appear to contain multiple sites of protein–DNA interaction.
2. Generally SAR are A/T-rich and contain recognizable motifs.
3. The fragments bind reversibly to scaffolds and the binding sites are saturable.

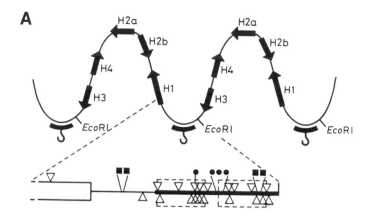

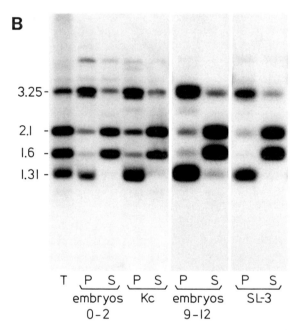

FIG. 3. Histone gene cluster: one repeat, one loop, invariant in all cell types tested. (A) Two repeats of the 5-kb DNA fragment containing the *Drosophila* histone genes. The 657-bp scaffold-attached region (SAR) occurs in the nontranscribed spacer between the *H1* and *H3* genes, and is indicated by a bar containing a hook (Mirkovitch *et al.*, 1984). Since there are roughly 100 tandem histone repeats in the genome, they would form a series of small loops. When LIS-extracted nuclei are digested with *Eco*RI, *Hin*DIII, and *Xho*I, the 657-bp SAR is contained within a 1.31-kb fragment, shown enlarged below the loops. Sequence motifs common to a number of SAR are indicated in the enlarged map. ▽, Sequences with >70% homology to the topoisomerase II consensus sequence (Sander and Hsieh, 1985), in either the Watson (top) or Crick (bottom) strand; ●, the 10-bp A-box; ■, the 10-bp T-box

4. The SAR are found exclusively in noncoding sequences. Generally they are also restricted to nontranscribed regions, with the exception of several mapping to introns.
5. The distances between SAR vary from <3 kb to 112 kb, and one or several genes can occur between two SAR. The largest region with no scaffold attachment sites mapped to date is 140 kb, surrounding the human α-globin locus.
6. In several instances SAR were found to flank one or a pair of coordinately regulated genes and to coincide with the boundaries of the nuclease-sensitive domain that arises upon activation of the gene.
7. Some SAR map very close to, or coincide with enhancerlike regulatory elements.

A. PRESENCE OF TOPOISOMERASE II CONSENSES AND A/T-RICH MOTIFS IN SAR

One of the best-characterized SAR is found on a 657-bp fragment upstream of the histone H1 gene in *Drosophila*. Exonuclease III digestion showed that this fragment contains two protein-binding regions of roughly 200 bp each, separated by ~100 bp (Gasser and Laemmli, 1986b) (Fig. 3A). Each half is able to bind the nuclear scaffold, although it is not known if the same proteins bind each site.

The presence of topoisomerase II in both nuclear and metaphase scaffolds prompted the screening of this and other SAR sequences for the putative cleavage consensus for topoisomerase II (Sander and Hsieh, 1985). Sequences that show >75% homology to the loose 15-bp consensus were found clustered in the SAR fragments relative to non-SAR regions (Fig. 3A) (Gasser and Laemmli, 1986a,b). Remarkably, a sequence highly homologous to this *Drosophila* cleavage site was also present twice in the matrix-attached region (MAR) of the immunoglobulin κ (*Igκ*) gene in mouse cells (Cockerill and Garrard, 1986a), and has since been recognized

(see text). Two 200-bp domains within the SAR (here encompassed by dashed lines) were resistant to exonuclease III digestion in intact scaffolds, indicating the presence of protein–DNA complexes (Gasser and Laemmli, 1986b). (B) The distribution of fragments from this locus between nuclear scaffolds (P) and the supernatant (S) after digestion of LIS-extracted *Drosophila* nuclei with *Eco*RI, *Hin*DIII, and *Xho*I. Equal amounts of total (T), pellet (P), and supernatant (S) DNA were analyzed on a 1% agarose gel and probed with a nick-translated clone containing all five histone genes. The bands of 2.1 and 1.6 kb contain the genes for histones H3 and H4, and H2A and H2B, respectively. The band at 3.25 kb contains the SAR sequence and derives from a variant repeat that lacks the *Eco*RI site. Nuclei were obtained from either Kc or SL-3 *Drosophila* tissue culture cells, or from embryos collected at 0–2 hours (0–2) or 9–12 hours (9–12) after egg deposition, as indicated.

also in SAR from yeast (Amati and Gasser, 1988), CHO (Käs and Chasin, 1987), and human cells (Jarman and Higgs, 1988). Udvardy *et al.* (1985) were able to show that topoisomerase II cleaves preferentially in the histone and heatshock SAR *in vitro,* which strengthens the significance of these sequence observations.

Clearly not all topoisomerase II cleavage sites fall within SAR fragments, and some human SAR do not contain this consensus (Jarman and Higgs, 1988). The SAR of the chick lysozyme gene were shown to bind scaffolds prepared from chick erythrocytes (Phi-van and Stratling, 1988), which have been reported to have <300 molecules of topoisomerase II per cell (Heck and Earnshaw, 1986). Thus while topoisomerase II may bind SAR, other proteins are likely to be involved in the scaffold–DNA interaction, particularly in interphase nuclei.

In general SAR are 70–75% A/T-rich, but are not highly repetitive elements. There is no cross-hybridization among the fragments, with the exception of a family of scaffold-attached fragments that all contain the middle repetitive X element of *Drosophila* (Lis *et al.,* 1981). The X element-containing restriction fragments found upstream of a number of heat shock genes are all scaffold-bound (Mirkovitch *et al.,* 1984).

A pairwise matrix analysis of the sequences of six *Drosophila* SAR revealed two conserved sequence motifs in addition to the topoisomerase II consensus (Gasser and Laemmli, 1986a), which again have been found in other systems. These are 10-bp sequences: $AATAAA^T/_CAAA$ called the A-box, and $TT^A/_TT^T/_ATT^T/_ATT$ called the T-box. The A-box is often found in several copies as a direct repeat, whereas the T-box occurs less frequently and is distinguished from the A-box by being a more contiguous run of T's. No proteins that bind specifically to these boxes have been purified to date, and the thymidine stretch may simply serve to exclude nucleosome formation. It is important to note, however, that A/T richness does not suffice to form a scaffold attachment site (Gasser and Laemmli, 1986a; Amati and Gasser, 1988; Jarman and Higgs, 1988).

B. Saturable and Reversible Binding to Scaffolds

The interaction between SAR and the nuclear scaffold meets the criteria one would expect from a specific DNA–protein interaction: it is both reversible (Cockerill and Garrard, 1986a; Gasser and Laemmli, 1986a; Izaurralde *et al.,* 1988) and saturable (Izaurralde *et al.,* 1988; Amati and Gasser, 1988). Different SAR show varying affinities for the scaffold, and high-affinity SAR can displace those of lower affinities. In *Drosophila* this was demonstrated by titrating in exogenous, cloned SAR DNA and showing that it displaces endogenous SAR fragments. Plasmid DNA (pBR322) does not displace SAR fragments at even higher concentrations. Exogenously added, end-labeled fragments generally bind with the same specificity as the SAR mapped in endogenous DNA, although minor exceptions are found (Jarman and Higgs, 1988; Dijkwel and Hamlin, 1988).

A comparison of the binding patterns of exogenous DNA with the binding of endogenous fragments has been useful in distinguishing the difference between the salt-extracted matrix and the LIS-extracted scaffold. Using a preparation of salt-extracted nuclei from various mouse cell lines, Cockerill and Garrard (1986a) showed that matrices bind an end-labeled MAR in the presence of nonspecific competitor DNA. The MAR also rebinds to LIS-extracted scaffolds, suggesting that scaffolds and matrices are equivalent at one level of complexity. *Drosophila* and chick SAR fragments also bind salt-extracted matrices specifically (Izaurralde *et al.*, 1988; Phi-van and Stratling, 1988). On the other hand, if endogenous sequences are probed in salt-extracted matrices, actively transcribed regions are found matrix-bound in a transcription-dependent manner (see, e.g., Cook *et al.*, 1982; Robinson *et al.*, 1982; Ciejek *et al.*, 1983; Jost and Sedron, 1984; Kirov *et al.*, 1984; Small and Vogelstein, 1985). In some cases nontranscribed regions that correspond to scaffold-bound regions in LIS-extracted nuclei were also bound (Small *et al.*, 1985). Razin *et al.* (1985) have made similar observations in NaCl-extracted matrices from chick erythrocytes and erythroblasts, and could distinguish experimentally between "transcription-dependent" and "permanent" attachment at the α-globin locus. The permanent attachment sites may correspond functionally to the SAR mapped in LIS-extracted nuclei (Razin *et al.*, 1986; see also Section VI). These results are consistent with the observations of Roberge *et al.* (1988), who suggest that RNA polymerase is precipitated in high salt-extracted matrices, and find it absent from LIS-extracted scaffolds.

C. PROXIMITY OF SOME SAR TO PROMOTERS AND/OR ENHANCERS

For three highly expressed, developmentally regulated loci in *Drosophila* (*Adh, ftz,* and *Sgs-4;* Gasser and Laemmli, 1986a), the 5' SAR map to regions that contain sequences that function as enhancers for high-level expression *in vivo,* as determined by P-element transformation studies (Hiromi *et al.*, 1985; Posakony *et al.*, 1985; McNabb and Beckendorf, 1986). For *Adh,* from which two transcripts are made, two enhancerlike regulatory regions were identified and each remarkably maps to a 5' SAR fragment. The tissue-specific regulatory elements of *Adh* and *ftz,* on the other hand, are not scaffold-attached (Gasser and Laemmli, 1986a). This cohabitation of SAR and enhancerlike regulatory sequences was also observed in the human β-globin gene, in which a SAR that maps to an intron coincides closely with a regulatory element (Jarmen and Higgs, 1988).

In the case of the mouse Igκ and λ chain genes, scaffold-attached domains flanked, but were clearly separable from, the well-characterized enhancer elements (Cockerill and Garrard, 1986a; Cockerill et al., 1987). Similarly, for the Drosophila hsp70 gene a SAR is located immediately upstream of what is apparently necessary and sufficient to allow the heat shock response, including the region with heat shock-specific DNase 1-hypersensitive sites (Wu, 1980; Pelham, 1982; Bienz and Pelhem, 1986). It has been proposed that the SAR (or MAR) might modulate the enhancer activity.

Not all enhancers are scaffold-bound. For example, in the chick lysozyme gene the known enhancer is not at or near a scaffold-binding site (Phi-van and Stratling, 1988), nor is the enhancer for flight muscle actin (locus Act-88F) in Drosophila (Mirkovitch, 1987).

D. ARE SAR TISSUE- OR CELL TYPE-SPECIFIC?

The close proximity of SAR to tissue-specific enhancers raises the question whether these fragments only bind the scaffold when the enhancer is functional. In this case one would expect to find cell type-specific SAR. In Drosophila, mouse, human, and chick systems, cell type-specific differences have been sought but were never observed. The κ light-chain gene MAR was mapped in six mouse cell types that had either the rearranged or the nonrearranged gene, yet no differences were observed (Cockerill and Garrard, 1986a). In Drosophila three different developmentally regulated genes (Adh, ftz, Sgs-4) have been studied in cells from three different stages of development as well as in tissue culture cells. All identified SAR showed nearly identical binding specificity for the various nuclear scaffolds, whether the adjacent gene was expressed or not (Gasser and Laemmli, 1986a).

Despite the consistency of these negative results, it is still possible that in vivo loop organization shows cell type specificity, which the present assays are unable to detect. If histones are necessary to regulate loop organization by forming nucleosomes that block a scaffold-binding site, the current assay for detecting SAR would exclude the possibility of seeing cell type differences. The assays currently in use detect all DNA sequences with potential to bind the scaffold. Along the same line one might argue that perhaps not all the observed attachment sites are actually used in a given cell type, particularly where numerous SAR map within a small region (e.g., human β-globin: Jarman and Higgs, 1988). In vivo footprint data might help to resolve this question.

E. Mapping of Some SAR to the Boundaries of Active Domains

A number of the *Drosophila* genes studied have SAR fragments both 5′ and 3′ of the gene, suggesting that the SAR physically define a functional unit. In general these are highly expressed genes and would be contained in small loops ranging from 5 to 15 kb (Gasser and Laemmli, 1986a; Mirkovitch *et al.*, 1984). In the case of *ftz* and *Sgs-4*, there is corroborative evidence from P-element transformation studies suggesting that transformants containing both the 5′ and 3′ SAR show less position effect on the expression of the transfected gene, than those in which only one SAR is present (Hiromi *et al.*, 1985; McNabb and Beckendorf, 1986). More convincingly, work on the chick lysozyme gene (Phi-van and Stratling, 1988), the human β-interferon (IFN-β) gene (Bode and Maass, 1988), and the human β-globin genes (Jarman and Higgs, 1988) present a clear correlation of scaffold attachment with the boundaries of chromatin domains.

At the chick lysozyme locus a 19-kb domain is bordered by SAR. This region was previously characterized as a "classic" chromatin domain featuring enhanced nuclease sensitivity in cells that express the gene (Stratling *et al.*, 1986). Similarly, SAR are found flanking the human IFN-β and β-globin domains. In the latter case eight SAR have been mapped within the large 90-kB domain, but again both borders coincide with SAR. Moreover, these flanking fragments allow a consistently high-level expression of the β-globin gene upon integration into heterochromatic regions in transgenic mice (Grosveld *et al.*, 1987). A final example consistent with these is that of the silent mating-type loci *HML* and *HMR* in *Saccharomyces cerevisiae*, in which a coordinately regulated pair of genes are flanked by SAR (Hofmann *et al.*, 1989; discussed later).

Nontranscribed regions at the borders of actively transcribed regions are often associated with highly ordered, or static, nucleosome positioning. In several cases mentioned here (IFN-β, β-globin) the SAR elements map to regions of static nucleosome formation, as do the SAR located 5′ of the *hsp*70 genes (loci *87C1* and *87A7*) and in the nontranscribed spacer between the genes of histones H1 and H3 in *Drosophila*. Taken together these results implicate at least a subset of SAR in the definition of chromatin domains.

Not all SAR flank highly expressed and coordinately regulated genes: in the 320 kb surrounding the *Ace* and *rosy* loci of *Drosophila*, there are four attachment sites spaced at distances from 26 to 112 kb (Mirkovitch *et al.*, 1986). From one to seven unrelated transcription units are contained within these larger loops, the majority of which encode low-abundance mRNAs. The most abundant transcript from this region, the *Ace* locus, is closely positioned downstream of a SAR. These results suggest a correlation between potential levels of expression and closely positioned

SAR: the relatively small (5–18 kb) loops seem to be characteristic of highly expressed genes that in fact make up a small portion of the genome. Regions of low expression appear to be organized in larger looped domains.

F. Binding of Interphase SAR by the Metaphase Scaffold

A systematic study of scaffold-associated fragments through the cell cycle has not been done. In *Drosophila*, however, one study has asked whether the SAR fragments identified by association with the interphase scaffold were bound to the scaffold isolated from metaphase chromosomes. Despite a significantly simpler protein pattern in the metaphase scaffold, no major differences in the binding specificity was detected among the seven SAR tested (Mirkovitch *et al.*, 1988). The results with metaphase scaffolds suggest that certain SAR maintain loop organization through the mitotic cell cycle (Mirkovitch *et al.*, 1988). Since the metaphase scaffolds from *Drosophila* contained small amounts of the nuclear lamin and a pore-associated glycoprotein, the study was extended to ask whether the *Drosophila* SAR could bind HeLa mitotic scaffolds. These were devoid of lamin and yet bound three of the six *Drosophila* SAR tested. These results, in conjunction with the observation that type II (unstabilized) nuclear matrices or scaffolds do not bind either the *Drosophila* SAR (Izaurralde *et al.*, 1988) or those at the chicken lysozyme gene (Phi-van and Stratling, 1988), strongly suggests that the nuclear lamina are not directly involved in the observed interaction.

V. SAR Mapping at Functional Elements in *S. cerevisiae*

The budding yeast *S. cerevisiae* offers particularly powerful tools to investigate the structural and functional aspects of chromatin folding, since the major chromosomal elements involved in chromosomal replication, maintenance, and segregation have been identified and cloned (for reviews see Williamson, 1985; Blackburn and Szostack, 1984). These include origins of replication known as autonomously replicating sequences (ARS: Brewer and Fangman, 1987; Huberman *et al.*, 1987), centromeres (CEN), and telomeres (TEL). A considerable number of transcriptional regulatory elements (enhancers and silencers) have been isolated and characterized in this organism. The method for mapping SAR outlined earlier has been adapted to yeast nuclei (Amati and Gasser, 1988). In contrast to higher eukaryotic systems in which the function of SAR is largely unclear, the SAR identified to date in yeast coincide with well-characterized functional elements (summarized in Table I). These include

TABLE I

SUMMARY OF SCAFFOLD-ATTACHED FRAGMENTS (SAR) IDENTIFIED IN
Saccharomyces cerevisiae[a]

Locus[b]	SAR Fragment[c]	Comments[d]	Reference
ARS1	311 bp	3' of *TRP1* gene	Amati and Gasser (1988)
Distal	293 bp	3' of *ARS1* fragment	Amati and Gasser (1988)
HO ARS	1.25 kb**	3' of *HO* gene	B. B. Amati and S. M. Gasser (unpublished)
H4 ARS	374 bp	3' of *H4* gene	B. B. Amati (unpublished)
2-μm Circle	1.54 kb*	Contains origin and *STB* locus	Amati and Gasser (1988)
HMR-E	379 bp	Entire E-silencer	B. B. Amati (unpublished)
Distal	~630 bp	Adjacent to *HMR*-E	B. B. Amati (unpublished)
HMR-I	1.23 kb**	Contains I element	S. M. Gasser (unpublished)
HML-E	395 bp	Partial E-silencer	Hofmann *et al.* (1989)
Distal	540 bp	Rest of E-silencer	Hofmann *et al.* (1989)
HML-I	420 bp	Distal I region and ARS consensus	Hofmann *et al.* (1989)
$\alpha1/\alpha2$ Promoter	474 bp	Intragenic region at *HMLα*	Hofmann *et al.* (1989)
CEN3	320 bp	Minimal centromere	Amati and Gasser (1988)
CEN4	1.3 kb**	*CEN4* and flanking regions	S. M. Gasser (unpublished)

[a] Abbreviations: ARS, autonomously replicating sequence; CEN, centromere; *HML* or *HMR*, the silent mating-type loci; *STB*, stability locus of the 2-μm plasmid.

[b] The scaffold-attached sites mapped in *S. cerevisiae* to date are listed by the name given the locus by other authors. The word distal indicates that the fragment is 3' of the locus in the preceding entry.

[c] The size of the smallest restriction fragment to which the SAR was mapped. * Fragment may not be the minimal size of the SAR, since it was not mapped more closely. ** Fragments were only mapped in the *in vitro*-rebinding assay, and not by Southern hybridization to genomic DNA.

[d] A short description of the element in a functional or positional sense.

ARS sequences, centromeres (Amati and Gasser, 1988), and the transcriptional silencers that flank the silent mating-type loci (Abraham et al., 1984; Hofmann et al., 1989).

A. ARS ELEMENTS

Autonomously replicating sequences were initially isolated from yeast DNA by their ability to increase the transformation efficiency of yeast plasmids, and to allow the autonomous replication of those plasmids in the cell (reviewed in Williamson, 1985; Campbell, 1986; Kearsey, 1986). The most convincing evidence that ARS elements are bona fide origins of replication came from two-dimensional electrophoretic analysis that allows one to distinguish between the linear, branched, or "bubbled" replication intermediates of replicating DNA (Brewer and Fangman, 1987). These ARS elements consist of an 11-bp consensus ($^A/_T$TTTAT$^A/_G$TTT$^A/_T$) necessary for ARS activity, and A/T-rich flanking sequences, which contribute to ARS function and mitotic stability.

Seven ARS elements and the origin of the 2-μm circle have been shown to be scaffold-bound (Amati and Gasser, 1988; Hofmann et al., 1989). A fine-mapping study of the scaffold-binding sites in the ARS1 element has been done to relate the scaffold-binding sites to the regions necessary for ARS function. Two short subregions were essential for scaffold association: the first is a stretch of around 100 bp, which spans the ARS consensus sequence and contains part of the flanking A/T-rich regions. The second was immediately downstream of the ARS1 locus, and has no known function.

Several factors have been identified that bind to ARS1. The binding site for the ARS binding factor 1 (ABF-1 or SBF-B: Shore et al., 1987; Buchman et al., 1988a; Diffley and Stillman, 1988; Sweder et al., 1988), was not sufficient to mediate scaffold association when separated from the ARS consensus by restriction digestion. Also, ABF-1 binds to the HMR-E silencer region (Shore et al., 1987), yet could not be detected by band-shift assay with an HMR-E probe and the soluble scaffold protein extract described earlier (Hofmann et al., 1989). Thus, either ABF-1 is not a component of the yeast scaffold, or it resists renaturation after LIS extraction. The stretch of bent DNA or ARS1 that spans the ABF-1 site (Snyder et al., 1986) is not sufficient to bind the scaffold. It is still possible, however, that ABF-1 or the bent DNA contributes indirectly to the stability of the scaffold–DNA interaction of the intact ARS1.

Another structural feature of ARS elements is an increased propensity to unwind the DNA helix: loss of the ease of unwinding and loss of replication activity correlate in mutant versions of the H4 ARS, and restoration of the unwinding activity by an unrelated sequence restores ARS activity (Umek and Kowalski, 1988). A/T richness is not a sufficient determinant of this feature (Umek and Kowalski, 1987). Although unwinding occurs more readily in circular, torsionally stressed molecules, we cannot rule out at present that the binding of SAR-containing restriction fragments might be mediated by such a structural property confined to a portion of the fragment.

A study by Tsutsui *et al.* (1988) supports the idea that supercoiled or single-stranded stretches of DNA might be important for specific DNA–scaffold interaction. These authors show that LIS-extracted rat liver nuclei bind supercoiled pBR322 preferentially, while having little or no affinity for linear or relaxed circular plasmid. Denatured DNA, on the other hand, was able to compete for binding of the supercoiled plasmid. We have also observed that single-stranded DNA competes for the binding of certain yeast SAR to scaffold proteins *in vitro* (J. F-X. Hofmann and S. M. Gasser, unpublished observations).

B. CENTROMERES

Centromeres have been isolated by their ability to confer proper mitotic and meiotic segregation to the mitotically unstable ARS plasmids (Clarke and Carbon, 1980; Hsiao and Carbon, 1981), or by selection for the decreased copy number of ARS plasmids (Hieter *et al.*, 1985). Centromere elements have been shown to consist of three distinct sequence elements: the central element is 78–86 bp of highly A/T-rich DNA (>90%), flanked on one side by an 8-bp element ($^A/_G$TCAC$^A/_G$TG) and on the other by a 25-bp element III (TGTTT$^T/_A$TGNTTTCCGAAANNNN$^A/_T$$^A/_T$$^A/_T$: Hieter *et al.*, 1985; Fitzgerald-Hayes *et al.*, 1982; reviewed in Blackburn and Szostack, 1984). These sequences contribute to different extents to the mitotic and meiotic segregation functions of centromeres. The scaffold-binding region was mapped to a 320-bp fragment that contains all of these elements. It is not yet known which, if any, of these mediates scaffold association in yeast, although element I can be deleted without loss of scaffold binding *in vitro* (S. M. Gasser, unpublished observations).

To determine whether ARS and CEN elements were binding the same scaffold sites, the ability of a particular SAR fragment added exogenously to compete for the binding of another to the scaffold was tested. Such experiments performed with *ARS1* and *CEN3* fragments demonstrated that while *CEN3* can compete for *ARS1* binding, the opposite is not true. While *CEN3* has been found to provide weak ARS activity on a plasmid (Kearsey and Edwards, 1987), *ARS1* has no centromere activity. Thus one explanation may be that *CEN3* recognizes the binding site for *ARS1* in addition to or in conjunction with other CEN-specific components that are not bound by *ARS1*.

The specificity and reversibility of binding to the yeast scaffold was also demonstrated by competition experiments: excess unlabeled *ARS1* added exogenously displaced a trace amount of end-labeled *ARS1,* while excess amounts of unlabeled pBR322 did not. Testing the ability of other fragments with high A+T content to bind the scaffold showed that A/T richness was not sufficient to promote binding (Amati and Gasser, 1988). This has also been shown in other systems: the A/T-rich *CEN3* fragment does not bind scaffolds from *S. pombe, Drosophila* (S. M. G., unpublished), or mammalian cells (Sykes *et al.,* 1988).

C. SILENT MATING-TYPE LOCI

The genes that control mating type are found in three copies in the yeast genome; one is active and two are, under normal circumstances, transcriptionally silent. The repressed loci, *HML* and *HMR,* are silenced by cis-acting sequences that flank the two genes, and by the effects of a number of trans-acting factors (e.g., SIR 1–4, RAP-1, and possibly ABF-1: reviewed in Nasmyth, 1982a; Shore *et al.,* 1987). The cis-acting silencer fragments have been shown to have ARS activity in yeast (Abraham *et al.,* 1984), although it is not known whether these are active as origins in the genome. In view of their spatial organization, the silent mating-type loci can each be considered as a coordinately regulated domain bounded by the cis-acting silencer–ARS sequences.

Hofmann *et al.* (1989) have identifed the silencers that flank the silent mating-type casettes *HML* and *HMR* as SAR. In addition, a fragment contining the $\alpha1/\alpha2$ promoter at the *HML* locus is scaffold-bound, albeit with decreased affinity. By gel retardation studies using a silencer DNA probe and the soluble scaffold extract described earlier, it was shown that RAP-1 is the only scaffold protein that binds the minimal 132-bp *HMR*-E silencer fragment. In addition, RAP-1 binds both of the cis-acting silencer fragments at *HML* and the $\alpha1/\alpha2$ promoter, but not to other ARS elements (Hofmann *et al.,* 1989; Shore and Nasmyth, 1987; Buchman *et al.,* 1988b). If both the RAP-1 binding site and either the ARS consensus or the ABF-1 binding sites are deleted, silencing activity is lost (Brand *et al.,* 1987).

The available data thus appear to distinguish three classes of SAR elements in yeast, coinciding with CEN, ARS, and silencer elements. Two other SAR elements have been described in regions that have no obvious function assigned (see Table I); these may bind to yet different sites.

D. SAR and Nucleosomal Organization

The variation in sensitivity of chromatin and naked DNA to nuclease- and chemical-induced cleavage also contributes to our understanding of higher-order chromatin structure (reviewed in Igo-Kemenes *et al.*, 1982). Early analyses led to the conclusion that the organization of DNA into nucleosomes is found in yeast as well as higher eukaryotes (Thomas and Furber, 1976). Chromatin structure at the *TRP1 ARS1* locus has been studied both on yeast minichromosomes (i.e., plasmids: Thoma *et al.*, 1984; Long *et al.*, 1985; Thoma, 1986) and in the chromosome (Lohr and Torchia, 1988). Although discrepancies exist between these studies, the *ARS1* element appears to be organized into a distinct, nonnucleosomal configuration, while the adjacent *TRP1* gene is packaged into nucleosomes.

The centromere sequences at *CEN3* and *CEN11* are embedded in a 220–250 bp nuclease-protected region, which is markedly distinct from the protection provided by the nucleosomal core (Bloom and Carbon, 1982). This region appears to be organize a highly static, phased nucleosome pattern on either side of itself, for at least 12 nucleosomal subunits, similar to the higher eukaryotic SAR mentioned earlier.

The yeast mating-type loci have also been studied by classic techniques to detect nuclease-sensitive sites (Nasmyth, 1982b). Transcriptional repression and derepression of the silent mating-type loci correlate with changes in chromatin structure within these loci. These changes are bounded on either side by the silencer elements E and I, although the nuclease-sensitive sites in the silencers themselves do not change. Such observations support the idea that the scaffold-attached silencers define a coordinately regulated domain.

E. Is Scaffold Binding Essential for ARS, CEN, or Silencer Activity?

It is clear that affinity for the scaffold alone is not sufficient for either ARS, CEN, or silencer function. The important question is, however, whether the association of these cis-acting sequences with the scaffold contributes to their activity. All functional ARS elements tested so far are able to bind the scaffold, and at *ARS1* and a heterologous ARS (see Section VI) the minimal essential binding sites appear to coincide with the sequences necessary for ARS activity (Amati and Gasser, 1988; Amati *et al.*, 1990). If loss of binding were to correlate with loss of activity in short deletions or point mutations of these elements, then the nuclear scaffold that is operationally defined in these studies would gain much in physiological significance. Such studies are now feasible in yeast, with the availability of extensive mutations within ARS, CEN, and silencer elements, and the knowledge of the specific sequence requirements of these elements.

VI. Are Scaffold Attachment Sites Origins of Replication in Higher Cells?

A. STUDIES ON HIGHER EUKARYOTIC ORIGINS

The findings in yeast suggest that SAR identified in higher eukaryotic cells may be good candidates for origins of replication. Indeed, it has been proposed for many years that origins of replication may be associated with a nuclear structure in eukaryotic cells (see, e.g., Dingman, 1974; Pardoll *et al.*, 1980; van der Velden and Wanka, 1987). On the other hand, apart from viral systems, it remains to be proven that there are sequence-specific origins of replication in eukaryotic nuclear DNA (reviewed in Laskey and Harland, 1981). In the following paragraphs we will discuss several attempts to map specific chromosomal origins in different eukaryotic cells, particularly those in which a correlation of the putative origin with a SAR was found.

At the α-globin gene domain of chicken, origin DNA was mapped by Razin *et al.* (1986) using a specific origin-DNA preparation method developed by Zannis-Hadjopoulos *et al.* (1981). At that locus the putative origin was found within a fragment enriched in a fraction they define as "permanently attached" matrix DNA; the association of this fraction with the high-salt matrix is independent of transcription (Razin *et al.*, 1985, 1986).

At the amplified *DHFR* locus (*DHFR* amplicon) in a methotrexate-resistant CHO cell line, the replication origin was mapped initially in a 28-kb region (Heintz and Hamlin, 1982). It has since been narrowed down to two initiation sites present within ~21 kb. At this amplicon, scaffold binding was mapped to a 3.4-kb fragment situated between the two initiation sites (Dijkwell and Hamlin, 1988). It should be noted, however, that in a previous study of this locus the site was not found to be scaffold-bound (Käs and Chasin, 1987).

Finally, a cloned 2.5-kb mouse chromosomal DNA fragment that hybridizes to part of the simian virus 40 (SV40)-origin palindrome has been shown to have ARS activity in mouse and human cells (Ariga *et al.*, 1987). Cotransfection with anti-c myc antibodies showed that the c-myc protein was necessary for the replication mediated by this fragment, and other similar elements could be isolated by virtue of their binding to c-myc

(Iguchi-Ariga *et al.*, 1987). These observations correlate with the presence of the c-myc protein in the nuclear matrix fraction as described by several authors (Evan and Hancock, 1985; Van Straaten and Rabbitts, 1987).

Not only has it been proposed that DNA replication initiates at a nuclear substructure, but it has also been widely presumed that the replication complex is matrix-bound. These models were based on the finding that newly synthesized DNA is enriched in the DNA bound to the high-salt matrix (for reviews see Zehnbauer and Vogelstein, 1985; Nelson *et al.*, 1986; Jackson, 1986; Razin, 1987; van der Velden and Wanka, 1987) and on a correlation of the size of the topologically constrained, looped domains in nuclei, and the size of replicons (Cook and Brazell, 1976; Benyajati and Worcel, 1976; Igo-Kemenes and Zachau, 1978; Vogelstein *et al.*, 1980; Paulson and Laemmli, 1977; Huberman and Riggs, 1968; Buongiorno-Nardelli *et al.*, 1982; Vogelstein *et al.*, 1982). It is important to note here that localization of replication to a nuclear substructure conflicts to a certain degree with the mapping of sequence-specific scaffold attachment in LIS-extracted nuclei. If association with the matrix is mediated by the DNA polymerase, then the population of attached DNA fragments should be random in S-phase, representing any region being replicated. With LIS-extracted scaffolds we find specific attachment sites that appear to be cell cycle-independent. This discrepancy again reflects the different methods used in the extraction of histones from nuclei or chromosomes: high-salt extraction apparently stabilizes (or precipitates) replication and transcription complexes, while these are dissociated by LIS.

B. HETEROLOGOUS ARS AND SAR

Ever since the discovery of ARS elements in yeast, the assumption that heterologous ARS (e.g., human DNA fragments with ARS activity in yeast) might represent origins of replication in the respective organisms has been a tempting one. Heterologous ARS elements have been described from a variety of DNA sources (partially reviewed in Williamson, 1985), yet none, to our knowledge, has been identified as a replication origin or as an ARS in the homologous organism, with the exception of some of the elements from the fission yeast *S. pombe* (Maundrell *et al.*, 1985). Only a few of the *S. pombe* ARS were functional ARS elements in *S. cerevisiae*, and even these appear to have different sequence requirements in the two organisms.

Contradictory observations have been reported concerning the relationship between heterologous ARS activity and attachment to the nuclear matrix or scaffold in human and animal cells. Cook and Lang (1984) did not detect any enrichment of heterologous ARS in a human matrix-bound

DNA fraction with respect to total chromosomal DNA, whereas Aguinaga *et al.* (1987) isolated a matrix-bound DNA fraction from synchronized HeLa cells consistently enriched in DNA that functioned as ARS elements in yeast. Similarly, a heterologous ARS element from the human hypoxanthine phosphoribosyltransferase (*HPRT*) locus was found to bind tightly to nuclear matrices from several mammalian cell types (Sykes *et al.*, 1988). This fragment was also shown to contain bent DNA, a characteristic that is shared with the *ARS1* element (Snyder *et al.*, 1986).

In our laboratory, we have observed ARS activity for two of four *Drosophila* SAR fragments tested in *S. cerevisiae* (Amati *et al.*, 1990). The same SAR fragments, plus an additional one, have ARS function in *S. pombe,* while none of four non-SAR fragments is able to replicate in either yeast species.

Such results suggest that perhaps association with the scaffold is a feature of replication origins conserved from yeast to humans. Indirect support of this comes from the large number of instances in which cross-species conservation of SAR-binding sites has been observed. This has been assayed by allowing end-labeled DNA fragments from one species to bind scaffolds from another, and assaying whether the pattern of scaffold-bound and released fragments is the same as that observed in the original organism. Reports include the binding of *Drosophila* SAR to rat liver (Izaurralde *et al.,* 1988), HeLa (Mirkovitch *et al.,* 1988), mouse (Cockerill and Garrard, 1986a), and chick nuclear scaffolds (Phi-van and Stratling, 1988), as well as the binding of the mouse Ig MAR to matrices from yeast (Cockerill and Garrard, 1986b) and scaffolds from chicken cells (Phi-van and Stratling, 1988). We have observed the specific binding of *Drosophila* SAR to scaffolds from both *S. cerevisiae* and *S. pombe,* and conversely the binding of ARS elements from *S. cerevisiae* and *S. pombe* to the *Drosophila* nuclear scaffold (Amati and Gasser, 1988; Amati *et al.*, 1990). While *S. pombe* ARS also bind to HeLa nuclear scaffolds, the *S. cerevisiae* ARS1 element failed to do so (S. M. Gasser, unpublished observations). Nonetheless, to establish that scaffold association has functional significance, SAR will have to be tested for origin activity in a homologous system.

VII. Reconstitution Studies

We feel that two approaches will be the most fruitful in the future for the characterization of higher-order chromatin structure. One is the cloning of genes for the various scaffold components, followed by the creation of null or conditional mutants in these genes. For this approach yeast is clearly the organism of choice. The second approach will be the *in vitro* reconstitution of various levels of chromatin organization that one hopes will exhibit properties similar to those observed with isolated nuclei. This section will discuss studies that have begun to address these problems.

A. Loop Formation *in Vitro* Using the *HML* Locus of *S. cerevisiae*

To address the question of whether interactions of scaffold proteins with SAR DNA mediates the formation of a loop, Hofmann *et al.* (1989) have reconstituted a DNA loop from solubilized components. A 5.6-kb fragment from *S. cerevisiae* that contains the entire silent mating-type locus *HMLα* flanked by its two known silencer elements was mixed with either the solubilized scaffold extract or purified RAP-1 protein, and assayed for *in vitro* loop formation by quantitative electron microscopy. Results showed loop formation between any two of the three scaffold-binding sites that had been previously mapped, that is, the *HML*-E silencer, the *HML*-I silencer, and the α1/α2 promoter fragment (see Fig. 4). Protein–protein interaction (and thus loop formation) increased in a sigmoidal fashion with protein concentration, suggesting some form of cooperative interaction. The fact that a highly purified RAP-1 allowed reconstitution of the same loops and that all loops were eliminated by addition of a competing RAP-1-binding site in competition experiments, suggests that the abundant nuclear factor RAP-1 mediates formation of the observed loops.

Although this only shows reconstitution at a single locus, it lends credence to the interpretation that DNA sequences that bind the scaffold actually can interact via proteins to form loops. Moreover, this loop coincides with the domain of transcriptional repression *in vivo* (reviewed in Nasmyth, 1982a). Other evidence suggesting that repression is mediated at least in part by the organization of the chromatin domain includes the nuclease sensitivity studies cited earlier (Nasmyth, 1982b) and the observation that deletion of the conserved N terminus of histone H4 derepresses the silent mating-type loci (Kayne *et al.*, 1988). The ability to reconstitute a single, coordinately regulated loop may allow reconstitution of an *in vitro* "silencing" assay, in which a loss of transcriptional activity is dependent on the addition of purified factors that reconstitute the chromatin loop.

B. Nuclear Assembly and Disassembly

Over the last several years a number of studies have demonstrated that nuclei can be assembled (Burke and Gerace, 1986; Blow and Laskey, 1986; Newmeyer *et al.*, 1986; Newport, 1987) and disassembled (Lohka and Maller, 1985; Maike-Lye and Kirschner, 1985; Suprynowicz and Gerace, 1986; Newport and Spann, 1987) in cell-free extracts derived either from

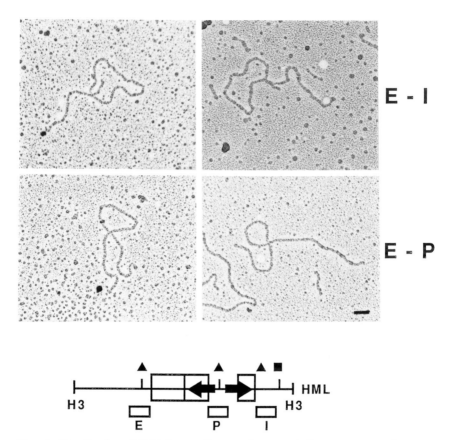

FIG. 4. Visualization of DNA loops: silencer–silencer and silencer–promoter loops are formed at the *HML* locus by interaction with scaffold proteins. Shown are several examples of DNA loops reconstituted by incubation of a 5.6-kb fragment from the yeast *HML* locus with a soluble scaffold extract. Below the micrographs is shown a map of the *Hin*DIII fragment used in the reconstitution assay. Black arrows represent the two coding regions for α1 and α2, and open boxes below the map show the promoter region (P) and the E and I silencer regions. ▲, Approximate binding sites for RAP-1 *in vitro;* ■, binding site of the nuclear factor ABF-1 (see text). The upper two micrographs represent interactions between the two silencer domains, E and I; and the lower two interactions between E and P. The regions used to define domains E and I come from Abraham *et al.* (1984). A loop was considered to be of a given class if its point of departure fell within 150 bp of the E, I, or P domains drawn here. Loop formation was not seen in the absence of scaffold extract nor with other DNA fragments such as pBR322 (Hofmann *et al.*, 1989). Bar = 100 nm.

Xenopus oocytes or mitotic tissue culture cells (reviewed in Newport and Forbes, 1987). The reassembly can occur around protein-free, noneukaryotic DNA as well as metaphase chromosomes, and appears to have two steps: an ATP-dependent step of chromatin condensation and an ATP-independent step of membrane re-formation (Newport, 1987). Remarkably the reconstituted nuclei are able to import nuclear proteins specifically (see e.g., Newmeyer *et al.*, 1986) and to replicate DNA (see e.g., Blow and Laskey, 1986).

Blow and Laskey (1988) have shown that membrane integrity is essential for properly limiting DNA replication to a single round per cell cycle. If the nuclear membrane is allowed to break down and re-form, a putative "licensing factor" is able to rebind to the chromatin and reinitiate a round of DNA synthesis.

Although the cell-free system used for these studies is still relatively crude, identification and purification of the individual factors involved will undoubtedly follow soon. Senior and Gerace (1988) have purified three proteins from rat liver nuclei that may serve to anchor the nuclear lamina to the inner nuclear membrane. Similarly, Worman *et al.* (1988) identify a 55-kDa "lamin B receptor" in the inner nuclear membrane of chick cells. Wilson and Newport (1988), moreover, report that digestion of nuclear membrane vesicles with protease prior to nuclear reconstitution blocks vesicle interaction with chromatin, suggesting the presence of a nonlamin chromatin receptor. Characterization of these elements and confirmation of their role in nuclear organization will help a great deal in our understanding of chromosomal structure in the interphase nucleus.

VIII. Concluding Remarks and Future Prospects

The organization of DNA beyond the nucleosomal level, within the nucleus and in metaphase chromosomes is still largely unsolved. Electron microscopy of histone-depleted scaffolds and SAR-mapping studies strongly favors a model in which the 30-nm solenoid fiber is constrained in looped domains. We have discussed the evidence suggesting that a residual protein skeleton might provide this constraint, yet the biological importance of the nuclear and metaphase scaffold and its role in dynamic functions such as DNA replication and transcription is much less clear on a molecular level.

At the very least nuclear matrices and scaffolds can be considered as useful tools for subfractionation and purification of various nuclear components. The field is burdened by the fact that small variations in the conditions of nuclear isolation and extraction (especially exposure to met-

als, heat shock temperatures, SH-oxidizing conditions, or high salt concentrations) can result in different, though most likely related, structures.
The specificity of the DNA–scaffold interaction and the conservation of
these binding sites (SARs) between higher eukaryotes and yeast is encour-
aging. The apparent association of replication origins with the nuclear
matrix of chick and the scaffold from yeast, and the ability of some
Drosophila SAR to replicate in yeast, justify a closer investigation of the
biological properties of these regions. If SAR (or a subset of them) can be
proven to be true replication origins of eukaryotic chromosomes, and if
sequences and protein factors involved can be characterized in greater
detail, we might begin to understand the conserved, basic features of
chromosome organization that underlie the regulated expression and
transmission of genetic information.

A few words of caution should be aired, however. First, it remains to be
proven that SAR are actually the bases of loops in intact, or extracted but
nondigested, nuclei. There is as yet no correlation between "SAR-defined
loops" and the loops observed in polytene chromosomal puffs or in lamp-
brush chromosomes. The mapping of SAR is performed on histone-
depleted nuclei, and obviously, if histones are involved in regulating the
association of DNA with nonhistone proteins, one level of regulation has
been eliminated. Finally, even though the reconstitution studies discussed
in Section VII suggest that two SAR *can* interact to form a loop, it remains
to be demonstrated that this occurs in intact cells.

The only way to prove that the "stabilized" scaffold or matrix structure
reflects an important level of nuclear organization is to ascertain from *in
vitro* studies what function the structure might influence, to make deletions
in the DNA sequences or mutants in the proteins that bind them, and test
for altered function *in vivo*. Yeast provides an ideal system for such
manipulations, yet yeast scaffolds and matrices are less easily characterized *in vitro*. For reasons not yet clear, the previously described "size
effect" (i.e., the preferential but nonspecific association of large DNA
fragments with the scaffold: Mirkovitch *et al.*, 1986, 1988; Mirkovitch,
1987) is more pronounced in yeast and must be controlled for (Amati and
Gasser, 1988; Conrad and Zakian, 1988). Still there is accumulating evidence that yeast nuclei and their skeletons resemble those of higher eukaryotes, notably with respect to conserved antigens (Hurt, 1988), motifs
within specific scaffold-binding sequences (Amati and Gasser, 1988), and
conserved features of the nuclear pore complex (Aris and Blobel, 1989).
Certainly the factors regulating transcription in yeast are in many cases
interchangeable with those functioning in higher plants and animals
(Ptashne, 1988).

In view of the recent interest in the formation of small loops that facilitate the interaction of distant promoter elements (Ptashne, 1986, 1988), it is perhaps important to distinguish two potential types of DNA loops in the nucleus: those created by the interaction of enhancer-binding factors with proteins at a more-or-less distant promoter, and those defining large chromatin domains. One would expect the latter not to depend on the binding of transcription factors nor on active transcription, and to represent landmark attachment points in the nucleus. Cohabitation of enhancer and SAR elements is found in several instances, but is likely not to be a general feature of enhancers. The instances observed might represent sites at which chromatin loops and transcription-regulating loops coincide.

Speculation on the function of DNA loops has been published before (Zehnbauer and Vogelstein, 1985; Jackson, 1986; Nelson *et al.*, 1986; Razin, 1987; Gasser and Laemmli, 1987). The presence of topologically fixed points in the genome raises the possibility that a modulation of supercoiling density regulates loop extension and retraction, and may similarly allow for the maintenance of gene activity. While there is no direct evidence implicating torsional stress in the induction of eukaryotic genes, it is quite clear that the transcription process itself induces localized supercoiling on a circular plasmid (Liu and Wang, 1987; Brill and Sternglanz, 1988). The accumulation of torsional stress might occur whenever an active gene is bounded by fixed points of attachment to the nuclear scaffold. This might explain the presence of the topoisomerase II cleavage consenses on SAR, since the enzyme could relieve the accumulating supercoils that arise from the transcription process. Thus topoisomerase II, rather than helping in the initiation of transcription, might maintain a high transcriptional level once a gene is activated.

An important task for the future, which can be approached in both yeast and mammalian cells, is to show unambiguously that SAR influence replication or transcription. In addition, the characterization of the SAR-binding scaffolding proteins will be necessary before we can evaluate what role this level of nuclear architecture plays in promoting and coordinating various nuclear activities during the cell cycle.

ACKNOWLEDGMENTS

We would like to acknowledge the critical reading of Dr. Viesturs Simanis and Robin Walter. Research was supported by grants to S. M. G. from the Swiss National Science Foundation and the Swiss Cancer League. M. E. C. is recipient of a fellowship from the Roche Foundation.

REFERENCES

Abraham, J., Feldman, J., Nasmyth, K. A., Strathern, J. N., Klar, A. J. S., Broach, J. R., and Hicks, J. B. (1984). *Cold Spring Harbor Symp. Quant. Biol.* **49,** 989–997.

Adolph, K. W. (1980). *J. Cell Sci.* **42,** 291–304.

Adolph, K. W., Cheng, S. M., and Laemmli, U. K. (1977). *Cell* **12,** 805–816.

Aguinaga, M. P., Kiper, C. E., and Valenzuela, M. S. (1987). *Biochem. Biophys. Res. Commun.* **144,** 1018–1024..

Amati, B. B., and Gasser, S. M. (1988). *Cell* **54,** 967–978.

Amati, B. B., Pick, L., and Gasser, S. M. (1990). *EMBO J.* (submitted).

Ariga, H., Itani, T., and Iguchi-Ariga, S. M. M. (1987). *Mol. Cell. Biol.* **7,** 1–6.

Aris, J. P. and Blobel, G. (1989). *J. Cell Biol.* **108,** 2059–2067.

Benyajati, C., and Worcel, A. (1976). *Cell* **9,** 393–407.

Berezney, R., and Coffey, D. (1974). *Biochem. Biophys. Res. Commun.* **60,** 1410–1419.

Berg, J. M. (1986). *Science* **232,** 485–486.

Berrios, M., Osheroff, N., and Fisher, P. A. (1985). *Proc. Natl. Acad. Sci. U.S.A.* **82,** 4142–4146.

Berrios, S., and Fisher, P. A. (1988). *Mol. Cell. Biol.* **8,** 4573–4575.

Bienz, M., and Pelham, H. R. B. (1986). *Cell* **45,** 753–760.

Blackburn, E. H., and Szostack, J. (1984). *Annu. Rev. Biochem.* **53,** 163–194.

Bloom, K. S., and Carbon, J. (1982). *Cell* **29,** 305–317.

Blow, J. J., and Laskey, R. A. (1986). *Cell* **47,** 577–587.

Blow, J. J., and Laskey, R. A. (1988). *Nature (London)* **332,** 546–548.

Bode, J., and Maass, K. (1988). *Biochemistry* **27,** 4706–4711.

Bouvier, D., Hubert, J., Seve, A. P., and Bouteille, M. (1982). *Biol. Cell.* **43,** 143–146.

Brand, A. H., Micklem, G., and Nasmyth, K. (1987). *Cell* **51,** 709–719.

Brewer, B. J., and Fangman, W. L. (1987). *Cell* **51,** 463–471.

Brill, S. J., and Sternglanz, R. (1988). *Cell* **54,** 403–411.

Buchman, A. R., Kimmerly, W. J., Rine, J., and Kornberg, R. D. (1988a). *Mol. Cell. Biol.* **8,** 210–225.

Buchman, A. R., Lue, N. F., and Kornberg, R. D. (1988b). *Mol. Cell. Biol.* **8,** 5086–5099.

Buongiorno-Nardelli, M., Micheli, G., Carri, M. T., and Marilley, M. (1982). *Nature (London)* **298,** 100–102.

Burdon, R. H., Qureschi, M., and Adams, R. L. P. (1985). *Biochim. Biophys. Acta* **825,** 70–79.

Burke, B., and Gerace, L. (1986). *Cell* **44,** 639–652.

Campbell, J. L. (1986). *Annu. Rev. Biochem.* **55,** 733–771.

Ciejek, E. M., Tsai, M., and O'Malley, B. W. (1983). *Nature (London)* **306,** 607–609.

Clarke, L., and Carbon, J. (1980). *Nature (London)* **287,** 504–509.

Cockerill, P. N., and Garrard, W. T. (1986a). *Cell* **44,** 273–282.

Cockerill, P. N., and Garrard, W. T. (1986b). *FEBS Lett* **204,** 5–7.

Cockerill, P. N., Yuen, M.-H., and Garrard, W. T. (1987). *J. Biol. Chem.* **262,** 5394–5397.

Conrad, M. N., and Zakian, V. A. (1988). *Curr. Genet.* **13,** 291–297.

Cook, P. R., and Brazell, I. A. (1975). *J. Cell Sci.* **19,** 261–279.

Cook, P. R., and Brazell, I. A. (1976). *J. Cell Sci.* **22,** 287–302.

Cook, P. R., and Brazell, I. A. (1980). *Nucleic Acids Res.* **8,** 2895–2907.

Cook, P. R., and Lang, J. (1984). *Nucleic Acids Res.* **14,** 1069–1075.

Cook, P. R., Lang, J., Hayday, A., Lanca, L., Fried, M., Chirwell, D. J., and Wyke, J. A. (1982). *EMBO J.* **1,** 447–452.

Davis, L. I., and Blobel, G. (1986). *Cell* **45,** 699–709.

Diffley, J. F. X., and Stillman, B. (1988). *Proc. Natl. Acad. Sci. U.S.A.* **85,** 2120–2124.

Dijkwel, P., and Hamlin, J. L. (1988). *Mol. Cell. Biol.* **8,** 5398–5409.

Dijkwel, P. A., and Wenink, P. W. (1986). *J. Cell Sci.* **84,** 53–67.

DiNardo, S., Voelkel, K., and Sternglanz, R. (1984). *Proc. Natl. Acad. Sci. U.S.A.* **82,** 2616–2620.

Dingman, C. W. (1974). *J. Theor. Biol.* **43,** 187–195.

Earnshaw, W. C., and Heck, M. M. S. (1985). *J. Cell Biol.* **100,** 1716–1725.

Earnshaw, W. C., Halligan, N., Cooke, C., and Rothfield, N. (1984). *J. Cell Biol.* **98,** 352–357.

Earnshaw, W. C., Halligan, B., Cooke, C. A., Heck, M. M. S., and Liu, L. F. (1985). *J. Cell Biol.* **100,** 1706–1715.

Evan, G. I., and Hancock, D. C. (1985). *Cell* **43,** 253–261.

Fields, A. P., and Shaper, J. H. (1988). *J. Cell Biol.* **107,** 1–8.

Finch, J. T., and Klug, A. (1976). *Proc. Natl. Acad. Sci. U.S.A.* **73,** 1897–1901.

Fisher, P. A., Berrios, M., and Blobel, G. (1982). *J. Cell Biol.* **92,** 674–686.

Fisher, P. A., Lin, L., McConnell, M., Greenleaf, A., Lee, J.-M., and Smith, D. E. (1989). *J. Biol. Chem.* **264,** 3464–3469.

Fitzgerald-Hayes, M., Clarke, L., and Carbon, J. (1982). *Cell* **29,** 235–244.

Foster, K. A., and Collins, J. M. (1985). *J. Biol. Chem.* **260,** 4229–4235.

Franke, W. W. (1987). *Cell* **48,** 3–44.

Fürst, P., Hu, S., Hackett, R., and Hamer, D. (1988). *Cell* **55,** 705–717.

Gasser, S. M., and Laemmli, U. K. (1986a). *Cell* **46,** 521–530.

Gasser, S. M., and Laemmli, U. K. (1986b). *EMBO J.* **5,** 511–518.

Gasser, S. M., and Laemmli, U. K. (1987). *Trends Genet.* **3,** 16–22.

Gasser, S. M., Laroche, T., Falquet, J., Boy de la Tour, E., and Laemmli, U. K. (1986). *J. Mol. Biol.* **188,** 613–629.

Gerace, L., and Burke, B. (1988). *Annu. Rev. Cell Biol.* **4,** 335–374.

Grosveld, F., Van Assendelft, G. B., Greaves, D. R., and Kollias, G. (1987). *Cell* **51,** 975–985.

Hancock, R. (1982). *Biol. Cell.* **46,** 105–122.

Hancock, R., and Hughes, M. E. (1982). *Biol. Cell.* **44,** 201–212.

Heck, M. M., and Earnshaw, W. C. (1986). *J. Cell Biol.* **103,** 2569–2581.

Heintz, H. H., and Hamlin, J. L. (1982). *Proc. Natl. Acad. Sci. U.S.A.* **69,** 4083–4087.

Hieter, P., Pridmore, D., Hegemann, J. H., Thomas, M., Davis, R. W., and Philippsen, P. (1985). *Cell* **42,** 913–921.

Hirano, T., Hiraoka, Y., and Yanagida, M. (1988). *J. Cell Biol.* **106,** 1171–1183.

Hiromi, Y., Kuriowa, A., and Gehring, W. (1985). *Cell* **43,** 603–613.

Hofmann, J. F. X., Laroche, T., Brand, A., and Gasser, S. M. (1989). *Cell* **57,** 725–737.

Holm, C., Goto, T., Wang, J. C., and Botstein, D. (1985). *Cell* **41,** 553–563.

Hsiao, C. L., and Carbon, J. (1981). *Proc. Natl. Acad. Sci. U.S.A.* **78,** 3760–3764.

Huberman, J. A., and Riggs, A. D. (1968). *J. Mol. Biol.* **32,** 327–341.

Huberman, J. A., Spotila, L. D., Nawotka, K. A., El-Assouli, S. M., and Davis, L. R. (1987). *Cell* **51,** 473–481.

Huet, J., Cottrelle, P., Cool, M., Vignais, M. L., Thiele, D., Marck, C., Buhler, J.-M., Sentenac, A., and Fromageot, P. (1985). *EMBO J.* **4,** 3539–3547.

Hurt, E. C. (1988). *EMBO J.* **7,** 4323–4334.

Igo-Kemenes, T., and Zachau, H. G. (1978). *Cold Spring Harbor Symp. Quant. Biol.* **42,** 108–118.

Igo-Kemenes, T., Höre, N., and Zachau, H. G. (1982). *Annu. Rev. Biochem.* **51,** 89–121.

Iguchi-Ariga, S. M. M., Itani, T., Kiji, Y., and Ariga H. (1987). *EMBO J.* **6,** 2365–2371.

Izaurralde, E., Mirkovitch, J., and Laemmli, U. K. (1988). *J. Mol. Biol.* **200**, 111–126.

Jackson, D. A. (1986). *TIBS* **11**, 249–252.

Jackson, D. A., and Cook, P. R. (1988). *EMBO J.* **7**, 3667–3677.

Jarman, A. P., and Higgs, D. R. (1988). *EMBO J.* **7**, 3337–3344.

Jost, J.-P., and Sedron, M. (1984). *EMBO J.* **3**, 2005–2008.

Käs, E., and Chasin, L. A. (1987). *J. Mol. Biol.* **198**, 677–692.

Kaufmann, S. H., and Shaper, J. H. (1984). *Exp. Cell Res.* **155**, 477–495.

Kaufmann, S. H., Fields, A. P., and Shaper, J. H. (1986a). *Methods Achiev. Exp. Pathol.* **12**, 141–171.

Kaufmann, S. H., Okret, S., Wikström, A. C., Gustafsson, J. A., and Shaper, J. H. (1986b). *J. Biol. Chem.* **261**, 11962–11967.

Kayne, P. S., Ung-Jin, K., Han, M., Mullen, R. R., Yoshizaki, F., and Grunstein, M. (1988). *Cell* **55**, 27–39.

Kearsey, S. (1986). *BioEssays* **4**, 157–161.

Kearsey, S. E., and Edwards J. (1987). *Mol. Gen. Genet.* **210**, 509–517.

Kirov, N., Djondjurov, L., and Tsanev, R. (1984). *J. Mol. Biol.* **180**, 601–614.

Laskey, R. A., and Harland, R. M. (1981). *Cell* **24**, 283–284.

Lebkowski, J. S., and Laemmli, U. K. (1982a). *J. Mol. Biol.* **156**, 309–324.

Lebkowski, J. S., and Laemmli, U. K. (1982b). *J. Mol. Biol.* **156**, 325–344.

Lewis, C. D., and Laemmli, U. K. (1982). *Cell* **29**, 171–181.

Lewis, C. D., Lebkowski, J. S., Daly, A. K., and Laemmli, U. K. (1984). *J. Cell Sci, Suppl.* No. 1, 103–122.

Lis, J., Neckmayer, W., Mirault, M.-E., Artvanis-Tsakonas, S., Lall, P. Martin, G., and Schedl, P. (1981). *Dev. Biol.* **83**, 291–300.

Littlewood, T. D., Hancock, D. C., and Evan, G. I. (1987). *J. Cell Sci.* **88**, 65–72.

Liu, L. F., and Wang, J. C. (1987). *Proc. Natl. Acad. Sci. U.S.A.* **84**, 7024–7027.

Lohka, M. J., and Maller, J. L. (1985). *J. Cell Biol.* **101**, 518–523.

Lohr, D., and Torchia, T. (1988). *Biochemistry* **27**, 3961–3965.

Long, C. M., Brajkovich, C. M., and Scott, J. F. (1985). *Mol. Cell. Biol.* **5**, 3124–3130.

McConnell, M., Whalen, A. M., Smith, D. E., and Fisher, P. A. (1987). *J. Cell Biol.* **105**, 1087–1098.

McNabb, S. L., and Beckendorf, S. K. (1986). *EMBO J.* **5**, 2331–2340.

Marchesi, V. T., and Andrews, E. P. (1971). *Science* **174**, 1247–1248.

Marsden, M. P. F., and Laemmli, U. K. (1979). *Cell* **17**, 849–858.

Maundrell, K., Wright, A. P. H., Piper, M., and Shall, S. (1985). *Nucleic Acids Res.* **13**, 3711–3722.

Miake-Lye, R., and Kirschner, M. W. (1985). *Cell* **41**, 165–175.

Miller, T. W., Huang, C.-Y., and Pogo, A. O. (1978). *J. Cell Biol.* **76**, 675–691.

Mirkovitch, J. (1987). Ph.D. Thesis, Univ. of Geneva.

Mirkovitch, J., Mirault, M.-E., and Laemmli, U. K. (1984). *Cell* **39**, 223–232.

Mirkovitch, J., Spierer, P., and Laemmli, U. K. (1986). *J. Mol. Biol.* **190**, 255–258.

Mirkovitch, J., Gasser, S. M., and Laemmli, U. K. (1988). *J. Mol. Biol.* **200**, 101–110.

Nasmyth, K. A. (1982a). *Annu. Rev. Genet.* **16**, 439–500.

Nasmyth, K. A. (1982b). *Cell* **30**, 567–578.

Nelson, W. G., Pienta, K. J., Barrack, E. R., and Coffey, D. S. (1986). *Annu. Rev. Biophys. Chem.* **15**, 457–475.

Newmeyer, D. D., Lucocq, J. M., Büglin, T. R., and De Robertis, E. M. (1986). *EMBO J.* **5**, 501–510.

Newport, J. W. (1987). *Cell* **48**, 205–217.

Newport, J. W., and Forbes, D. J. (1987). *Annu. Rev. Biochem.* **56**, 535–565.

Newport, J. W., and Spann, T. (1987). *Cell* **48**, 219–230.

Nigg, E. A. (1989). *Curr. Opinion Cell Biol.* **1**, 435–440.

Nishizawa, M., Tanabe, K., and Takahasi, T. (1984). *Biochem. Biophys. Res. Commun.* **124**, 917–924.

Pardoll, D. M., Vogelstein, B., and Coffey, D. S. (1980). *Cell* **19**, 527–536.

Paulson, J. R. (1988). *In* "Chromosomes and Chromatin" (K. W. Adolph, ed), Vol. 3, pp. 3–36. CRC Press, Boca Raton, Florida.

Paulson, J. R., and Laemmli, U. K. (1977). *Cell* **12**, 817–828.

Pelham, H. R. B. (1982). *Cell* **36**, 357–367.

Phi-van, L., and Stratling, W. H. (1988). *EMBO J.* **7**, 655–664.

Posakony, J. W., Fischer, J. A., Corbin, V., and Maniatis, T. (1985). *In* "Current Communications," pp. 194–200. Cold Spring Harbor Lab., Cold Spring Harbor, New York.

Ptashne, M. (1986). *Nature (London)* **322**, 697–701.

Ptashne, M. (1988). *Nature (London)* **335**, 683–689.

Rattner, J. B., and Lin, C. C. (1985). *Cell* **42**, 291–296.

Razin, S. V. (1987). *BioEssays* **6**, 19–23.

Razin, S. V., Yarovaya, O. V., and Georgiev, G. P. (1985). *Nucleic Acids Res.* **13**, 7427–7444.

Razin, S. V., Kekelidze, M. G., Lukanidin, E. M., Scherrer K., and Georgiev, G. P. (1986). *Nucleic Acids Res.* **14**, 8189–8207.

Roberge, M., Dahmus, M. E., and Bradbury, E. M. (1988). *J. Mol. Biol.* **201**, 545–555.

Robinson, D. R., and Jencks, N. P. (1965). *J. Am. Chem. Soc.* **87**, 2470–2480.

Robinson, S. I., Nelkin, B., and Vogelstein, B. (1982). *Cell* **28**, 99–106.

Rowe, T., Wang, J. C., and Liu, L. F. (1986). *Mol. Cell. Biol.* **6**, 985–992.

Sander, M., and Hsieh, T. S. (1985). *Nucleic Acids Res.* **13**, 1057–1072.

Senior, A., and Gerace, L. (1988). *J. Cell Biol.* **107**, 2029–2036.

Shore, D., and Nasmyth, K. (1987). *Cell* **51**, 721–732.

Shore, D., Stillman, D. J., Brand, A. H., and Nasmyth, K. A. (1987). *EMBO J.* **6**, 461–467.

Small, D., and Vogelstein, B. (1985). *Nucleic Acids Res.* **13**, 7703–7713.

Small, D., Nelkin, B., and Vogelstein, B. (1985). *Nucleic Acids Res.* **13**, 2413–2431.

Smith, H. C., Puvion, E., Buchholtz, L. A., and Berezney, R. (1984). *J. Cell Biol.* **99**, 1794–1802.

Snyder, M., Buchman, A., and Davis, R. W. (1986). *Nature (London)* **324**, 87–89.

Spelsberg, T. L., and Hnilica, L. S. (1971). *Biochem. Biophys Acta* **228**, 202–211.

Stratling, W. H., Dolle, A., and Sippel, A. E. (1986). *Biochemistry* **25**, 495–502.

Suprynowicz, F. A., and Gerace, L. (1986). *J. Cell Biol.* **103**, 2073–2081.

Sweder, K. S., Rhode, P. R., and Campbell, J. L. (1988). *J. Biol. Chem.* **263**, 17270–17277.

Sykes, R. C., Lin, D., Hwang, S. J., Framson, P. E., and Chinault, A. C. (1988). *Mol. Gen. Genet.* **212**, 301–309.

Thoma, F. (1986). *Mol. Biol.* **190**, 177–190.

Thoma, F., Bergman, L. W., and Simpson, R. T. (1984). *J. Mol. Biol.* **177**, 715–733.

Thomas, J. O., and Furber, V. (1976). *FEBS Lett.* **66**, 274–280.

Tsutsui, K., Tsutsui, K., and Muller, M. T. (1988). *J. Biol. Chem.* **263**, 7235–7241.

Tubo, R. A., and Berezney, R. (1987a). *J. Biol. Chem.* **262**, 1148–1154.

Tubo, R. A., and Berezney, R. (1987b). *J. Biol. Chem.* **262**, 5857–5865.

Tubo, R. A., and Berezney, R. (1987c). *J. Biol. Chem.* **262**, 6637–6642.

Udvardy, A., Schedl, P., Sander, M., and Hsieh, T.-S. (1985). *Cell* **40**, 933–941.

Udvardy, A., Schedl, P., Sander, M., and Hsieh, T.-S. (1986). *J. Mol. Biol.* **191**, 231–246.

Uemura, T., and Yanagida, M. (1984). *EMBO J.* **3**, 1737–1744.

Uemura, T., and Yanagida, M. (1986). *EMBO J.* **5**, 1003–1010.

Uemura, T., Ohkura, H., Adachi, Y., Morino, K., Shiozaki, K., aqnd Yanagida, M. (1987). *Cell* **50,** 917–925.

Umek, R. M., and Kowalski, D. (1987). *Nucleic Acids Res.* **15,** 4467–4480.

Umek, R. M., and Kowalski, D. (1988). *Cell* **52,** 559–567.

van der Velden, H. M. W., and Wanka, F. (1987). *Mol. Biol. Rep.* **12,** 69–77.

Van Straaten, J. P., and Rabbitts, T. H. (1987). *Oncog. Res.* **1,** 221–228.

Verheijen, R., Van Venrooij, W., and Ramaekers, F. (1988). *J. Cell Sci.* **90,** 11–36.

Vogelstein, B., Pardoll, D. M., and Coffey, D. S. (1980). *Cell* **22,** 79–85.

Vogelstein, B., Nelkin, B. D., Pardoll, D. M., Robinson, S. I., and Small, D. (1982). *ICN–UCLA Symp. Gene Regul.* pp. 105–119.

Wang, J. C. (1985). *Annu. Rev. Biochem.* **54,** 665–697.

Williamson, D. H. (1985). *Yeast* **1,** 1–14.

Wilson, K. L., and Newport, J. (1988). *J. Cell Biol.* **107,** 57–68.

Wood, S. H., and Collins, J. M. (1986). *J. Biol. Mol.* **261,** 7119–7122.

Worman, H. J., Yuan, J., Blobel, G., and Georgatos, S. D. (1988). *Proc. Natl. Acad. Sci. USA* **85,** 8531–8534.

Wu, C. (1980). *Nature (London)* **286,** 854–860.

Zannis-Hadjopoulos, M., Persico, M., and Martin, R. G. (1981). *Cell* **27,** 155–163.

Zehnbauer, B. A., and Vogelstein, B. (1985). *BioEssays* **2,** 52–54.

INTERNATIONAL REVIEW OF CYTOLOGY, VOL. 119

Biochemistry and Cell Biology of Amphibian Metamorphosis with a Special Emphasis on the Mechanism of Removal of Larval Organs

KATSUTOSHI YOSHIZATO

Developmental Biology Laboratory, Department of Biology, Faculty of Science, Tokyo Metropolitan University, Fukazawa 2-1-1, Setagaya-ku, Tokyo 158, Japan

I. Introduction

Many invertebrate and some vertebrate animals go through a special developmental phase between embryo and adulthood, referred to as the larval phase. The evolution from embryo to larva proceeds relatively slowly and continuously in the morphology and physiology of the individual. Embryo-to-larva development is usually characterized by the appearance of new structures, increased complexity of body systems, and an increase in body volume and weight. By contrast, developmental changes in the transition from larva to adult are more qualitative. The change is rapid and discontinuous: some larval organs cease to function and are destroyed, while others survive and transform into adult organs. This complicated process of development from larva to adult has been called metamorphosis. In this review, metamorphosis of amphibia is mainly considered. It is natural that many biologists have long been attracted by amphibian metamorphosis because it includes many basic biological processes, including cell differentiation, endocrinology, cell death, cell growth, and histolysis.

All the tissues of tadpoles transform from the larval to the adult type under pressure of metamorphosis. The extent and nature of the transformation appear to be diverse and to depend on the particular organs concerned. A tadpole is composed of three types of organs: larva-specific, larva-to-adult, and adult-specific. The larva-specific organs exist and function only in a larva (e.g., gill and tail). The larva-to-adult organs exist and function throughout the larval phase to adulthood; however, their structures and functions in the larva are different from those in the adult. This organ type includes the liver, gut, and body skin. An example of the third type of organ is the forelimbs, which do not exist in a larva but appear during metamorphosis and continue to function through adult life.

It is interesting to think about the mechanism of changes in these organs

97

during metamorphosis at the cellular level. I will summarize this metamorphic transformation of larval organs into adult ones from the standpoints of
cell differentiation and the mechanism of action of thyroid hormone (TH).
With this goal this review deals with three topics. First, an aspect of cell
differentiation of metamorphic transformation of larval organs is discussed. Second, I focus on the larva-specific organ, the tail. The larva-
specific organ is destined to disappear eventually at metamorphosis, as
mentioned before: the process of tail breakdown is described morphologically and biochemically, and the mechanism of the breakdown is considered in relation to both epithelium–mesenchyme interactions and TH-
induced cell death. Third, studies on larva-to-adult organs are surveyed,
targeting the skin, intestine, and liver. The mechanism of action of TH in
metamorphosis is also discussed, with special reference to receptors for
this hormone.

Throughout the review, an emphasis is placed on the removal of larval
organs, formulating such questions as these: What is the cellular and
biochemical basis for the death of cells and breakdown of tissues? How do
the larval cells change to adult cells? Is this change a transformation or a
replacement? Are there any processes of this change that are common to
various organs? How is TH involved in determining the fate of larval cells
and adult cells during metamorphosis?

For convenience, developmental stages of tadpoles are schematically
given in Fig. 1 (modified from Broyles, 1981). Tadpoles of *Xenopus laevis*
are staged according to Nieuwkoop and Faber (1967; NF stages, and those
of *Rana pipiens* and *Rana catesbeiana* are according to Taylor and Koll-

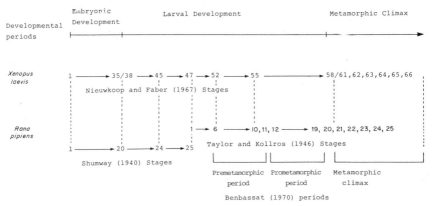

FIG. 1. Staging of tadpole development from embryos to tadpoles at metamorphic climax. Tadpoles of *Rana catesbeiana* are staged according to Taylor and Kollros (1946).
Modified from Broyles (1981).

ros (1946; TK stages). Benbassat (1970) periods of metamorphosis are also commonly used in the literature.

II. Metamorphic Transformation of Larval Organs and Cell Differentiation

As mentioned in the previous section, organs of an anuran larva can be grouped into three types according to their fate during metamorphosis: larva-specific organs like the gill and tail; larva-to-adult organs such as body skin, digestive organs, and blood circulation; and adult-specific organs like forelimbs. In order to understand the mechanism of transformation of larval organs during metamorphosis, it is apparently important to consider the transformation at the cellular level. In general, organs contain two types of tissues (i.e., epithelial and mesenchymal), the cells of which are called epithelial cells and mesenchymal cells, respectively.

The cells of each type of tissue are thought to comprise two kinds of cell populations: (1) undifferentiated and proliferative cells (germinative cells or stem cells), and (2) differentiated and nonproliferative cells (mature cells). Cells in an organ have their own life span and are engaged in a physiological turnover. When they come to the end of the span, they are lost from the tissue, the loss being compensated for by offspring of germinative cells. In this way the apparent cell number is kept relatively constant in a tissue (a state of dynamic equilibrium). In other words, a tissue, in general, contains a specific population of cells (germinative cells) that can enter the cycle of cell division to replace the mature cells that die.

In the following paragraphs the aim is to correlate the changes that larval organs undergo during metamorphosis with changes in the two kinds of cell populations mentioned previously. For this it should be kept in mind that TH initiates and governs all these changes. Possibilities of changes in the cells of larval organs during metamorphosis are discussed here, and are summarized in Table I.

A. LARVA-SPECIFIC ORGANS

It is natural to suppose that all the cells in this type of organ are subjected to cell death and are destined to be removed from the body during spontaneous metamorphosis. Before metamorphosis, larva-type germinative cells proliferate and produce progeny that go on to differentiate into larval-type mature cells. At metamorphosis, it can be expected that both the germinative and mature cells undergo cell death, because these larva-specific organs have no counterpart in the adult body. It should be empha-

TABLE I
POSSIBLE CHANGES IN CELLS OF TADPOLE TAIL DURING METAMORPHOSIS

Organs	Larva	Metamorphosis	Adult
Larva-specific organ	Larval germinative cells ↓	→ Metamorphic cell death	
	Larval mature cells ↓	→ Metamorphic or physiological cell death	
	Physiological cell death		
Larva-to-adult organ	Larval germinative cells ↓	→ Transformation	→Adult germinative cells ↓
	Larval mature cells ↓	→ Metamorphic or physiological cell death	Adult mature cells ↓
	Physiological cell death		Physiological cell death
	Adult germinative cells (dormant)[a]	→ Activation	→Adult germinative cells ↓
			Adult mature cells ↓
			Physiological cell death
Adult-specific organ	Adult germinative cells (dormant)[a]	→ Activation	→Adult germinative cells ↓
			Adult mature cells ↓
			Physiological cell death
	Larval germinative cells ↓	→ Transformation	→Adult germinative cells ↓
	Larval mature cells ↓	→ Metamorphic or physiological cell death	Adult mature cells ↓
	Physiological cell death		Physiological cell death

[a] Dormant adult germinative cells could also be called "progenitor cells for adult germinative cells."

sized that there seem to be two types of cell death: (1) the physiological death of mature cells at the end of their life span and (2) a metamorphosis-related death of cells (metamorphic cell death). The latter type of death occurs in cells that are still proliferation-competent and may or may not be near to physiological death. Even germinative cells could be subject to metamorphic cell death. Physiological cell death is considered to be TH-independent, but metamorphic cell death appears to be TH-dependent.

B. LARVA-TO-ADULT ORGANS

There are two possible pathways of cell transformation in this case. Either there is only one population of germinative cells or there are two. In the first case, larval-type germinative cells survive the metamorphosis and transform into the adult type. Mature larval-type cells would terminate their life as a result of either physiological or metamorphic death. In the other case, two types of germinative cells coexist in the larval organ— some larva-specific and others adult-specific. The larva-specific germinative cells would function only before metamorphosis and undergo metamorphic cell death, whereas the adult-specific germinative cells (progenitor cells for adult germinative cells) do not function and remain inactive or dormant until metamorphosis. At metamorphosis they are activated and start to function as germinative cells for adult life.

C. ADULT-SPECIFIC ORGANS

Theoretically, two possibilities can be considered. In the first, adult-type germinative cells are nonfunctional and dormant in premetamorphic tadpoles, becoming active only in prometamorphic or metamorphosing tadpoles. The other possibility is that the germinative cell for a larva-to-adult type organ, which has been functioning as a larval germinative cell, transforms into a germinative cell for the adult-specific organ during metamorphosis.

There is no doubt that thyroid hormones play leading roles in these changes of larval cells. Quantification of TH in serum of tadpole and young froglets has been published by investigators for bullfrog (Miyauchi et al., 1977; Regard et al., 1978; Suzuki and Suzuki, 1981) and for X. laevis (Leloup and Buscaglia, 1977). There is agreement that the level of TH in prometamorphic animals is very low or undetectable by radioimmunoassay, and that the TH surge starts at the metamorphic climax—indicating that TH is effective only at the climax of metamorphosis. However, we have very limited knowledge about the pathways of cellular changes, and we cannot yet say which is the actual pathway among possibilities previously mentioned in each case. The present review deals with the first and second types of organs.

III. Removal of a Larva-Specific Organ: Tail

The larva-specific organ may be the simplest type to study in considering the metamorphic changes at the cellular level, because this organ type appears to contain only a population of germinative cells for the larval organ that is to be removed during metamorphosis and therefore does not require the presence of progenitor germinative cells for adult life. The critical questions regarding the larva-specific organs are how TH terminates the life of cells that make up the organ, and how the materials other than cells are destroyed during metamorphosis. To answer these questions, we have made cell biological and biochemical studies targeting the tadpole tail as a typical larva-specific organ.

A. OVERVIEWS OF THE TADPOLE TAIL

The tail constitutes a large proportion of the tadpole's body—almost two-thirds of total body length in the case of bullfrog tadpoles. Tadpole tail has been one of the most frequently studied organs in experimental work on amphibian metamorphosis, for several reasons. The tail is easily recognizable; as an organ that is visible to the naked eye, without surgical intrusion, it shows dramatic changes during natural metamorphosis and also in response to exogenously administered TH (i.e., induced metamorphosis). It is known that the tail is one of the organs most sensitive to stimulation by TH. In addition, the tail can be cultured with relative ease for >1 week in a simple salt solution; furthermore, under these conditions the tail undergoes regressive changes in response to TH at physiological concentrations (references are cited later). Many studies on histolysis of the tail during spontaneous and TH-induced metamorphosis have accumulated since the last century (Guieysse, 1905; Brown, 1946; Weber, 1967; Frieden and Just, 1970; Atkinson, 1981).

The tail has a relatively simple architecture but does contain several tissues: nerve tissues, blood vessels, notochord, and muscle tissues (Fig. 2). Skeletal muscles are most voluminous. Highly hydrated gelatinous connective tissues surround the muscles and support the structure of the tailfins. The tail is covered by a thin epidermis. The cells proper to this organ therefore include epidermal cells, fibroblastic cells, macrophages, neurons, glial cells, endothelial cells and smooth muscle cells of blood

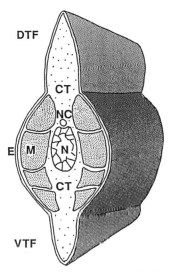

FIG. 2. A schematic representation of the tadpole tail. DTF, Dorsal tailfin; CT, connective tissue; NC, nerve cord; E, epidermis; M, muscle; N, notochord; VTF, ventral tailfin.

vessels, chondrocytes, muscle cells. These are cells that reside in the tail; needless to say, the tail also contains cells that circulate through the entire body: red blood cells, leukocytes, and lymphocytes. Naturally, we focus mostly on the resident cells in this review, because we are interested in the fate of the tail-proper cells during metamorphosis.

At the final stage of metamorphosis the whole structure of the tail disappears, as do the various types of cells just mentioned as well as substances of the extracellular matrices (ECM). How do they disappear? Do they die (in the case of the cells), or are they completely hydrolyzed to low molecular weight substances (in the case of ECM)? Do they survive the metamorphosis? Are they reused as cells or components of the adult body?

B. MORPHOLOGICAL DESCRIPTIONS OF BREAKDOWN OF THE TAIL

1. *Overall Features of Tail Regression*

The tail regression is one of the most conspicuous morphological changes during anuran metamorphosis. All the materials of tail tissues are subject to destruction. For example, in tadpoles of *R. catesbeiana,* the rate of protein degradation increases ~7-fold between TK stages 20 and 21: the rate at prometamorphic stages is 3×10^{-6} mg protein/mg tissue per minute and 22×10^{-6} at TK stage 22 (Little *et al.,* 1973). There seems to be a

distal–proximal gradient of the rate of regression according to the quantitative studies by Dmytrenko and Kirby (1981). They marked off with silk
sutures the tails of *R. catesbeiana* tadpoles into three segments along a
longitudinal axis (i.e., distal, middle, and proximal parts). The length of
each segment was recorded during $3,3',5$-triiodothyronine (T_3)-induced
tail regression. Their results clearly show that a regressive area is not
confined to the distal tip of the tail, and that reductions in length are
observed concurrently in proximal, middle, and distal segments.
However, the rate of shortening varies among the three segments, the
highest rate is observed in the distal segment, which regresses about twice
as rapidly as the proximal part.

Regression of the tadpole tail seems to include two different processes:
condensation and histolysis. Several workers have reported that the predominant initial response of the tissue to the metamorphic stimulus is
water loss, resulting in an apparent tissue regression by the condensation
of cells and ECM. For example, Lapiere and Gross (1963) found that
T_3-treated tadpoles lose 50% of their tail water during 6 days; thereby the
collagen and DNA content of the tail increasees 2-fold. The many studies
on the loss in tissue water have been reviewed by Frieden (1961). As
detailed later, the breakdown of tail requires cell–cell interactions and
biochemical (enzymatic) reactions. The condensation might facilitate
these cellular and biochemical processes. Another component of tail regression is a histolysis of the tissue. The condensation and the histolysis
progress concomitantly from the beginning to the completion of the tail
breakdown, which might be the basis of a well-organized process of degeneration.

The condensation of tail tissues during metamorphosis might help to
explain the mechanism of tissue breakdown, but this has received little
attention. The mechanism of condensation or contraction of tissues has
been largely unknown, as attested to by the work of Coulombre and
Coulombre (1964). The cornea contains multilayered structures of collagen in a highly organized orthogonal arrangement, which resembles
structures of collagen layers in the basement lamella of tadpole skin as
detailed later. Coulombre and Coulombre showed that dehydration occurs
at the beginning of corneal development; it is interesting to note that this
corneal dehydration is dependent on the presence of TH.

Bell *et al.,* (1979) have developed an experimental model in which
fibroblasts are cultured three-dimensionally in hydrated collagen lattices.
The model shows an extensive contraction of collagen fibrils resulting in a
rearrangement of the fibrils. This cell-mediated condensation of the model
is partly dependent on the presence of TH (Taira and Yoshizato, 1987).
These observations strongly suggest that tail dehydration during metamor-

phosis might also be induced by the action of TH. Thus condensation appears to be an important phenomenon in the mechanism of tissue regression and is worthy of further intensive study.

There have been many terms used to refer to tail regression in the literature: breakdown, dissolution, autolysis, histolysis, phagocytosis, heterolysis, degeneration, regression, contraction, absorption, atrophy, and necrosis. Kerr *et al.* (1974) suggest that the terms atrophy, necrosis, and degeneration are not suitable to describe the tail breakdown, because this process is composed of developmentally programmed cell death and organized removal of the dead cells and ECM by phagocytosis and enzymatic degradation. Apoptosis might be a suitable term for the tissue destruction of metamorphosing tadpoles. Cell death in higher animals is classified into two distinct types: necrosis and apoptosis (Kerr *et al.*, 1972). The former is characterized by "marked swelling of mitochondria followed by dissolution of internal and plasma membranes," which is widely recognized in pathological cell death. The latter, by contrast, is characterized by "rapid condensation of the cell with convolution of its surface, followed by separation of the surface protuberances to form membrane-bounded globules in which organelle integrity is initially maintained" (Kerr *et al.*, 1987). The proposal of Kerr *et al.* (1974) is supported by the electron-microscopic (EM) observation of tail muscle destruction by Sasaki *et al.* (1985). Even when the junctional folds of the nerve endings disappear at the beginning of metamorphosis, acetylcholinesterase activity is still observed in the synaptic cleft. This observation supports the idea that muscle breakdown at metamorphosis is not a denervation atrophy. Earlier investigators had already demonstrated that the tail degeneration is not due initially to a failure of blood supply (Helff, 1930; Brown, 1946; Atkinson, 1981) or of innervation (Brown, 1946).

2. *Breakdown of Skin*

The skin is composed of two tissues of developmentally different origin: epidermis and dermis (mesenchymal tissue). These two tissues are separated by basement membrane. As discussed later in detail, epidermis of the tail is different in cellular composition from that of the body skin. This difference is of biological significance in that the tail skin is a larva-specific tissue while the body skin is a larva-to-adult tissue: The presence in this case of two types of an organ (i.e., skin) that have different metamorphic fates provides a unique and interesting opportunity to gain an insight into a mechanism of metamorphic transformation at the cellular level. Epidermis of the tail is three cell layers in depth; the outermost layer is made up of apical cells and the remaining two layers contain cells with figures of Eberth. The figures of Eberth, discovered by Rudneff (1865) and described

by Eberth (1866) independently, are threadlike structures in epidermal
cells of frog tadpoles (see Singer and Salpeter, 1961, for historical description of figures of Eberth). Robinson and Heintzelman (1987) call these cells
skein cells; others call them the cells with bobbins (Weiss and Ferris, 1954;
Usuku and Gross, 1965) or tufts of tonofilaments (Gona, 1969). The figures
of Eberth, which are now regarded as huge bundles of intermediate filaments (11 nm in diameter), are specific to tadpole epidermal cells. Some
skein cells attach to basement membrane with hemidesmosomes, and, in
this case, Eberth's figures are associated with the hemidesmosomes.

The metamorphic changes of tail epidermal cells were surveyed
electron-microscopically using tadpoles of *Rana japonica* (Kinoshita *et
al.*, 1985). The earliest change is noticed in the outermost cells, where
vacuoles with acid phosphatase activity appear and the cell membranes
are destroyed. These changes spread toward the inner layers as metamorphosis progressses. The next changes observed are ruptures of structures
of desmosomes, and the pyknosis of nuclei. Lymphocytes, neutrophils,
and especially macrophages become conspicuous around TK stages
23–24. Macrophages are often observed to phagocytose the dead cells.
From these observations Kinoshita *et al.* propose that the destruction of
tail skin cells proceeds in two sequential steps: autolysis and heterolysis.
The autolysis is the death of epidermal cells, which progresses inward
from the apical layer. The heterolysis is the phagocytosis of the autolysed
cells by macrophages. Kerr *et al.* (1974) made extensive EM observations
of autolysis of tail epidermal cells of *Litoria glauerti*. The first change in
the cells is a condensation of chromatins on nuclear membranes, followed
by breakdown of nuclei and condensation of cytoplasm. As a result many
fragmented cell bodies with cell membranes are produced. Organelles in
the cell body such as mitochondria appear to be intact. These authors
claim these changes are not specific to the metamorphosis but are generally observed in the process of programmed cell death in vertebrate embryogenesis and physiological turnover of cells. The fragmented cell bodies are named apoptotic bodies. The process leading to the formation of
apoptotic bodies revealed by Kerr and Searle corresponds to autolysis as
described by Kinoshita *et al.* (1985). Apoptotic bodies are then phagocytosed by macrophages (heterolysis). Based on these studies, the process of
removal of tail epidermal cells is schematically summarized in Fig. 3.

Thyroid hormone-induced metamorphosis causes changes in tail
skin, which are similar to the spontaneous metamorphosis mentioned
earlier (Usuku and Gross, 1965). The effect of TH is noticed at day 5 of the
treatment, when intracellular structures start to be disordered and the
figures of Eberth become less prominent. At day 8–10 after treatment, the
epidermis thickens and cornified cells appear, suggesting that TH acceler-

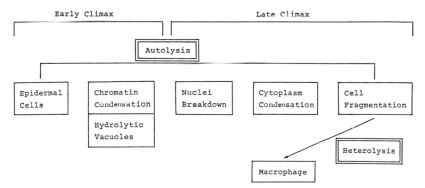

FIG. 3. Metamorphic changes in tail epidermal cells.

ates both breakdown and differentiation of the epidermal cells. *In vitro* metamorphosis of the epidermis was thoroughly examined electron-microscopically by Gona (1969), who cultured tailfin tissues of *R. pipiens* in the presence of TH and the identical changes were observed: the epidermis shows hormonal effects at day 4 and thickens by day 6 from two to three, to five or six cell layers thick.

Dermis (mesenchyma) of tadpole tail has unique structures of collagen fibers named basement lamella, in which each layer of collagen fibers is orthogonally arranged, with 20 layers of thickness at TK stage 4 in case of *R. pipiens* (Weiss and Ferris, 1954). This characteristic structure of collagen lamellae appears as early as at Shumway stage 22 in tadpoles of *R. pipiens* (Kemp, 1959). The basement lamella reaches to 3–6 μm in thickness in *R. catesbeiana* before metamorphosis. Beneath this thick collagen layer there exists an amorphous gelatinous core of mesenchyme, where mesenchymal cells scatter.

Usuku and Gross (1965) gave detailed EM observations on the TH-induced breakdown of the tail mesenchyme of *R. catesbeiana*. Swelling of basement lamellae is the most outstanding feature in the dermis of TH-treated tadpoles. The lamella swells to 6–9 μm thick on day 5. Swelling continues thereafter to 9–12 μm at day 8, 14–17 μm at day 10, and 30–40 μm at day 14, when the tail is almost absorbed. Concomitantly with the lamella thickening, collagen layers are frayed out and fragmented without any participation of mesenchymal cells in the initial phase (autolytic phase), but later phagocytic mesenchymal cells (macrophages) became prominent, the cytoplasm of which is often shown to contain fragmented collagen bundles (heterolytic phase). The almost identical process of mesenchymal degradation of the tail of *R. catesbeiana* was described electron-microscopically by Gona (1969) for tailfin cultured in the presence of TH.

The origin of the macrophages often observed in the metamorphosing tail is controversial. Usuku and Gross (1965) characterize the macrophages as rich in granules and organelles like mitochondria. Many researchers have described macrophages or macrophagelike mesenchymal cells (see review in Dodd and Dodd, 1976). Weber (1964) proposed the possibility of transformation of connective-tissue mesenchymal cells to macrophages. Others have pointed out that macrophages originating in blood capillaries as well as resident macrophages could function in the heterolytic phase of tail destruction (Lehman, 1953; Kinoshita *et al.,* 1985). We have succeeded in isolating an almost pure population of macrophages and culturing them as discussed later. Determination of the origin and transformation of macrophages promises to be a fascinating and important subject for researchers in modern cell biology and developmental biology of metamorphosis.

3. *Breakdown of Muscle*

Several authors have provided morphological descriptions of the breakdown of tail skeletal muscles. The study by Watanabe and Sasaki (1974) can be summarized as follows. The degradation does not occur in a simultaneous and concomitant fashion in all parts of tail muscles; rather, it occurs apparently in a random fashion, so that intensively fragmented and degenerated muscles are in proximity to intact and apparently active muscle tissues. There seem to be two patterns of muscle degradation, although both patterns are initially similar, characterized by the degradation and disappearance of Z disks. After this, some are fragmented into the sarcolytes and the others frayed and fanned out at the region previously occupied by Z-disks. The red (or slow) muscles that occupy superficial layers of muscle tissue are subjected to the former degradation pattern, whereas white (rapid) muscles found in deeper layers undergo the latter pattern.

There seems to be some difference of opinion in the literature with respect to the role of mesenchymal cells in muscle degradation. Watanabe and Sasaki (1974) emphasize the participation and key roles of mesenchymal cells at the beginning of the breakdown; their invasion into the muscle fibers by extending long cytoplasmic protrusions is an inevitable process for the ensuing fraying out of the muscle and disappearance of Z disks. Kerr *et al.* (1974) do not describe this initial invasion of the mesenchymal cells. Both authors reached the same conclusion for the role of macrophages in the final stage of degeneration of muscles; that is, macrophages with intense lysosomal enzymes like acid phosphatase phagocytose sarcolytes. Kerr *et al.* (1974) emphasize the concept of apoptosis for the degenerative process of muscles, because they noticed electronmicroscopically the presence of condensation of chromatins and the remarkable intactness of intracellular structures of sarcolytes.

C. Epithelium–Mesenchyme Interactions in Tail Regression

There are many examples of biological phenomena that require normal interactions between tissues, among them inductive interactions in embryogenesis and epithelium–mesenchyme interactions in both organogenesis and maintenance of structures and functions of mature organs. Several studies have demonstrated the importance of these kinds of interactions in the phenomena of metamorphic breakdown of tadpole tail as well.

1. Determination of Epidermis as "Tail" Epidermis by the "Tail" Mesenchyme

The tail, as a larva-specific organ, is characterized by its self-destruction during metamorphosis. There remains the question of how this character of the "tail" is determined. Kinoshita *et al.* (1986a) tried to answer this question with a classical organ transplantation technique. As detailed in Section II, body skin is a larva-to-adult organ, while tail skin a larva-specific one. It is interesting to look for the mechanism by which this apparent difference in metamorphic fate originates in the two skin types. Dorsal skin autotransplanted into the tail of a premetamorphic tadpole of *R. japonica* maintains its original character; that is, the skin is not subjected to self-destruction during metamorphosis, but survives and transforms into adult-type skin. On the other hand, the fate of tail skin transplanted into the back is quite interesting in that it does change its original character—that of tail skin. The transplanted tail skin is not destroyed at the metamorphic stage, but instead survives and transforms into adult-type skin. This study shows that not only back skin but also tail skin is capable of transforming into adult-type skin.

In surveying electron-microscopically this unexpected response of the transplanted tail skin, it was noticed that the tail dermis undergoes self-destruction as its original fate in the transplanted site: the tail mesenchymal collagens are phagocytosed by macrophages. Then, fibroblasts in the surrounding host tissues migrate into the degenerating transplanted tail dermis and secrete collagens, resulting in a transient coexistence of degenerating collagens of the transplant and newly formed host collagens. The tail dermis of the transplant is completely replaced by host back skin dermis. On the other hand, the dermis in the transplanted back skin grows and remains intact during metamorphisis. In summary, transplanted dermis of tadpole undergoes self-determination irrespective of the transplantation site, as has been shown by other investigators (Rand and Pierce, 1932; Jacobson and Baker, 1968; Baker and Jacobson, 1970). However, the transplanted epidermis does not undergo self-determination; the tail epidermis transdifferentiates into back epidermis.

To answer the question on this apparent difference between epidermis and dermis. Kinoshita *et al.* (1986b) carried out transplantation experiments using recombinant skins. Skin was removed from both sites and treated with proteases; epidermis and dermis were then separated. Homotypically and heterotypically recombined skins were reconstructed from the two separated tissues and were homografted to the tail region from which the skin had been excised. The fate of the grafts was monitored electron-microscopically until the subject tadpoles became young froglets. Among the recombinant skin types studied, the epidermis combined with tail dermis undergoes degenerative changes, irrespective of the origin of the epidermis—tail or back. This study emphasizes the determining role played by the dermis in changing the character of tail epidermis into that of back epidermis. It can be concluded from these results that the information that leads the tail skin to degenerate at metamorphosis under the influence of TH exists in dermis, not in epidermis. What is the chemical nature of this information that resides in dermis? This is an intriguing question that requires further intensive study to be answered.

The factor in the dermis that determines it to be "tail dermis" at some developmental stage induces not only the degeneration of tail epidermis but also the transdetermination of epidermis that was predetermined to be the epidermis of back skin, into "tail epidermis." The determination of dermis as the "tail" type is suggested to occur at some time between the late-neurula and tailbud embryo stages; this was estimated using the same procedure of skin autotransplantation between regions of presumptive back and tail. With additional confirmational experiments, Kinoshita *et al.* (1989) conclude that mesoderm of the presumptive tail region of a tailbud embryo contains the information previously proposed. The epithelium–mesenchymel interactions in the degeneration of tail skin are summarized schematically in Fig. 4. The mesenchyme-dependent survival of tail epidermis beyond metamorphosis has also been reported by Naughten and Kollros (1971).

2. An Epidermal Factor That Is Required for TH-Induced Mesenchymal Regression

Another type of epithelium–mesenchyme interaction in the degeneration of tail skin has been reported by Yoshizato and associates. Niki *et al.* (1982, 1984) tried to culture epidermis-free mesenchymal blocks from tail skin of a tadpole of *R. catesbeiana*. Mesenchymal blocks were prepared

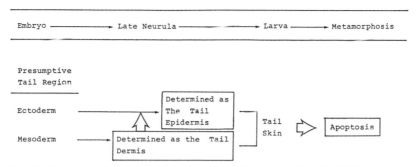

FIG. 4. Determination of dermis as the tail dermis and the fate of tail skin during metamorphosis.

either by a mechanical separation with fine forceps or by chemical (EDTA, ethylenediaminetetracetic acid) and trypsin treatments of tailfins. Epidermis-deprived mesenchymes were cultured in media containing TH. They failed to regress in response to TH, while normal fins and the mesenchymes cocultured with epidermis showed TH-dependent regression. The TH insensitivity was also found for epidermis-deprived muscle tissue. Therefore, it is strongly suggested that some factor(s) released from epidermis is required for the TH-dependent regression of mesenchymes or muscle tissues. This suggestion was further supported by the following observations. Skin from tadpole tails was cultured for 2 days. Skin-conditioned media (SCM) were recovered, centrifuged, filtered, and used as media for culturing mesenchymal blocks. The mesenchymes regressed in SCM when TH was added, but not without TH; thus, SCM should contain the factor suggested earlier. Only limited amounts of information on the chemical nature of this factor are available at present. Niki and Yoshizato (1986) characterize it as follows: the factor is not inactivated by boiling or by treating with trypsin or pronase, is dialyzable, and is taken up by Sephadex G-10 resin, indicating a nonpeptide substance of low molecular weight. However, we have noticed irreproducibility of TH dependence of the regression induced by SCM. Sometimes SCM induces mesenchymal regression without the cooperation of TH. There seems to be a critical concentration of the putative factor in SCM, beyond which SCM induces the regression without TH.

There should be a discussion on the relationship between the active substance in SCM and collagenase, because this enzyme has been well known as a key enzyme for the destruction of collagens rich in mesenchymal tissues (Section III,D,1). According to Gross and Lapiere (1962), SCM should contain relatively high amounts of collagenase; we have confirmed this observation. There are no reports available that demonstrate the direct effect of purified collagenase on the mesenchymal regression. The active substance is heat-stable and of low molecular weight; this excludes the possibility that the factor is collagenase. Eisen and Gross (1965) showed that mesenchyme of the tailfin does not contain collagenase activity. However, we were able to show the presence of collagenase activity in this tissue using a more sensitive assay method, although the activity is much less than that found in epidermis (K. Yoshizato, unpublished observations). At present, there is a possibility that the factor in the epidermis is an activator of collagenase in mesenchyme.

Kinoshita *et al.* (1986a,b) emphasize the principal role of mesenchyme in regression of the tail skin; for them, the epidermis is just a passive tissue, the fate of which depends on the dermis to which it adheres. Niki *et al.* (1982, 1984; Niki and Yoshizato, 1986), in contrast, emphasize the important role of the epidermis in the regression of dermis. A detailed explanation of the relationship between the two observations should become possible as soon as the chemical nature of the information transmitted from dermis to epidermis claimed by Kinoshita *et al.* and of the factor in epidermis claimed by Niki *et al.* have been clarified. There seems to be no doubt about the principal importance of the mesenchyme in the TH-dependent regression of epidermis. However, it is not clear whether the tail mesenchyme can independently respond to TH and regress autonomously during metamorphosis, when transplanted to the back. As described earlier, we repeatedly noticed that epidermis-deprived tail dermis does not regress in response to TH *in vitro*. However, the tail dermis transplanted with epidermis to the back regresses in response to TH (Kinoshita *et al.*, 1986a), suggesting that tail dermis is capable of responding to TH and regresses autonomously. This discrepancy between the two observations remains to be resolved.

D. BIOCHEMISTRY OF BREAKDOWN OF LARVA-SPECIFIC TISSUES

In Section III,B,2 and 3, the morphological aspects of tissue breakdown were summarized, focusing on the skin and muscle tissues of the tadpole tail. In this section we review proteolytic enzymes, which function in removing major chemical constituents of the two tail tissues, mesenchyme and muscle.

1. *Mesenchyme*

As has been mentioned, the tail mesenchyme is characterized by the presence of thick collagen lamellae (basement lamella) beneath the epidermis and of gelatinous amorphous matrices in the core. Tail mesenchyme contains exclusively type I collagen (K. Yoshizato, unpublished observations), which is a major protein constituent of basement lamellae and gelatinous-core matrices. As is well known, animal collagenase was detected first in media in which tadpole skin had been cultured (Gross and Lapiere, 1962), and was purified by Nagai et al. (1966). Highly purified preparations of tadpole collagenase were characterized in several laboratories (Hori and Nagai, 1979; Bicsak and Harper, 1984, 1985): the molecular weight was ~49K and the isoelectric point 5.0. Bicsak and Harper (1984) hypothesized the existence of two distinct types of this enzyme, sharing similar physical properties but with different specificities. It has been shown that tissues of back skin and tailfin of tadpoles produce procollagenase and an activator that converts the zymogen into active collagenase (Harper et al., 1971; Harper and Gross, 1972). Eisen and Gross (1965) claimed that epidermis releases collagenase into media when cultured, but mesenchyme does not. However, we detected collagenase activity in the mesenchyme of tailfin as well, although the activity was less than one-tenth that of the epidermis (K. Yoshizato, unpublished observations).

According to Lipson and Silbert (1965), tadpole skin contains hyaluronic acid (HA) as a major component of extracellular glycosaminoglycans (GAG). At least 96% of the GAG is HA and the remaining 4% has not been fully characterized. Evidently, hyaluronidase plays key roles in removing ECM of tadpole tail. Gross and his associates (Silbert et al., 1965) purified this enzyme from the tadpole tail and characterized it as endo-β-N acetyl-hexosaminidase with activity similar to that of testicular hyaluronidase. Eisen and Gross (1965) demonstrated that hyaluronidase is produced by mesenchyme but not by epidermis.

There is no doubt that these two enzymes have important roles in destroying and removing ECM of the larva-specific tissue during metamorphosis. However, little is known about the metamorphic regulation of these enzyme activities. The activity of collagenase has been reported to increase in the tail skin during TH-induced metamorphosis (Lapiere and Gross, 1963). Tadpoles of R. catesbeiana were induced to metamorphose by placing them in water containing 10^{-6} M thyroxine (T$_4$). Little increase in activity was found at day 2. Collagenolytic activity of T$_4$-treated animals increased 4-fold as compared to that of controls at day 6, when the tail had shortened by ~30%. Induction of collagenase by TH was reported also in tailfins cultured in vitro. (Davis et al., 1975). Explants of tailfin tissues of R. catesbeiana were cultured by the method of Derby (1968). Reepithelialized and healed explants respond to TH and release collagenase into the medium, whereas they do not in the absence of TH. These studies indicate the regulation of collagenase activity, most likely at the level of transcriptional expression of collagenase genes by TH, because regression of the explants by TH is known to be inhibited when actinomycin D or puromy-cin is present as discussed later. Molecular biological approaches are essential to further our understanding of the mechanism of collagenase induction by TH.

There has been little evidence that hyaluronidase is also induced by TH in the tail tissues. In the back skin, where extensive removal of HA occurs at metamorphosis as described later, Polansky and Toole (1976) showed that TH increases the enzyme activity 2.5-fold. Eisen and Gross (1965) suggested that the enzyme is stored in the mesenchymal cells and probably released when needed, because the production of hyaluronidase from the cultured mesenchyme is not inhibited by puromycin.

There has been little experimental work attempting to correlate the morphological changes in dermis during metamorphosis, described in Sec-tion III,B,2, with the biochemical studies described in this section. Weiss and Ferris (1954) postulated that hyaluronidase secreted by mesenchymal cells is responsible for the initial fraying out of collagen lamellae. No data are yet available regarding the question whether collagenase is directly responsible for the ensuing changes in collagen lamellae, the fragmentation of collagen bundles. As discussed in Section III,B,2, the mesenchyme of tadpole tail is characterized by the presence of thick collagenous layers (basement lamella) with a remarkable degree of architectural order (Weiss and Ferris, 1954). Therefore, the most prominent event occurring in the mesenchyme is a degradation of basement lamella. Little is known about the biochemical aspect of this breakdown. There is no doubt that col-lagenase functions *in vivo* to destroy collagen fibers, because Dresden (1971) was able to demonstrate the presence of collagenase degradation products in tailfin from tadpoles undergoing TH-induced metamorphosis. Whether collagenase plays a principal role in the destruction of the base-ment lamella could be determined by biochemically and electron-microscopically examining changes in the lamella after the lamella isolated from tail is digested by highly purified tadpole collagenase. Such an experi-ment has been published for human rheumatoid synovial collagenase: Woolley *et al.* (1975) demonstrated that the enzyme can degrade highly crosslinked insoluble collagen without intervention of nonspecific proteases.

There have been no experimental attempts to gain an insight into the biochemical mechanisms of phagocytosis of collagen bundles by macro-phages as described in Section III,B,2. Cathepsin D might be involved in a still-undetermined process in the breakdown of mesenchyme, because Seshimo *et al.* (1977) reported that pepstatin, a specific inhibitor of the enzyme, suppresses the TH-dependent regression of cultured tailfins. The possibility that an activator of collagenase is the TH-dependent mesenchyme regression-inducing factor of epidermis (Niki *et al.,* 1982, 1984; Niki and Yoshizato, 1986) is discussed in Section III,C,2.

2. Muscle Tissues

The enzymes to which investigators have given intensive attention in relation to degradation of muscle proteins have been cathepsins. It has been well known since Weber's report in 1957 that cathepsin activity increased markedly and rapidly right after the initiation of metamorphosis (Eeckhout, 1969; Weber, 1977). Horiuchi and associate (Sakai and Horiuchi, 1979a,b) confirmed this for cathepsin D and characterized this enzyme. Nanbu *et al.* (1988) have purified cathepsin D and characterized it as follows: the molecular weight is 38K and the pH optimum when hemoglobin is used as a substrate is 3.5; this enzyme hydrolyzes muscle proteins when incubated with myosin B prepared from tail muscle.

Yoshizato and Nakajima (1982) demonstrated the presence of a new and unique enzyme (protease T_1) that preferentially attacks actin. This enzyme was accidentally discovered when patterns of protein banding of tail muscle tissues of *R. catesbeiana* were compared at various stages of metamorphosis on polyacrylamide gels electrophoresed in the presence of sodium dodecyl sulfate (SDS). It was noticed that some tadpoles at TK stages 18–20 and all tadpoles at climax stages examined lack the band corresponding to actin. We postulated the presence of a proteinase that preferentially attacks actin, because the band of actin was normally observed when proteins were detected from the tissue boiled just after removal from the animals.

This enzyme was purified as follows (Motobayashi and Yoshizato, 1986): precipitation with acetone (35–55%), gel filtration chromatography with Sephadex G-75, affinity chromatography with Affi-Gel 501, and cation-exchange chromatography with CM-cellulose. The enzyme was purified to homogeneity, giving a single band on SDS–PAGE (polyacrylamide gel electrophoresis). The enzyme has a molecular weight of 26K, being composed of two subunits each having an apparent molecular weight of 13K. Until the gel filtration chromatography step, protease T_1 does not attack actin in the absence of SDS, suggesting the presence of a specific inhibitor for this enzyme. This suggestion has been substantiated, because the inhibitor was removed from the enzyme at the step of CM-cellulose chromatography. Inhibitor-free protease T_1 digests actin in the absence of SDS, giving major degradation products with molecular weights of 38K, 35K, and 28K, while in the presence of SDS, only the MW 28K polypeptide is prominent. The enzyme requires thiol compounds for the expression of the activity. Studies with artificial synthetic peptides reveal that protease T_1 preferentially splits a peptide bond at the carboxyl side of the lysyl residue of a Val-Leu-Lys-X-sequence. The substrate specificity of protease T_1 has been checked among several proteins: actin, myosin H chains, and troponin I and C are degraded; the enzyme does not hydrolyze myosin L chain, tropomyosin, troponin T, or bovine serum albumin. Some characterization of the inhibitor has been done, revealing that it is a peptide with a relatively low molecular weight around a few thousand. The inhibitor is released from the enzyme–inhibitor complex at an acidic pH.

As mentioned earlier, *in vitro* tests of the enzyme show that protease T_1 is not a protease specific to actin. However, this enzyme preferentially attacks actin filaments, when myofibrils or muscle tissues of tadpole tail are used as its substrate. (Motobayashi *et al.,* 1986; Yoshizato, 1986). Protease T_1 converts myofibrils into small fragments. Analyses of the digest by SDS–PAGE reveal the presence of a major degradation product with a molecular weight of 28,000, which is a specific polypeptide produced when actin molecules are hydrolyzed by this enzyme; this indicates a physiologically significant role for protease T_1 in breakdown of muscle tissues. At several time points during the digestion, the tissue specimens were subjected to EM observations. This enzyme preferentially attacks the I band of muscle, which is disorganized as early as 5 minutes of digestion; actin filaments are frizzled, while the A band is intact and the Z disks are discernible. With a 4-hour digestion the I band and Z disks completely disappear, while the A bands are still intact and are overlapped, resulting in the appearance of electron-dense zones between them. The M line and the H band remain intact. In some photographs, fanlike structures of frayed myosin filaments are visible. These degradation patterns of muscle fibers by purified protease T_1 strikingly resemble the patterns of *in vivo* degradation during spontaneous metamorphosis as described in Section III,B,3. The mode of action of protease T_1 toward muscle fibers is schematically demonstrated in Fig. 5.

This resemblance suggests the following hypothesis: (1) protease T_1 plays a key role in the metamorphosis-induced degradation of muscle tissues, and (2) protease T_1 might be useful for determining whether the process of degradation of tail muscle tissues is heterolysis or the autolysis, or both (Section III,B,3). In autolysis, proteases synthesized by muscle cells play leading roles, while in heterolysis macrophages have a principal function in removing muscle fibers. Mesenchymal cells are thought to play some role at the beginning of muscle regression in fraying out the fibers. Protease T_1 induces the changes in muscle from the initial to late stages of degeneration. Presently, it is not known which cells in muscle tissues produce protease T_1. If muscle cells, but not mesenchymal cells do it, then autolysis is the more likely mechanism. If, however, it turns out that mesenchymal cells, but not muscle cells produce protease T_1, then the

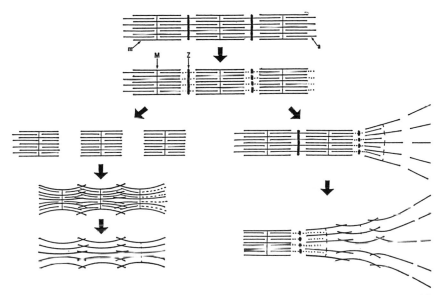

FIG. 5. A schematic representation of the process of muscle degradation induced by protease T_1 a, Actin filaments; m, myosin filaments; M, M lines; Z, Z disks. Dotted lines show the structures split by the enzyme. From Yoshizato (1986).

heterolysis theory would be better. It remains to be studied whether TH regulates the synthesis of protease T_1.

As mentioned earlier, protease T_1 should be inactive *in vivo* because the enzyme exists as a complex with the inhibitor, the chemical nature of which has not been understood completely yet. Even at the climax stages of muscle degradation most protease T_1 is complexed with the inhibitor, because the enzyme is known to be recovered as a complex form from tissues undergoing massive degradation. It is not known at present what mechanism is reponsible for the dissociation of the inhibitor from the complex. As mentioned earlier, the solution pH has an influence on the association–dissociation balance between the enzyme and the inhibitor. It has been generally accepted (Weber, 1967; Frieden, 1967) since the report of Aleschin (1926) that degenerating tissues become acidic. Therefore it is probable that the enzyme is easily activated only in degenerating tissues, which show acidic pH.

E. CELL DEATH

1. *Epidermal Cells*

Every cell in the body has it own life span and eventually proceeds to die. As discussed in Section II, tissues usually contain two populations of cells, germinative (stem cells) and differentiated—the latter of which are progenies of the germinative cells. Differentiated cells undergo cell death and are replaced by new progenies from germinative cells (physiological turnover of cells). Epidermis is a unique tissue for considering the physiological cell turnover: germinative cells and differentiated cells are easily distinguishable, because the two populations exist in different layers in the tissue. Proliferative cells locate basally and usually attach to the basement membrane or are close to it. Basal cells are in the stratum basale and prickle cells in the stratum spinosum; the two layers are called collectively the germinative layer or Malpighian layer. The differentiated progeny moves outward and becomes cornified cells at the terminal state, just prior to death. Thus we can observe every population of cells with varying extents of proliferation and differentiation as a distinguishable population, within the same tissue and at the same time. In other words, epidermis is an ideal tissue to employ in investigating the physiological turnover of cells.

It is well known that in mammalian epidermis germinative cells are smaller than differentiated cells, which suggests there may be some correlation between proliferative activity and cell size. Barrandon and Green (1985) verified this: small epidermal cells have higher colony-forming activity than large cells. A similar positional hierarchy of epidermal cells is found in the epidermis of adult frogs: there are layers of germinative cells, granular cells, and cornified cells.

As illustrated in Fig. 15, the epidermis of the tadpole tail is relatively simple in structural organization. It contains three types of cells known as apical, skein, and basal cells (Section III,B,2; Robinson and Heintzelman, 1987). What is the relationship among the three? Are the cells in a deeper layer more proliferative as compared to upper ones, as is the case in mammals or adult frogs? Are they in a state of physiological cell turnover in the cycle of basal cells to skein cells to apical cells?

According to data currently available, the larval epidermis seems to be quite unique in lacking the expected hierarchy found in the typical epidermis. There is no population of germinative epidermal cells in tadpole skin (discussed in Section IV,A), which does not mean that tadpole epidermis contains no population of proliferative cells.

Epidermal cells were isolated from the tail by successive treatments with EDTA and trypsin in that order (Nishikawa and Yoshizato, 1985). All apical cells and some skein cells were removed by EDTA. Most skein cells and all basal cells were obtained by the trypsin digestion. There is no significant difference in size among apical, skein, and basal cells; the cells from the EDTA treatment are 14–15 μm in diameter and those from the trypsin digestion 13–14 μm. This indicates that there is no gradient of cell size distribution between inner and outer cells of epidermis, as in the case of mammalian skin.

A major population of cells obtained by the trypsin treatment consists of skein cells with the figures of Eberth. Robinson and Heintzelman (1987) described the lack of basal cells in tail epidermis, a conclusion they reached after light-microscopic observations of fixed and sectioned specimens of tail. The cells obtained were cultured on a plastic surface coated with fibronectin. We estimated from the appearance of cells under a phase-contrast microscope that 95% of the cells are skein cells and the remainder basal cells (Nishikawa et al., 1989). Both cells show mitotic figures, indicating they are proliferative. Indeed they can grow in vitro, increasing cell number 2.5-fold in 3 days (Nishikawa and Yoshizato, 1985).

The fate of tail epidermal cells during metamorphosis is inferred from the response of the cells to TH in culture. For the culture of tail cells, it has been noted that TH-deprived serum should be used, because the cells are sensitive to TH present in fetal calf serum, which is commonly used for cell culture: calf serum usually contains 10^{-8} M T_4 and 10^{-9} M T_3 (Nishikawa and Yoshizato, 1985). Therefore, serum was treated with charcoal to remove TH. The epidermal cells remain healthy and at steady state for at least 10 days in culture when TH is absent. It was also shown that the presence of cortisol improves the maintenance of epidermal cells. In the presence of TH as 10^{-8} M T_3, the same cells cannot survive more than a week. The survival time of the cell is dependent on the concentration of TH in the medium (Fig. 6). Apparently, TH hastens the time of death, shortening the life span of the cell (Nishikawa and Yoshizato, 1986a).

The following experiment demonstrates that TH induces epidermal cell death by suppressing DNA synthesis (Nishikawa et al., 1989). Epidermal cells were cultured for 4 days and the rate of DNA synthesis was measured by the incorporation of [^{3}H]thymidine into cells. Thyroid hormone markedly reduced DNA synthesis (~15% of the control; Fig. 7). This has been confirmed by studies with bromodeoxyuridine (BrdU), a thymidine analog. The cells that enter S phase of the cell cycle can be distinguished by labeling with the analog and by detecting this with anti-BrdU antibody. It was found that TH decreased the rate of entry to 50% of the control culture. When the rate of DNA synthesis is calculated on the basis of "per cells in S phase," there is no difference in the rate between experimentals and controls. Therefore, it is suggested that TH suppresses the entry of epidermal cells into S phase and thereby the DNA synthesis, and induces the eventual death of the cells. We call this TH death or metamorphic death to distinguish it from the physiological death that results from normal turnover of epidermal cells.

We have conducted experiments to try to determine whether TH exerts

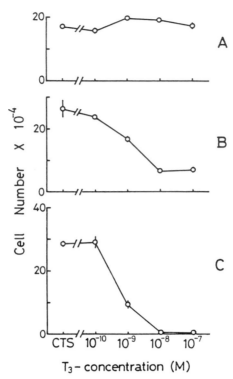

FIG. 6. Thyroid hormone-dependent death of epidermal cells. Epidermal cells of tadpole tail were cultured for ≤8 days in the presence of various concentrations of T_3. (A) Day 0; (B) day 5; (C) day 8. CTS, Cultures without T_3. From Nishikawa and Yoshizato (1986a).

its effect also on the physiological death of tail epidermal cells. For this, morphological and biochemical characters were examined, which are associated with terminal differentiation of epidermal cells (Nishikawa et al., 1989). Transmission EM observations revealed that the epidermal cells cultured in the presence of TH have thickened cell membranes and are significantly keratinized. The activity of transglutaminase, a marker enzyme of keratinization, is markedly elevated—as high as ~10-fold (Fig. 8). The keratinized epidermal cells were also determined quantitatively by the method of Miyazaki et al. (1982) and were found to increase markedly in the presence of TH.

These results lead to the conclusion that TH accelerates the terminal differentiation of tail epidermal cells; this appears to be of biological significance, because it suggests the occurrence of adult-type skin even in the tail epidermal cells. This suggestion is discussed later, and in Section IV,A in detail.

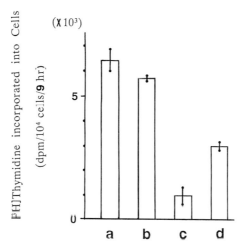

Fɪɢ. 7. Effect of TH on the rate of DNA synthesis of tail epidermal cells. Tail epidermal cells were cultured for 4 days in the presence and absence of T_3, and the rate of $[^3H]$thymidine incorporation was determined. a, Control; b, $5 \times 10^{-7} M$ cortisol; c, $10^{-8} M T_3$; d, $5 \times 10^{-7} M$ cortisol and $10^{-8} M T_3$. From Nishikawa *et al.* (1989).

As mentioned briefly before, there should be at least two types of death in cells of tail tissue during metamorphosis. First, cell death occurs as a terminal phase of cell turnover; this type can be called physiological cell death and would be TH-independent. The second type of cell death is specific to metamorphosis; this would be a TH-dependent type of cell

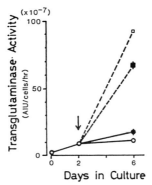

Fɪɢ. 8. Effect of TH on transglutaminase activity of cultured tail epidermis. Epidermal cells were cultured in the presence and absence of cortisol and TH for 4 days and the cells were assayed for the enzyme activity. The arrow indicates the time of addition of the hormones. ○ No hormones; ●, $5 \times 10^{-7} M$ cortisol; □, $10^{-8} M T_3$; ■, $5 \times 10^{-7} M$ cortisol and $10^{-8} M T_3$. From Nishikawa *et al.* (1989).

death. As will be described in Section IV,A, two of the three epidermal cell types—apical and skein cells—are larva-specific, while the basal cells are thought to be precursors of germinative epidermal cells in the adult skin (Robinson and Heintzelman, 1987). There is evidence that the two larva-specific kinds of cells have mitotic potentials but they do not enter the pathway to differentiation. Basal cells, in contrast, can differentiate through granular cells to cornified cells during metamorphosis under the influence of TH (Nishikawa and Yoshizato, 1989; see also Section IV,A). Considering these observations, it is reasonable to postulate the action of TH during metamorphosis as follows: TH-dependent cell death (metamorphic cell death) occurs in the larva-specific apical and skein cells. They terminate their life as a result of suppression of DNA synthesis by TH. The basal cells proceed toward physiological cell death, which may or may not directly involve TH; however, the basal cell requires TH to metamorphose into the germinative cell, the progeny of which is the differentiated epidermal cell, which is destined to die after terminal differentiation. It appears that TH accelerates the rate of differentiation of the stem cell to the granular cell to cornified cell. In summary, TH accelerates both types of cell death. This seems to provide a cellular basis for rapid degradation of the tail during metamorphosis. As mentioned before, basal cells in tail epidermis are a minor population ($<5\%$); this indicates that the metamorphic cell death greatly contributes to the breakdown of tail epidermis.

2. Mesenchymal Cells

We succeeded in isolating mesenchymal cells from the tail of prometamorphic tadpole by digesting EDTA and trypsin-treated tailfins with collagenase and hyaluronidase (Yoshizato and Nishikawa, 1985). Two major cell populations are obtained when the digests are centrifuged on a Percoll gradient. One is fibroblasts with densities of 1.030–1.050; the other is macrophages with densities of 1.045–1.050. Fibroblasts do not proliferate when cultured and keep a steady and healthy-looking appearance for at least 10 days. They actively synthesize collagen *in vitro*. Macrophages do not proliferate either. Macrophages adhere to the plastic surface more rapidly than do fibroblasts. Once macrophages have adhered to plastic, it is difficult to detach them by conventional techniques. Macrophages protrude from the cytoplasm and take on bizarre appearances. They take in latex beads and show intense activity of acid phosphatase around nuclei and latex beads (Nishikawa and Yoshizato, 1986b). When cultured >1 week they fuse with each other, forming giant multinuclear cells. These characteristics are typical also in mammals. The two types of tail mesenchymal cells show quite different responses to TH. Upon exposure to TH, fibroblasts manifest changes identical to those of epidermal cells (Yoshizato and Nishikawa, 1985). They detach themselves from the plastic surface, probably as a result of cell death induced by TH, as in the case of epidermal cells. The rate of death of fibroblasts is clearly correlated with the concentration of TH (Fig. 9).

In contrast to fibroblasts, macrophages require the presence of TH for *in vitro* maintenance. Without TH, they detach themselves from the plastic surface—exactly the opposite of the response shown by fibroblasts. This might seem logical considering the proposed functions of macrophages during metamorphosis. They should function actively as phagocytic cells during metamorphosis, when the concentration of TH in the body fluid is high (Section II). It remains to be determined whether macrophages isolated and cultured as mentioned earlier phagocytose myofibrils and collagen fibrils as they do *in vivo,* and if so, whether or not *in vitro* phagocytosis requires the presence of TH.

F. MODE OF ACTION OF TH IN TISSUE BREAKDOWN

As often mentioned in the previous sections, the breakdown of larva-specific tissues completely depends on TH. As Kaltenbach (1959) clearly demonstrated by implanting cholesterol–T_4 pellets into the tailfin, where the local absorption of tissues was observed around the pellet, the action of TH is a direct one without intervention of any mediator. This was confirmed further by Weber's experiment (1962), in which he showed that the isolated and cultured tail of *X. laevis* shrinks in response to TH in the medium. Since then, it has been widely accepted that TH acts directly on

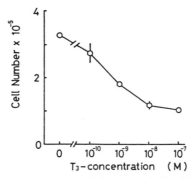

FIG. 9. Thyroid hormone-dependent cell death of tail fibroblasts. Fibroblasts were cultured for 6 days in the presence of various amounts of T_3 and 5×10^{-7} M cortisol. From Yoshizato and Nishikawa (1985).

the larval tissue and induces tissue breakdown (Shaffer, 1963; Derby, 1968).

However, there have been a few studies that show the possibility of the presence of a mediator or an activator of TH action. One of these studies is discussed in detail in Section III,C,2. Niki *et al.* (1982, 1984; Niki and Yoshizato, 1986) postulate the presence of some active factor in the tail skin, which is required for TH to induce breakdown of mesenchyme. Another possibility has been put forward by Hamburgh and colleagues (1981; Kim *et al.*, 1977), who adopted unique techniques. They made a triplet of tailfin blocks by sandwiching one block with the other two blocks and fusing them. The triplets were cultured in the presence of TH and the changes in size of the sandwiched block were followed during culture. The block starts to regress with a lag period of 3–4 days. This lag phase shortens to 24 hours when the block is sandwiched by blocks that have been immersed in TH-containing solution for 3 days and washed extensively with the hormone-free solution. The authors interpret this observation by postulating some active intermediate that is synthesized in the lag phase and is required for the action of TH. Hamburgh and colleagues (1981) admit that the active intermediate might be nothing more than hydrolytic enzymes such as collagenase. No studies on the identification of this substance have been reported since then.

The precise and detailed mechanism of action of TH in the degradation of larva-specific organs remains largely uncertain at the molecular level. It has been well recognized since Weber's report (1965) that actinomycin D completely inhibits the action of TH on the regression of the cultured tissues of tail (Eeckhout, 1965; Tata, 1966). Therefore, it seems likely that TH induces or stimulates some mRNAs, the translations of which are needed for the breakdown of larval tissues. What is this specific protein? This is an important question for understanding the complex phenomenon of TH-dependent tissue breakdown from the viewpoint of modern molecular biology. However, there has been no significant progress in this intriguing area of research.

Overall synthesis of proteins in tadpole tail appears to be rather depressed by TH (Little *et al.*, 1973; Kistler *et al.*, 1975); this contradicts the expectation described previously. There remains one possibility to be examined: that the synthesis of some minor specific proteins is induced or enhanced by TH, although the synthesis of major proteins in tail is suppressed by TH because the tissue is destined for destruction. Smith and Tata (1976) tried to examine this possibility by comparing fluorographic patterns of newly synthesized proteins on SDS–polyacrylamide gels between tail tissues cultured in the absence and presence of TH. However, no significant difference was noted.

Nishikawa and Yoshizato (1989) analyzed newly synthesized proteins by two-dimensional gel electrophoresis. Epidermal cells from the tail of a prometamorphic tadpole of *R. catesbeiana* were cultured in the presence or absence of 10^{-8} *M* T_3 for 4 days. The cells were labeled with [^{35}S]methionine for 12 hours and cellular proteins were subjected to two-dimensional gel electrophoresis. The same analyses were performed for epidermal cells of body skin, which is the larva-to-adult organ, in order to contrast the TH-induced changes occurring in the larva-specific organ. The results are interesting and are worthy of further study (Table II). About 30 spots were identified and were numbered sequentially from 1 to 30 in order of their molecular weights. There is no tail-specific protein the synthesis of which is induced by TH. Two proteins (numbers 4 and 30) are tail-specific proteins; their syntheses are completely abolished by the hormone. There are at least nine proteins, the syntheses of which are stimulated by TH. Three of them (numbers 24, 26, and 27) are marked in their extent of stimulation. Thyroid hormone induces the synthesis of several proteins (numbers 2, 7, 8, 9, 10, 11, 13, 22, 23, 29) that are common to tail and body skins. The rate of synthesis of proteins of spots number 2 and 11 of these are much more remarkable in the tail than in the back. The result, summarized in Table II, is a clear demonstration that TH acts directly on tail epidermal cells and induces or stimulates the synthesis of some proteins or completely inhibits the synthesis of other proteins, although the biological significance of these observations has not been elucidated.

The first step of TH action on cells is believed to be common in amphibia and mammals: binding of TH to the binding sites of the hormone-responsive cell. Among several possible binding sites—such as cell membrane, nucleus, cytosol, and mitochondrion—nuclear sites have received intensive attention. Presence of specific nuclear binding sites for TH (nuclear receptors) in cells of tadpole tail was first demonstrated by Yoshizato *et al.* (1975). Characterization of the receptors was performed during spontaneous metamorphosis of *R. catesbeiana* for the numbers of binding sites and the dissociation constant (K_d) (Yoshizato and Frieden, 1975) (Table III). The K_d does not show significant changes during spontaneous metamorphosis (TK stages 18–23) as compared to that for the premetamorphic animals. In contrast, maximum binding capacity (B_{max}) increases 2-fold at TK stage 22 as compared to TK stage 25. These characteristics of the receptors suggest that the responsiveness of the tailfin cells to TH are correlated with B_{max} of the nuclear receptors, but not with K_d.

A similar correlation between B_{max} and cell responsiveness to TH has been suggested by other studies. It has been well known that exogenously introduced glucocorticoid hormone accelerates the TH-induced metamor-

TABLE II

THYROID HORMONE-INDUCED CHANGES IN PROTEIN SYNTHESIS OF EPIDERMAL CELLS REVEALED
BY TWO-DIMENSIONAL POLYACRYLAMIDE GEL ELECTROPHORESIS[a]

Tail only[b]				Tail > back skin[c]				Tail = back skin[d]			
Spot Number	M_r	TH	Effect	Spot Number	M_r	TH	Effect	Spot Number	M_r	TH	Effect
1	270K	↑		2	190K*	⇧		5	105K	↑	
3	175K	↑		6	88K*	↑		12	75K	↑	
4	165K		⇩	7	87K*	⇧		13	70K	⇧	
14	60K	↑		8	87K*	⇧		17	60K	↑	
15	60K	↑		9	87K*	⇧		19	58K	↑	
16	60K*		↓	10	87K*	⇧		22	48K*	⇧	
18	59K	↑		11	84K*	⇧		23	48K*	⇧	
20	56K*		↓	21	52K	↑		29	31K	⇧	
24	45K*	↑									
25	37K	↑									
26	36K*	↑									
27	35K*	↑									
28	33K		↓								
30	23K		⇩								

[a] ↑ , Increased by T_3; ↓ , decreased by T_3; ⇧ , appearance associated with T_3; ⇩ , disappearance associated with T_3.

[b] Protein spots found only in tail.

[c] Protein spots found in both tissues, but TH effect is more marked in tail than back.

[d] Protein spots found in both tissues, and TH effect is evenly observed in the two tissues.

* Indicates a prominent change.

phosis of anuran tadpoles (Gasche, 1942; Frieden and Naile, 1955; Kobayashi, 1958; Kaltenbach, 1958). This effect of glucocorticoid on metamorphosis should have some physiological meaning, because Krug et al. (1983) showed that the concentration of glucocorticoid in the tadpole serum increases significantly during spontaneous metamorphosis of bullfrog tadpoles. The effect of the steroid hormone can be reproduced in vitro: the steroid accelerates the TH-dependent regression of cultured tailfin (Kikuyama et al., 1981). The precise mechanism of this acceleration is not yet known. There is a possibility that the steroid hormone increases the sensitivity of the cells to TH. The glucocorticoids were shown to change the binding characteristics of TH nuclear receptors: B_{max} increases, but K_d is unchanged by the action of the steroid (Niki et al., 1981; Suzuki and Kikuyama, 1983).

Moriya et al. (1984) studied the binding characteristics of TH receptors of red blood cells of tadpoles in relation to metamorphic transformation of larval erythrocytes to adult ones. The B_{max} for T_3 almost doubles at the

TABLE III
CHANGES IN B_{MAX} AND K_D FOR T_3 BY TAILFIN
NUCLEI DURING METAMORPHOSIS[a]

TK Stage[b]	B_{max} per Cell Nucleus	K_d (pM)
10	1300	112
15	1580	105
18–21	2830	155
20–21	2410	169

[a] From Yoshizato and Frieden (1975).
[b] Taylor and Kollros (1946) stages.

climax, as compared to that of premetamorphic stages, while K_d rather increases, supporting the idea described previously that sensitivity of tadpole tissues to the metamorphosis-inducing hormone is associated with B_{max} of the TH receptors in nuclei, but not with affinity constant.

As discussed before, tail epidermis, for example, contains three types of cells. These cells respond to the same hormone in different ways: apical and skein cells come out of the cell cycle, which leads them to death (metamorphic type); basal cells are transformed into germinative cells, which function in the skin of adult life. The studies on the TH receptors have been carried out for mixed populations of cells. It appears important that characterizations and roles of TH receptors are studied for a homogeneous population for each type of the cells. Considerable progress has been made in the study of TH receptors in mammalian cells. The receptor proteins have been suggested to be coded from the c-erb-A gene (Sap et al., 1986; Weinberger et al., 1986).

IV. Transformation of Larval Organs into Corresponding Adult Organs

Many tadpole organs are destined to transform into the corresponding adult organs during metamorphosis. In this section, back skin, small intestine, and liver are described as examples of the larva-to-adult organ for reviewing recent advances focusing on the changes in cell populations in a tissue during metamorphosis.

A. SKIN

The fate of skin in tail during metamorphosis is discussed in detail in Section III. The tail skin is destined to be destroyed, whereas the skin of body or head—though identical in appearance to that of tail—undergoes quite a different metamorphic fate, in that it survives the metamorphic period and transforms into adult-type skin. The transformation of larval skin to adult skin has been well characterized biochemically (Lipson and Silbert, 1965, 1968; Hata and Nagai, 1973). The larval back skin is quite different from the adult form in the chemical nature of GAG. More than 96% of GAG in tadpoles of *R. catesbeiana* is HA, while the back skin of adult frogs contains dermatan sulfate, HA, chondroitin sulfate A and/or C, with relative ratios of 45%, 25%, and 15%, respectively. The remaining 15% of the GAG may be a dermatanlike substance. This biochemical transformation of skin GAG starts at TK-stage 20 and terminates at TK stage 22 (Lipson *et al.*, 1971). For this transformation, hyaluronidase appears to play key roles in removing HA (Polansky and Toole, 1976).

These very different fates—that is, breakdown (tail skin) versus survival to transform (back skin)—are guided by the same hormone, TH. To try to determine the cellular basis for this apparent difference, we will next review the literature. This question is addressed briefly in Section III,E,1.

1. *Structures of Body Skin*

Robinson and Heintzelman (1987) have provided detailed morphological descriptions of the ventral body epidermis of tadpoles of *R. catesbeiana*. The epidermis is composed of three kinds of cells. The outermost are the apical cells, which are cuboidal or columnar in shape and have microvilli on the cell membrane. Nuclei are large and regular in shape, and are located at the bottom of the cells. The cytoplasm is dense. Beneath the apical cells are three to four layers of relatively large cells with lobulated nuclei located near the upper cell membrane and with clear cytoplasm. These are skein cells or cells with bobbins. Some skein cells are found to attach on basement membrane. The lowermost cells, attached to the basement membrane, are basal cells; they are columnar or triangular in shape with irregular-shaped large nuclei and dark-stained cytoplasm. Basal cells are electron-microscopically characterized as having a dense cytoplasm, less prominent organelles like mitochondria, Golgi vesicles and endoplasmic reticulum, and no granular structures. As contrasted with body epidermis, tail epidermis is characterized by its abundance of skein cells and paucity of basal cells (Section III,E,1).

2. *Metamorphic Changes in Epidermal Cells*

The changes observed by Robinson and Heintzelman (1987) are summarized in Table IV. Metamorphosis-related changes in epidermal cells start at TK stage 19 with the enucleation of apical cells and their detachment from epidermis, accompanied by the detachment of skein cells from the

TABLE IV
METAMORPHIC CHANGES OF TADPOLE VENTRAL EPIDERMAL CELLS[a]

	TK Stages[b]				
	Before 18	19	20	21	22
Apical cells		Enucleated Sloughing			
Skein cells	3–4 layers thick With figures of Eberth Some on BM[c]	Removed from BM	Moving upward Breakdown of figures of Eberth Degenerating	Disappeared	
Basal cells	1–2 layers thick		>2 layers thick		
Prominent events			Appearance of progranular cells	Appearance of germinative cells	Adult-type structure

[a] From Robinson and Heintzelman (1987).
[b] Taylor and Kollros (1946) stages.
[c] BM, Basement membrane.

basement membrane. This is probably the result of their being pushed up as the basal cells proliferate. Accordingly all the cells attached to basement membrane at this stage are basal cells. The most important events at TK stage 20 are the appearance of degenerating skein cells, proliferation of basal cells to occupy half the thickness of the tissue, and the appearance of progranular cells in two to three layers of thickness. The main events at TK stage 21 are the disappearance of most skein cells, appearance of cornified cells, and the appearance of germinative cells that resemble the corresponding cells in the adult. The latter event of the three just described should be biologically most important among the many changes in the epidermis during metamorphosis, because the appearance of germinative cells indicates that the larval-type epidermis has begun to transform into adult-type tissue. At TK stage 22, the overall structure of the epidermis, with four to five layers, resembles that of the adult. Distinct layers of hierarchy of differentiation of epidermal cells are established: the layer of cornification at the outermost surface, strata granulosum and spinosum at the inner layers, and stratum germinativum on the basement membrane.

Skein cells, one of the larva-specific cell types, are lost during metamorphosis, as shown quantitatively in Fig. 10 (Robinson and Heintzelman,

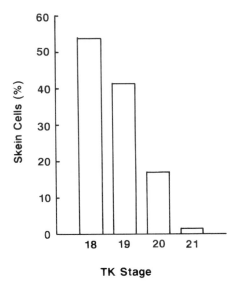

TK Stage

FIG. 10. Histogram showing the decrease in skein cells in ventral epidermal cells during metamorphosis. Modified from Robinson and Heintzelman (1987).

1987). Before the onset of metamorphosis, 54% of the epidermal cells are skein cells; at TK stage 19 they constitute 42%, 16–20% at TK stage 20, only 0.8% at TK stage 21, dropping to nothing at TK stage 22. The mechanism of loss of this cell should be identical to those described in Section III,E,1 for the skein cells of tail epidermis: inhibition of the cell cycle by the direct action of TH (Nishikawa *et al.*, 1989).

There remains one important question: What is the origin of germinative cells appearing at TK stage 21? Evidence that suggests the basal cell as the precursor of the stem cell has been presented by Robinson and Heintzelman (1987), who measured the labeling indices of the three types of larval epidermal cells after injecting [^{3}H]thymidine (Fig. 11). Before the onset of metamorphosis, the indices are very low. A drastic increase occurs in the labeling indices, as much as 10-fold as compared to the previous stages. This increase is ascribed to the increase in the basal cells. Similar results have been reported for the shank epidermis (Wright, 1973) and limb epidermis (Wright, 1977) of tadpoles of *R. pipiens*.

The assertion by Robinson and Heintzelman (1987) just described is supported by the fact that the labeling indices for germinative cells of adult epidermis are 7–8% (Jorgensen and Levi, 1974; Levi and Nielsen, 1982). It is most likely with our present state of knowledge that basal cells in tadpoles are inactive (dormant) germinative cells, which are activated or

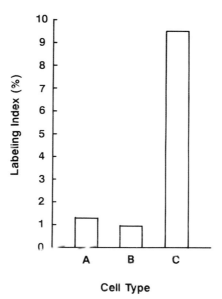

FIG. 11. Labeling-index histogram of the tadpole ventral epidermal cells. A, Basal cell; B, skein cell; c, apical cell. Modified from Robinson and Heintzelman (1987).

transformed into "mature" germinative cells for adult skin under the influence of TH. This hypothesis is intriguing and is worthy of verification. Nishikawa *et al.* (1989) isolated a population of back skin epidermal cells, the major cells of which are basal cells. The rate of DNA synthesis of the population is increased 2.5-fold by TH (Fig. 12), indicating that the activation of the basal cells is the result of direct action of TH. Cells in S phase of the cell cycle were labeled with BrdU and the rate of DNA synthesis was calculated as [^{3}H]thymidine incorporated per cells in S phase as described in tail epidermal cells at Section III,E,1. The same extent of stimulation by TH was obtained, suggesting the possibility that the basal cells in larval skin are in the cell cycle even before the onset of metamorphosis, although the rate of turnover is very low. The aforementioned study also supports the hypothesis with respect to basal-to-germinative cell transformation. The population of epidermal cells used contains skein cell and basal cells. There is no doubt that the activation in DNA synthesis induced by TH occurs in basal but not in skein cells, because as discussed in Section III,E,1, DNA synthesis of skein cells is suppressed by TH (Nishikawa *et al.*, 1989).

As mentioned in Section III,E,1, there is a hierarchy among typical epidermal cells in their extent of differentiation or proliferative activity. Is there also this kind of hierarchy in the tadpole epidermal cell? The labeling-

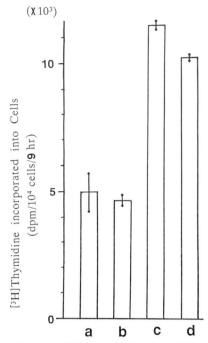

FIG. 12. Stimulation of the rate of DNA synthesis of tadpole back epidermal cells by TH. Epidermal cells were treated with hormones for 4 days and the rate of DNA synthesis was determined. a, Control; b, 5×10^{-7} M cortisol; c, 10^{-8} T_3; d, 5×10^{-7} cortisol and $10^{-8}M$ T_3. From Nishikawa *et al.* (1989).

indices experiment shown in Fig. 11 does not support the presence of such a hierarchy, because the highest indices are found in apical cells not in basal cells. This gives rise to another question: What is the relationship among these three types of cells? The data presently available suggest the following pathway of metamorphic changes of larval epidermal cells, which leads to the transformation of larval epidermis into adult epidermis: of the three types of cells constituting larval skin, apical and skein cells are intrinsic to larval life and undergo metamorphosis-associated cell death by the direct action of TH. Basal cells are dormant germinative cells that survive the metamorphosis and are destined to be activated to germinative cells for adult life during metamorphosis—also via the direct action of TH (Fig. 13).

3. Type A Antigens

Kaiho and Ishiyama (1987) found a new and unique marker for epidermal cells of adult frog skin. It has been known that blood cells and vascular

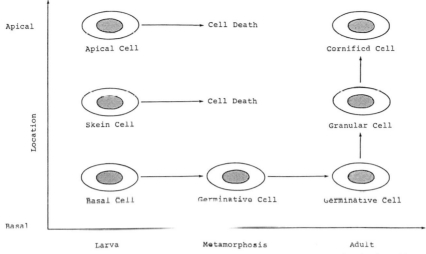

FIG. 13. Schematic representation of the fates of the three types of tadpole epidermal cells.

endothelial cells of frog have type B antigens of human blood cells on their surface (Yada and Yamazawa, 1962; Yada *et al.,* 1962). Kaiho and Ishiyama surveyed immunothistochemically the presence of the type A antigen of human blood cells and found that this antigen (A antigen) is localized on the cell surface only of acinar cells of pancreas and epidermal cells of adult bullfrog. It is interesting that the pancreatic exocrine cells are positive also when in tadpoles, but the epidermis is negative for this antigen. Germinative cells and cornified cells of adult epidermis are negative for the A antigen. Adult granular cells are positive. This finding provides us with a unique and valuable tool to visualize the appearance of adult-type epidermal cells in larval skin during metamorphosis.

Epidermis of tadpoles was surveyed for the presence of A antigens, and the results are schematically summarized in Fig. 14 (Nishikawa and Yoshizato, 1989). Epidermis of tadpoles before TK stage 19 is negative. At TK stage 20, positive cells appear. Apical, skein, and basal cells are negative. The positive cells are located above basal cells, which are small and have dark-stained cytoplasm. According to the study by Robinson and Heintzelman, progranular cells appear at TK stage 20 (Table IV). Naturally these cells are thought to be the A antigen-positive ones. In other words, granular cells that appear at the metamorphic climax are differentiated adult-type cells. This statement seems to be very important, because it logically means that there must exist adult-type germinative cells in the tadpole skin at TK stage 20, a period of early metamorphic climax. Con-

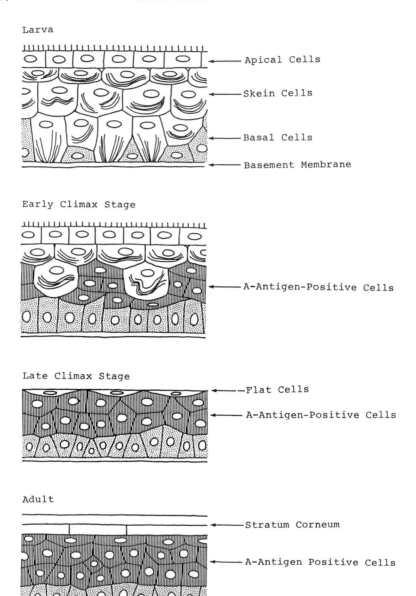

Fig. 14. Changes in epidermal cells of tadpole back skin during metamorphosis. A-antigen-positive cells are shown by vertical hatched lines.

sidering these observations together, it is highly plausible that larval basal cells start to be activated by TH and undergo "metamorphic transformation" into adult-type germinative cells at the very early metamorphic climax.

It appears that data are now available to answer the question raised at the beginning of this section: What is the cellular basis for the difference in metamorphic fate between tail epidermis (death) and body epidermis (transformation)? There is no difference between the two skin types in the kind of constituent epidermal cells; both contain apical, skein, and basal cells. This means that the difference in metamorphic fate could not be ascribed to the difference in the kind of the constituent cells.

It may be fruitful to approach the question in another way, by asking: Is it true that the tail epidermal cells are destined merely to be removed as a result of metamorphic cell death? There is a possibility that even larva-specific tissue could show adult-type characteristics under "metamorphic pressure," because the larva specific cells should contain identical sets of genetic information. Yoshizato and colleagues utilized the A antigen and surveyed its expression in the tail epidermis during metamorphosis (Nishikawa *et al.*, 1989). The results are schematically shown in Fig. 15. At the very late stage of metamorphic climax (TK stage 23), a few faintly positive

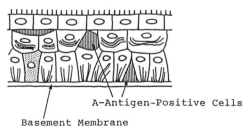

FIG. 15. Occurrence of A-antigen-positive cells in tail epidermal cells. The positive cells are shown by vertical hatched lines.

cells are detectable in the middle layer of epidermis or very rarely—if ever—in the basal layer. This convincingly shows that the even the cells in larva-specific tissue have the capacity to become adult-type cells and try to do so. These positive cells should be granular cells according to the results of studies on the body skin. It can then be logically inferred that basal cells are capable of being activated to "transform" into germinative cells for adult even in the tail, which has no counterpart in the adult.

The possibility just described has been further confirmed with the following experiment (Nishikawa and Yoshizato, 1989). Epidermal cells were obtained from the back and tail and were cultured in the presence or absence of TH. The cell preparations contained skein and basal cells. The A-antigen-positive cells appeared in both cultures only in the presence of TH. This experiment clearly demonstrates that (1) TH directly acts on larval cells and induces their "transformation" into adult cells, and (2) this TH-induced transformation also occurs in the tail.

Then, what is the explanation for the different metamorphic fates of the two types of epidermis (i.e., breakdown in tail versus transformation in body)? At present it is assumed that the difference is ascribed to the difference in the ratio of cell numbers between skein cells and basal cells. According to Robinson and Heintzelman (1987), ~50% of the entire epidermal cell population of the body skin is skein cells. Probably the majority of the remaining 50% are basal cells. We estimate that ~95% of epidermal cells in the tail are skein cells (Section III,E,1; the exact ratios of apical cells to skein cells to basal cells have not been determined). Basal cells are very few as compared to the body epidermis. As illustrated in Fig. 16, skein cells are larva-specific cells that perish through the action of TH; this indicates that 95% of the tail epidermal cells will be lost during metamorphosis. This could be the reason for the breakdown of the tail epidermis as a whole. Contrary to tail, body skin contains a much higher proportion of basal cells, which can actively proliferate under the influence of TH. This enables the body skin to survive metamorphosis.

4. *Expression of Keratin Genes*

Keratin is an epidermis-specific protein and is therefore a useful marker for investigating the TH-induced transformation of larval to adult skin at the molecular level. Reeves (1977) isolated the adult keratin from *X. laevis* frogs and prepared monospecific antibodies against this protein. As described in the following, the antibody is most likely to recognize the 63-kDa keratin of Miller and associate, which is a major and adult-type keratin (Mathisen and Miller, 1987). With this specific antibody, it was demonstrated that the synthesis of adult keratin is initiated at the onset of metamorphoses. Tadpoles are induced to synthesize the keratin when TH

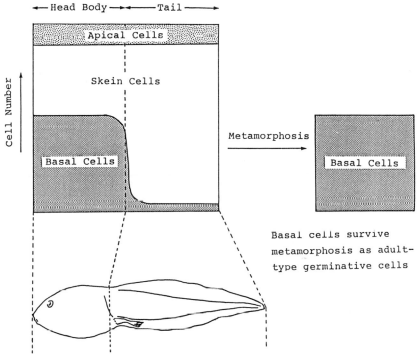

Fɪɢ. 16. Hypothetical scheme showing the cellular basis of disappearance of tadpole tail during metamorphosis.

is given after they reach NF stage 50–52, but not before this stage, suggesting that tadpoles are not able to respond to TH until they develop a sensitivity, although the nature of the sensitivity has not been clarified. Thyroid hormone also induces the synthesis of keratin by acting on the isolated and cultured back skin with a latent period of 2–3 days.

Miller and colleagues (Ellison *et al.*, 1985) have further advanced studies on TH-dependent gene expression of keratin proteins. They purified keratins from tadpoles with different developmental stages of *X. laevis* and classified them into three types (Table V). Nine keratin proteins were identified, four of which are the adult type, three the larval type, and the remaining two the embryo type. The genes for embryo-type keratins appear to be expressed until the onset of metamorphosis. Of these keratins, Ellison *et al.* focused on the 63-kDa keratin for investigating the mechanism of action of TH on the expression of keratin genes, because the 63-kDa keratin is most abundant in adult skin (Mathisen and Miller, 1987). Synthesis of the keratin is detected at as early as NF stage 18 by a Western

TABLE V

KERATINS OF *Xenopus laevis*[a]

Keratin Protein Type (kDa)	NF Stages[b]						
	26	33	45	52	59	63	Adult
Adult type							
63				+	+	+	+ +
56				+	+	+	+ +
53	+	+	+	+	+	+	+ +
49						+	+ +
Larval type							
60			+	+	+	+	+
48	+	+	+				
47	+	+	+				
Embryo type							
59	+	+	+	+	+		
58	+	+	+	+	+		

[a] From Ellison *et al.* (1985).
[b] Nieuwkoop and Faber (1967) Stages. +, Presence of a keratin; + +, a major keratin of the adult epidermis.

hybridization method. This low level of keratin synthesis does not require activation by TH. Messenger RNA for the 63-kDa keratin was quantified and is presented in Table VI. The gene is activated between NF stages 56 and 57. It has been confirmed in tadpoles injected with TH that TH activates the gene expression when the tadpoles have developed to NF stage 48, but earlier than this stage they do not respond to TH. It is interesting to point out that the 63-kDa keratin is also synthesized in the

TABLE VI

CONCENTRATION OF 63-kDa KERATIN mRNA
DURING METAMORPHOSIS[a]

NF Stages[b]	mRNA content (pg/μg of Total RNA)
52	1
53	4
54	11
55–56	13
57	300
62	3500
Adult	4500

[a] From Mathisen and Miller (1987).
[b] Nieuwkoop and Faber (1967) stages.

tail epidermis at the late-climax stage of metamorphosis, even though the level of synthesis is quite low as compared to body skin. This was also confirmed *in vitro,* where TH is found to induce a rather high level of 63-kDa keratin synthesis in cultured tail epidermal cells (Mathisen and Miller, 1989). These results support the observations on the appearance of A antigen in the tail epidermis described previously.

B. Small Intestine

Tadpoles are herbivorous; after metamorphosis the adult frog is carnivorous. Accompanying the transition in food habit, the structures and functions of the digestive organs drastically change. The small intestine loses its coiled structure and regresses; the bullfrog small intestine, for example, is 88% of its length in the tadpole (~45 cm at TK stage 18) (Sumiya and Horiuchi, 1980). The intestine of tadpoles of *X. laevis* shrinks from ~10 cm in length at NF stage 58 to 1 cm at NF stage 65 (Marshall and Dixon, 1978; Ishizuy-Oka and Shimozawa, 1987a). Prominent structural changes are as follow:

1. The larval intestine has no folded structures except in the region of the typhlosole, whereas the adult intestine is characterized by folded structures.
2. Connective tissues are very rare and localized in the typhlosole in tadpoles. During metamorphosis, connective tissues increase markedly.
3. Epithelial cells of intestine transform from the larval type to the adult type.

These changes are schematically illustrated in Fig. 17.

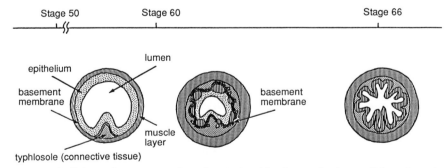

FIG. 17. Schematic representation of metamorphic changes in the small intestine of a *Xenopus laevis* tadpole. Dotted areas, larval epithelium; vertically hatched areas, adult epithelium. Stages are based on Nieuwkoop and Farber (1967).

Intestinal epithelial cells of the larval type are called the primary epithelial cells, whereas those of the adult type are called secondary epithelial cells (Smith-Gill and Carver, 1981). The primary epithelial cells undergo apoptosis during metamorphosis, which resembles the process in autolysis of epidermal cells in tail (Hourdry and Dauca, 1977; Dauca and Hourdry, 1983).

1. *Transformation of Larval Epithelial Cells to Adult Epithelial Cells*

As shown in Fig. 18, there is a regional difference in adult epithelial tissue due to folded structures forming troughs and crests. As described later, epithelial cells in fold troughs are undifferentiated and proliferative, while those in fold crests are differentiated and nonproliferative. Differentiating epithelial cells in the trough migrate toward the crest, forming a kind of hierarchy with respect to the extent of proliferation and differentiation, as in skin epidermis (Section III,E,1), with the cells at the crest sloughing off. It would be interesting also to determine the locations of populations of proliferative and nonproliferative cells in the tadpole intestine, which has no folded structures. Studies by Marshall and Dixon (1978) tried to answer this question. They histologically examined cells with mitotic figures and found that differentiated larval epithelial cells are capable of proliferation. There are no specific sites at which proliferative cells congregate. These observations indicate that the differentiated cells replace the cells that have been lost naturally through termination of their life span. There appear to be no stem cells in the tadpole epithelial tissue, which is found in the trough in the case of adult intestine. This is in contrast to the larval skin, which has dormant germinative basal cells, as discussed in Section IV,A.

Changes in proliferative activity were extensively studied during spontaneous metamorphosis with tadpoles of *X. laevis* (McAvoy and Dixon, 1977; Ishizuya-Oka and Shimozawa, 1987a, b) or of *R. pipiens* (Dournon and Chibon, 1974). Almost identical results have been reported by these authors. At the onset of metamorphosis (NF stage 60), there appear nests of cells that can be easily distinguished from neighboring larval epithelial cells. The same structures are called islets by Ishizuya-Oka and Shimozawa (1987a, b). Cells in the nest have relatively clear cytoplasm and frequently show mitotic figures. The nests increase their size and number as metamorphosis progresses. At NF stage 62, each nest coalesces and develops toward the side of connective tissue, thus pushing the larval epithelial cells over to the side of the lumen. Finally, all larval cells are replaced with cells from the nests, which are most likely the cells of adult-type epithelia. The new epithelia start to fold at NF stage 63. From this description, important roles of cells in the nest or islet in the metamorphic transformation of the intestine are clear, although the change in intestine from larva to adult type appears to be a replacement rather than a transformation.

The regional distributions of epithelial cells that had been in metaphase or anaphase were determined in the fold during metamorphosis (McAvoy and Dixon, 1977). The cells with mitotic figures are distributed evenly along the longitudinal axis of the fold at NF stage 63, but they are restricted to the troughs of the fold at NF stage 65, which is identical to the distribution in adult intestine. An adult frog was injected with [^{3}H]thymidine, and the upward migration of differentiating epithelial cells was autoradiographically confirmed, in that the labels first appeared in the "stem cells" localized in the troughs of folds.

The origin of "stem cells" of adult-type epithelium (secondary epithelium) has been the subject of studies on the metamorphosis of small intestine. Several investigators support the idea that undifferentiated cells in the nest or islet are the stem cells. According to Dauca and Hourdry (1983), the stem cells are characterized by their high nucleocytoplasmic ratio and their abundance of free ribosomes. However, there has been no generally accepted idea about the origin of the stem cell. Marshall and Dixon (1978) asserted that there are no undifferentiated stem cells in the primary epithelium, and that differentiated primary epithelial cells are capable of proliferation. Thus it is probable that primary epithelial cells differentiate and transform to the stem cells of nests, which develop into the secondary epithelium. This hypothesis is intriguing and requires further testing.

2. *Epithelium–Mesenchyme Interactions*

It is well known that connective tissues develop extensively in small intestine during spontaneous metamorphosis (Dauca and Hourdry, 1985; Smith-Gill and Carver, 1981). Studies by Ishizuya-Oka and Shimozawa (1987a,b) suggest some active roles of the connective tissues in the transformation of larval epithelium to the adult type. Before the onset of metamorphosis, connective tissues are not prominent, except in the typhlosole. The connective tissues start to grow as the metamorphosis begins; numbers of fibroblasts and quantity of collagen fibrils and periodic acid–schiff (PAS)-positive substances increase.

The reasons Ishizuya-Oka and Shimozawa postulate as active roles of the connective tissues in the metamorphic transformation of intestine can be summarized as follows:

1. Basement lamella of intestine markedly thickens from 40 nm on average at NF stage 56 to 2 μm at NF stage 60, the onset of metamorpho-

sis. As tadpoles undergo metamorphosis, the basement lamella become irregular in thickness; the domains in the lamella where larva-type cells attach remain as thick as before. In contrast, the domains where newly appearing adult-type cells attach become thin, suggesting a correlation of the change in the cell type with that in the thickness of basement lamella.

2. Transmission EM surveys reveal the contacts between epithelial cells and fibroblasts through gaps in the thickened basement lamella. These figures are often observed between NF stages 60 and 62 at the regions where the nests of secondary epithelial cells locate or in the troughs of newly developed intestinal folds, indicating spatiotemporal correlations of the figures of epithelium–mesenchyme contact with the active proliferation of stem cells for secondary epithelium. The changes in intestinal structures of epithelium and connective tissue during spontaneous metamorphosis are schematically illustrated in Fig. 18, placing emphasis on epithelium–meshenchyme relationships as asserted by the aforementioned authors.

The proposed role of mesenchymal cells in the appearance and proliferation of cells for adult-type epithelium is noteworthy and is worthy of further study. There has been wide acceptance of the concept that mesen-

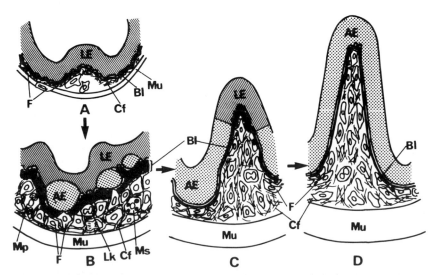

FIG. 18. Schematic illustration of changes in the intestinal connective tissue and epithelium during development of *Xenopus laevis* tadpoles. (A) Larva; (B) NF stage 60; (C) NF stage 63; (D) NF stage 66. AE, Adult epithelium; Bl, basal lamina; Cf, collagen fibril; F, fibroblast; LE, larval epithelium; Lk, leukocyte; Mp, macrophage; Ms, mastlike cell; Mu, muscular cell. From Ishizuya-Oka and Shimozawa (1987b).

chyme influences morphogenesis of epithelia, which includes proliferation and differentiation of epithelial cells, since the work of Grobstein (1956). Considerable literature is available in this still-developing area of research (Hahn *et al.*, 1986). Rheinwald and Green (1975) have established a cell culture in which clonal growth of epidermal cells is obtained by introducing fibroblasts as a feeder layer (Rheinwald and Green, 1975), suggesting that fibroblastic mesenchymal cells secrete a growth factor-like substance that influences proliferation and differentiation of epithelial cells.

C. LIVER

The liver has been one of the organs most frequently studied to obtain information on mainly biochemical aspects of metamorphosis and the mechanism of TH actions at molecular levels. There are several reasons for this specific situation: the liver is composed of relatively simple populations of cells, parenchymal liver cells (hepatocytes) as being the major population; in addition, the liver is a "metabolic center" of the body and is responsible for synthesis of proteins involved in metabolism. Many metabolic processes are known to change in tadpole liver during metamorphosis. The liver is easy to handle and prepare for biochemical and molecular biological analyses; cultures of hepatocytes have been developed in several laboratories as described later. These cultured hepatocytes can be maintained as cells that respond to hormonal stimuli. Liver undergoes multiple, well-characterized biochemical adaptations in association with metamorphosis. Excellent and detailed reviews on liver metamorphosis have been published (Frieden and Just, 1970; Smith-Gill and Carver, 1981). In this section a brief review is given of the research progress that has been made on the metamorphosis-related transformation of hepatocytes of larval type to those of adult type.

The cellular basis of this liver transformation has been one of the most important issues. As detailed in Section II and summarized in Table I, there are two possible pathways of transformation: (1) larval germinative cells transform into adult germinative cells, or (2) dormant (presumptive) adult germinative cells are activated to become adult germinative cells. Several studies have tried to determine whether changes accompanying the liver transformation occur in a fixed population of hepatocytes, or whether cell division accompanies or precedes the metamorphic transformation (Smith-Gill and Carver, 1981). An increase in DNA synthesis in liver has been shown at preclimax or early-climax stages for *R. catesbeiana* (Atkinson *et al.*, 1972; Smith-Gill, 1979; Smith-Gill *et al.*, 1979). Data on labeling indices of [³H]thymidine revealed by autoradiography support the biochemical-incorporation data (Smith-Gill and Carver, 1981).

With these and other results, Smith-Gill and Carver (1981) suggest that the proliferation and turnover of liver cells at climax may involve a replacement of larval by adult cells.

It is generally agreed that the synthesis of vitellogenin in *Xenopus* is induced in liver parenchymal cells by estrogen (Tata, 1976; Ryffel, 1978; Wahli *et al.*, 1981) and increases drastically at the metamorphic climax (May and Knowland, 1980; Skipper and Hamilton, 1979). The causal factor for this induction has not been clarified, although the most probable candidate seems to be TH. Weber and associates (Huber *et al.*, 1979) suggested that acquisition of competence to respond to estrogen by *Xenopus* liver cells is controlled by TH.

Cultured hepatocytes are apparently useful for determining the factor. Hepatocytes of anurans have been isolated and cultured by investigators (Stanchfield and Yager, 1978, 1980; Wangh *et al.*, 1979; Wahli *et al.*, 1978). Kawahara *et al.* (1981) developed further the method of hepatocyte culture and improved upon previous procedures. In searching for the aforementioned factor, Kawahara *et al.* (1989) succeeded in demonstrating the presence of progenitor cells of adult-type hepatocytes in larval liver, using cultures of hepatocytes.

A primary culture of *Xenopus* hepatocytes consisting of >90% parenchymal cells was obtained, which could remain healthy for at least 7 days with respect to responses to hormonal stimulation and to the rate of protein synthesis. Proliferation of hepatocytes in the culture was negligible, if not nil. Hepatocytes were obtained from tadpoles at different developmental stages and cultured in the presence of estrogen. Vitellogenesis was assayed with an enzyme immunostaining method using antivitellogenin antibodies (Kawahara *et al.*, 1987). Hepatocytes of tadpoles before NF stage 57 did not respond to TH for vitellogenesis. One of four tadpoles could be induced to synthesize the protein at NF stage 60. The cells of tadpoles NF stage 62 were positive for the immunostaining without exception, indicating that cells competent for the TH-dependent induction appear at this particular stage. Interestingly, only a rather small fraction of hepatocytes in the culture dish (<5%) became competent. The coexistence of competent and incompetent cells continues, even after metamorphosis, in 6-month-old frogs. No incompetent cells are found in 4-year-old male frogs, however. These observations strongly suggest that hepatocytes of tadpoles at prometamorphic and metamorphic climax stages contain at least two populations of cells: incompetent and competent. These appear to correspond to larval-type and adult-type hepatocytes, respectively; moreover, the two types of cells appear to have different cell lineages.

The direct action of TH in the presumed fraction of competent cells has been demonstrated by the same group (Kawahara *et al.*, 1989). Larval hepatocytes were obtained from NF stage 59 tadpoles and cultured in the presence of estrogen. In the culture without TH, ~1.2% of hepatocytes were immunostained by antivitellogenin antibody. With TH, the immuno-stainable cells increased to 20%. Vitellogenin-synthesizing cells are distributed as clusters of aggregates. The possibility that TH might stimulate the proliferation of preexisting competent cells has been excluded, because aphidicholin, an inhibitor of DNA synthesis, did not influence TH-dependent vitellogenin induction. These studies strongly support the concept mentioned earlier that prometamorphic tadpoles contain a small fraction of precursor cells for adult-type parenchymal hepatocytes. The characterization of this progenitor cell is important for understanding the metamorphic transformation of larval liver to adult liver from the viewpoint of cell differentiation. At present it appears that TH is not involved in the proliferation of the precursor cell, but involved in rendering the precursor cell responsive to estrogen.

V. Concluding Remarks

In closing this chapter, I would like to return to the questions raised in Section I.

1. What is the cellular and biochemical basis for the death of cells and breakdown of tissues? I have tried to correlate the morphological changes occurring in apoptotic tissues with the activities of hydrolytic enzymes, putting emphasis on collagenase and protease T_1. We are now able to see good correlations between them. However, more efforts are necessary for understanding the mechanism of activation or stimulation of the enzymes by TH at the level of gene expression. Thyroid hormone directly induces death of larva-specific cells by suppressing DNA synthesis. However, the mechanism of this cell death has been largely unknown.

2. How do larval cells change into adult cells? Is this change a transformation or a replacement? The key event occurring in larval skin appears to be the transformation of larval basal cells into adult germinative cells. It should be emphasized that the identical transformation seems to occur also in the larva-specific tail skin. The origin of stem cells of adult intestinal epithelium (secondary epithelium) has not been determined, although most investigators believe the transdifferentiation theory. Progress in studies on liver strongly suggests that the metamorphic change in hepatocytes is a replacement. Tadpole liver seems to contain both larval and adult hepatocytes, which have different cell lineages. It remains unresolved whether TH induces cell death of larval hepatocytes as in larval epidermal

cells. Cell culture techniques should be developed and elaborated for cells of tadpoles to obtain decisive evidence for either the replacement theory or the transformation theory.

3. How is TH involved in determining the fate of larval cells and adult cells during metamorphosis? This question is most important and fascinating. Large parts of the mechanism of action of TH remain unknown. There is no doubt that modern techniques of cell biology and molecular biology are now available for challenging this intriguing and unsettled problem of amphibian metamorphosis.

Acknowledgments

I wish to thank Dr. Masayoshi Kaiho, Tsutomu Kinoshita, and Atsuko Ishizuya-Oka for valuable discussions on the topics reviewed in this chapter. Figures 14 and 15 were originally drawn by Dr. Kaiho and Fig. 17 by Dr. Ishizuya-Oka. Special thanks are owed to Mr. Ken Ofusa and Ms. Yumi Izutsu for excellent assistance in preparation of the manuscript. The skillful typing of Ms. Ritsuko Aoki and Ms. Haruka Kikuchi and the expert drawing of Ms. Yuko Hata are gratefully acknowledged. The work discussed in this chapter was partly funded by a grant-in-aid for scientific research from the Ministry of Education, Science, and Culture of Japan.

References

Aleschin, B. (1926). *Biochem. Z.* **171**, 79–82.
Atkinson, B. G. (1981). *In* "Metamorphosis, A Problem in Developmental Biology" (L. I. Gilbert and E. Frieden, eds.), 2nd Ed., pp. 397–444. Plenum, New York.
Atkinson, B. G., Atkinson, K. H., Just, J. J., and Frieden, E. (1972). *Dev. Biol.,* **29**, 162–175.
Baker, R. E., and Jacobson, M. (1970). *Dev. Biol.* **22**, 476–494.
Barrandon, Y., and Green, H. (1985). *Proc. Natl. Acad. Sci. USA* **82**, 5390–5394.
Bell, E., Ivarsson, B., and Merrill, C. (1979). *Proc. Natl. Acad. Sci. U.S.A.* **76**, 1274–1278.
Benbassat, J. (1970). *Dev. Biol.* **21**, 557–583.
Bicsak, T. A., and Harper, E. (1984). *J. Biol. Chem.* **259**, 13145–13150.
Bicsak, T. A., and Harper, E. (1985). *Arch. Biochem. Biophys.* **242**, 256–262.
Brown, M. E. (1946). *Am. J. Anat.* **78**, 79–113.
Broyles, R. H. (1981). *In* "Metamorphosis, A Problem in Developmental Biology" (L. I. Gilbert and E. Frieden, eds.), 2nd Ed., pp. 461–490. Plenum, New York..
Coulombre, A. J., and Coulombre, J. L. (1964). *Exp. Eye Res.* **3**, 105–114.
Dauca, M., and Hourdry, J. (1985). *In* "Metamorphosis" (M. Balls and M. Bownes, eds.), pp. 36–58. Clarendon Press, Oxford.
Davis, B. P., Jeffrey, J. J., Eizen, A. Z., and Derby, A. (1975). *Dev. Biol.* **44**, 217–222.
Derby, A. (1968). *J. Exp. Zool.* **168**, 147–156.
Dmytrenko, G. M., and Kirby, G. S. (1981). *J. Exp. Zool.* **215**, 179–182.
Dodd, M. H. I., and Dodd, J. M. (1976). *In* "Physiology of the Amphibia" (B. Lofts, ed.), Vol 3, pp. 467–599. Academic Press, New York.
Dournon, C., and Chibon, P. (1974). *Wilhelm Roux' Arch. Entwicklungsmech. Org.* **175**, 27–47.
Dresden, M. H. (1971). *Nature (London) New Biol.* **231**, 55–56.
Eberth, C. J. (1866). *Arch. Microsk. Anat.* **2**, 490–506.

Eeckhout, Y. (1965). Thèse, Univ. Catholique de Louvain..
Eeckhout, Y. (1969). *Mem. Acad. R. Med. Belg.* **38**, 1–113.
Eisen, A. Z., and Gross, J. (1965). *Dev. Biol.* **12**, 408–418.
Ellison, T. R., Mathisen, P. M., and Miller, L. (1985). *Dev. Biol.* **112**, 329–337.
Frieden, E. (1961). *Am. Zool.* **1**, 115–149.
Frieden, F. (1967). *Recent Prog. Horm. Res.* **23**, 139–194.
Frieden, E., and Just, J. J. (1970). In "Biochemical Actions of Hormones" (G. Litwack, ed.), Vol. 1, pp. 1–52. Academic Press, New York.
Frieden, E., and Naile, B. (1955). *Science* **121**, 37–39.
Gasche, P. (1942). *Verhandl. Schweiz. Naturforsch. Ges.* **122**, 158–159.
Gona, A. G. (1969). *Z. Zellforsch. Mikrosk. Anat.* **95**, 483–494.
Grobstein, C. (1956). *Exp. Cell Res.* **10**, 424–440.
Gross, J., and Lapiere, C. (1962). *Proc. Natl. Acad. Sci. USA* **48**, 1014–1022.
Guieysse, P. A. (1905). *Arch. Anat. Microsc. Morphol. Exp.* **7**, 369–428.
Hahn, U., Schuppan, D., Hahn, E. G., Merker, H.-J., Riecken, E.-O., and Steglitz, K. (1986). In "Mesenchymal–Epithelial Interactions in Neural Development" (J. R. Wolff, J. Sievers, and M. Berry, eds.), pp. 111–117. Springer-Verlag, Berlin.
Hamburgh, M., Kim, Y., Tung, G., Crenshaw, R., and Mendoza, L. A. (1981). *Dev. Biol.* **81**, 392–398.
Harper, E., and Gross, J. (1972). *Biochem. Biophys. Res. Commun.* **48**, 1147–1152.
Harper, E., Bloch, K. J., and Gross, J. (1971). *Biochemistry* **10**, 3035–3041.
Hata, R., and Nagai, Y. (1973). *Biochim. Biophys. Acta* **304**, 408–412.
Helff, O. M. (1930). *Anat. Rec.* **47**, 177–186.
Hori, H., and Nagai, Y. (1979). *Biochim. Biophys. Acta* **566**, 211–221.
Hourdry, J., and Dauca, M. (1977). *Int. Rev. Cytol., Suppl.* **5**, 337–385.
Huber, S., Ryffel, G. U., and Weber, R. (1979). *Nature (London)* **278**, 65–67.
Ishizuya-Oka, A., and Shimozawa, A. (1987a). *Anat. Anz.* **164**, 81–93.
Ishizuya-Oka, A., and Shimozawa, A. (1987b). *J. Morphol.* **193**, 13–22.
Jacobson, M., and Baker, R. E. (1968). *Science* **160**, 543–545.
Jorgensen, C. B., and Levi, H. (1974). *Comp. Biochem. Physiol. A* **52A**, 55–58.
Kaiho, M., and Ishiyama, I. (1987). *Zool. Seien* **4**, 627–634.
Kaltenbach, J. C. (1958). *Anat. Rec.* **131**, 569–570.
Kaltenbach, J. C. (1959). *J. Exp. Zool.* **140**, 1–17.
Kawahara, A., Sato, K., and Amano, M. (1981). *Dev. Growth Differ.* **23**, 599–611.
Kawahara, A., Kohara, S., Sugimoto, Y., and Amano, M. (1987). *Dev. Biol.* **122**, 139–145.
Kawahara, A., Kohara, S., and Amano, M. (1989). *Dev. Biol.* **132**, 73–80.
Kemp, N. E. (1959). *Dev. Biol.* **1**, 459–476.
Kerr, J. F. R., Wyllie, A. H., and Currie, A. R. (1972). *Br. J. Cancer* **26**, 239–257.
Kerr, J. F. R., Harmon, B., and Searle, J. (1974). *J. Cell Sci.* **14**, 571–585.
Kerr, J. F. R., Searle, J., Harmon, B. V., and Bishop, C. J. (1987). In "Perspectives on Mammalian Cell Death" (C. S. Potten, ed.), pp. 93–128. Oxford Univ. Press, London..
Kikuyama, S., Yamamoto, K., Seki, T., Niki, K., and Yoshizato, K. (1981). *Abstr. Int. Symp. Comp. Endocrinol., 9th* p. 42.
Kim, Y., Hamburgh, M., Frankfort, H., and Etkin, W. (1977). *Dev. Biol.* **55**, 387–391.
Kinoshita, T., Sasaki, F., and Watanabe, K. (1985). *J. Morphol.* **185**, 269–275.
Kinoshita, T., Sasaki, F., and Watanabe, K. (1986a). *J. Exp. Zool.* **238**, 201–210.
Kinoshita, T., Sasaki, F., and Watanabe, K. (1986b). *Cell Tissue Res.* **245**, 297–304.
Kinoshita, T., Takahama, H., Sasaki, F., and Watanabe, K. (1989). *J. Exp. Zool.* **251**, 37–46.
Kistler, A., Yoshizato, K., and Frieden, E. (1975). *Dev. Biol.* **46**, 151–159.
Kobayashi, H. (1958). *Endocrinology* **62**, 371–377.

Krug, E. C., Honn, K. V., Batista, J., and Nicoll, C. S. (1983). *Gen. Comp. Endocrinol.* **52,** 232–241.

Lapiere, C. M., and Gross, J. (1963). *In* "Mechanisms of Hard Tissue Destruction" (R. F. Sognnaes, ed.), pp. 663–694. Am. Assoc. Adv. Sci., Washington, D.C..

Lehman, H. E. (1953). *Biol. Bull.* **105,** 490–495.

Leloup, J., and Buscaglia, M. (1977). *C.R. Hebd. Seances Acad. Sci., Ser. D* **284,** 2261–2263.

Levi, H., and Nielsen, A. (1982). *J. Invest. Dermatol.* **79,** 292–296.

Lipson, M. J., and Silbert, J. E. (1965). *Biochim. Biophys. Acta* **101,** 279–284.

Lipson, M. J., and Silbert, J. E. (1968). *Biochim. Biophys. Acta* **158,** 344–350.

Lipson, M. J., Cerskus, R. A., and Silbert, J. E. (1971). *Dev. Biol.* **25,** 198–208.

Little, G., Atkinson, B. G., and Frieden, E. (1973). *Dev. Biol.* **30,** 366–373.

McAvoy, J. W., and Dixon, K. E. (1977). *J. Exp. Zool.* **202,** 129–148.

Marshall, J. A., and Dixon, K. E. (1978). *J. Anat.* **126,** 134–144.

Mathisen, P. M., and Miller, L. (1987). *Genes Dev.* **1,** 1107–1117.

Mathisen, P. M., and Miller, L. (1989). *Mol. Cell. Biol.* **9,** 1823–1831.

May, F. E. B., and Knowland, J. (1980). *Dev. Biol.* **77,** 419–430.

Miyauchi, H., LaRochelle, F. T., Jr., Suzuki, M., Freeman, M., and Frieden, E. (1977). *Gen. Comp. Endocrinol.* **33,** 254–266.

Miyazaki, K., Masui, H., and Sato, G. H. (1982). *Cold Spring Harbor Conf. Cell Proliferation* **9,** 657–661.

Moriya, T., Thomas, C. R., and Frieden, E. (1984). *Endocrinology* **114,** 170–175.

Motobayashi, N. Y., and Yoshizato, K. (1986). *Zool. Sci.* **3,** 83–89.

Motobayashi, N. Y., Horiguchi, T., and Yoshizato, K. (1986). *Zool. Sci.* **3,** 91–96.

Nagai, Y., Lapiere, C. M., and Gross, J. (1966). *Biochemistry* **5,** 3123–3130.

Nanbu, M., Kobayashi, K., and Horiuch, S. (1988). *Comp. Biochem. Physiol. B* **89B,** 569–575.

Naughten, J. C., and Krollros, J. J. (1971). *Am. Zool.* **11,** 687. (Abstr.).

Nieuwkoop, P. D., and Faber, J. (1967). *In* "Normal Table of *Xenopus Laevis* (Daudin)," 2nd Ed. North-Holland Publ., Amsterdam.

Niki, K., and Yoshizato, K. (1986). *Dev. Biol.* **118,** 306–308.

Niki, K., Yoshizato, K., and Kikuyama, S. (1981). *Proc. Jpn. Acad., Ser. B* **57,** 271–275.

Niki, K., Namiki, H., Kikuyama, S., and Yoshizato, K. (1982). *Dev. Biol.* **94,** 116–120.

Niki, K., Yoshizato, K. Namiki, H., and Kikuyama, S. (1984). *Dev. Growth and Differ.* **26,** 329–338.

Nishikawa, A., and Yoshizato, K. (1985). *Zool. Sci.* **2,** 201–211.

Nishikawa, A., and Yoshizato, K. (1986a). *J. Exp. Zool.* **237,** 221–230.

Nishikawa, A., and Yoshizato, K. (1986b). *J. Exp. Zool.* **239,** 133–137.

Nishikawa, A., and Yoshizato, K. (1989). In preparation..

Nishikawa, A., Kaiho, M., and Yoshizato, K. (1989). *Dev. Biol.* **131,** 337–344.

Polansky, J. R., and Toole, B. P. (1976). *Dev. Biol.* **53,** 30–35.

Rand, H. W., and Pierce, M. E. (1932). *J. Exp. Zool.* **62,** 125–170.

Reeves, R. (1977). *Dev. Biol.* **60,** 163–179.

Regard, E., Taurog, L., and Nakajima, T. (1978). *Endocrinology* **102,** 674–684.

Rheinwald, J. G., and Green, H. (1975). *Cell* **6,** 331–343.

Robinson, D. H., and Heintzelman, M. B. (1987). *Anat. Rec.* **217,** 305–317.

Rudneff, M. (1865). *Arch. Mikrosk. Anat.* **1,** 295–298.

Ryffel, G. U. (1978). *Mol. Cell. Endocrinol.* **12,** 237–246.

Sakai, J., and Horiuchi, S. (1979a). *Comp. Biochem. Physiol. B* **62B,** 269–273.

Sakai, J., and Horiuchi, S. (1979b). *Zool. Mag.* **88,** 116–121.

Sap, J., Munoz, A., Damm, K., Goldberg, Y., Ghysdael, J., Leutz, A., Beug, H., and Vennström, B. (1986). *Nature (London)* **324,** 635–640.

Saskai, F., Grillo, B., Horiguchi, T., and Watanabe, K. (1985). Cell Tissue Res. 239, 511–517.
Seshimo, H., Ryuzaki, M., and Yoshizato, K. (1977). Dev. Biol. 59, 96–100.
Shaffer, B. M. (1963). J. Embryol. Exp. Morphol. 11, 77–90.
Silbert, J. E., Nagai, Y., and Gross, J. (1965). J. Biol. Chem. 240, 1509–1511.
Shumway, W. (1940). Anat. Rec. 78, 139–149.
Singer, M., and Salpeter, M. M. (1961). J. Exp. Zool. 147, 1–19.
Skipper, J. K., and Hamilton, T. H. (1979). Science 206, 693–695.
Smith, K. B., and Tata, J. R. (1976). Exp. Cell. Res. 100, 129–146.
Smith-Gill, S. J. (1979). Dev. Growth Differ. 21, 291–301.
Smith-Gill, S. J., and Carver, V. (1981). In "Metamorphosis, A Problem in Developmental Biology" (L. I. Gilbert and E. Frieden, eds.), 2nd Ed., pp. 491–544. Plenum, New York.
Smith-Gill, S. J., Reilly, J. G., and Weber, E. M. (1979). Dev. Growth Differ. 21, 281–290.
Stanchfield, J. E., and Yager, J. D. (1978). Exp. Cell Res. 116, 239–252.
Stanchfield, J. E., and Yager, J. D. (1980). J. Cell Biol. 84, 468–475.
Sumiya, M., and Horiuchi, S. (1980). Zool. Mag. 89, 176–182.
Suzuki, M. R., and Kikuyama, S. (1983). Gen. Comp. Endocrinol. 52, 272–278.
Suzuki, S., and Suzuki, M. (1981). Gen. Comp. Endocrinol. 45, 74–81.
Taira, T., and Yoshizato, K. (1987). In "Cytoprotection and Cytobiology" (F. Nagao et al., eds.), Vol. 4, pp. 318–326. Excerpta Med. Found. Amsterdam.
Tata, J. R. (1966). Dev. Biol. 13, 77–94.
Tata, J. R. (1976). Cell 9, 1–14.
Taylor, A. C., and Kollros, J. J. (1946). Anat. Rec. 94, 7–23.
Usuku, G., and Gross, J. (1965). Dev. Biol. 11, 352–370.
Wahli, W., Abraham, I., and Weber, R. (1978). Wilhelm Roux's Arch. Dev. Biol. 185, 235–248.
Wahli, W., David, I. B., Ryffel, G. U., and Weber, R. (1981). Science 212, 298–304.
Wangh, L. J., Osborne, J. A., Hentchel, C. C., and Tilly, R. (1979). Dev. Biol. 70, 479–499.
Watanabe, K., and Sasaki, F. (1974). Cell Tissue Res. 155, 321–336.
Weber, R. (1957). Experientia 13, 153–155.
Weber, R. (1962). Experientia 18, 84–85.
Weber, R. (1964). J. Cell Biol. 22, 481–487.
Weber, R. (1965). Experientia 21, 665–666.
Weber, R. (1967). In "The Biochemistry of Animal Development" (R. Weber, ed.), Vol. 2, pp. 227–301. Academic Press, New York.
Weber, R. (1977). Collog. Int. C. N. R. S. 266, 137–146.
Weinberger, C., Thompson, C. C., Ong, E. S., Lebo, R., Gruol, D. J., and Evans, R. M. (1986). Nature (London) 324, 641–646.
Weiss, P., and Ferris, W. (1954). Proc. Natl. Acad. Sci. USA 40, 528–540.
Woolley, D. E., Granville, R. W., Crossley, M. J., and Evanson, J. M. (1975). Eur. J. Biochem. 54, 611–622.
Wright, M. L. (1973). J. Exp. Zool. 186, 237–256.
Wright, M. L. (1977). J. Exp. Zool. 202, 223–234.
Yada, S., and Yamazawa, K. (1962). Jpn. J. Legal Med. 16, 62–64.
Yada, S., Yamazawa, K., and Mori, S. (1962). Jpn. J. Legal Med. 16, 233–237.
Yoshizato, K. (1986). Zool. Sci. 3, 219–226.
Yoshizato, K., and Frieden, E. (1975). Nature (London) 254, 705–707.
Yoshizato, K., and Nakajima, Y. (1982). Dev. Growth Differ. 24, 553–562.
Yoshizato, K., and Nishikawa, A. (1985). Dev. Growth Differ. 27, 621–631.
Yoshizato, K., Kistler, A., and Frieden, E. (1975). Endocrinololy 97, 1030–1035.

Localized mRNA and the Egg Cytoskeleton

William R. Jeffery

Center for Developmental Biology, Department of Zoology, University of Texas at Austin, Austin, Texas 78712

I. Introduction

For many years after its discovery mRNA was thought to be distributed uniformly in the cytoplasm of eukaryotic cells. During the last decade, however, our understanding of mRNA distribution has changed radically. It is now known that mRNAs coding for secretory proteins contain a signal sequence that targets polysomes to the endoplasmic reticulum, or ER (Walter *et al.*, 1984). This has raised the possibility that other mRNAs may be localized in the cytoplasm. Until recently, it was almost impossible to determine the spatial distribution of mRNA. With the advent of gene cloning and *in situ*-hybridization techniques, however, it is now possible to define patterns of mRNA distribution in reasonable detail, providing appropriate probes are available. The application of these methods to a variety of somatic cells has shown that the distribution of mRNA and its nuclear precursors (Lawrence *et al.*, 1989) can be nonuniform. For example, actin, tubulin, and vimentin mRNAs are localized in different regions of chick myoblasts and fibroblasts (Lawrence and Singer, 1986), microtubule-associated protein 2 mRNA is localized in the dendrites and tubulin mRNA in the cell bodies of rat brain neurites (Garner *et al.*, 1988), and acetylcholine receptor mRNA is localized in the neuromuscular junction in chick striated muscle (Merlie and Sanes, 1985; Fontaine *et al.*, 1988; Fontaine and Changeux, 1989). Other mRNAs, such as those coding for histones, appear to be distributed evenly in the cytoplasm (Lawrence *et al.*, 1988). However, even in prokaryotic cells, there is evidence for uneven distribution of mRNA. During *Caulobacter* cell division, flagellin mRNA segregates specifically to the swarmer cell progeny, in which it directs the synthesis of a new flagellum (Milhausen and Agabian, 1983).

Most of the localized messages described in various somatic cells are not associated with intracellular membranes. Instead, their nonrandom distribution may be based on an association with the cytoskeleton. When somatic cells are treated with nonionic detergents, most of the mRNA is retained in a detergent-insoluble residue (Lenk *et al.*, 1977; Lenk and Penman, 1979; Cervera *et al.*, 1981; van Venrooij *et al.*, 1981; Jeffery,

151

1982; Pramanik *et al.*, 1986; Bagchi *et al.*, 1987; Brodeur and Jeffery, 1987; Davis *et al.*, 1987). This detergent-insoluble residue contains cytoskeletal elements (Brown *et al.*, 1976; Lenk *et al.*, 1977; Small and Celis, 1978; Ben-Ze'ev *et al.*, 1979; Pudney and Singer, 1980; Schliwa *et al.*, 1981; Bravo *et al.*, 1982), including microtubules (Mt), actin filaments, and intermediate filaments. Electron microscopy (EM) confirms the results of detergent extraction, showing that polysomes are associated with cytoskeletal networks in various somatic cells (Brökelmann, 1977; Wolosewick and Porter, 1979; Ramaekers *et al.*, 1983), and nuclear mRNA precursors are also associated with a fibrillar matrix (Miller *et al.*, 1978; Herman *et al.*, 1978; van Eekelen and van Venrooij, 1981; Mariman *et al.*, 1982; Maundrell *et al.*, 1981; Ciejek *et al.*, 1982; Nakayasu and Ueda, 1985). The existence of mRNA–cytoskeleton interactions has been more difficult to demonstrate *in vivo,* particularly since messages that distribute specifically to the cytoskeletal fraction have not been reported (van Venrooij *et al.*, 1981; Cervera *et al.*, 1981). Support for the reality of this phenomenon, however, has been provided by the observation that cellular mRNA is released from its association with the cytoskeleton during viral infection and replaced by viral mRNA (Lenk and Penman, 1979; van Venrooij *et al.*, 1981; Bonneau *et al.*, 1985). Although the nature of mRNA–cytoskeleton interactions is still unresolved (Heuijerjans *et al.*, 1989), there is some indication that the 3' poly(A) tail may be a potential cytoskeletal binding site (Pramanik *et al.*, 1986). Likewise, the function of mRNA localization is still obscure in somatic cells, although it is conceivable that this localization reflects the coordinated translation and assembly of proteins into cellular architecture (Fulton *et al.*, 1980; Fulton and Wan, 1983).

Eggs are specialized cells that accumulate many different components to be used during subsequent embryogenesis. Among these components is mRNA, which is stored in relatively large quantities in the egg cytoplasm (Davidson, 1986). The fact that many eggs contain distinct cytoplasmic domains raises the possibility that egg mRNAs may be localized in these regions and that mRNA localization may have a role in embryonic development. In this article, we review the evidence for localized mRNAs in eggs and the possibility that mRNA localization may be mediated by the egg cytoskeleton.

II. mRNA Localization in Egg Cytoplasmic Domains

It has been known for >100 years that some eggs contain distinct cytoplasmic domains. Electron microscopy has subsequently shown that these domains contain large cytoplasmic organelles such as yolk platelets, mitochondria, and pigment granules (see, e.g., Berg and Humphreys, 1960). The early descriptive work on egg cytoplasmic domains was reviewed in E. B. Wilson's classic volume "The Cell in Development and Heredity" (Wilson, 1925), and will not be repeated here. What we will do, however, is describe recent work on egg cytoplasmic domains that contain localized mRNA. We shall also examine the role of these cytoplasmic domains and localized mRNAs in development.

A. LOCALIZED mRNA IN *Chaetopterus* EGGS

Chaetopterus is a polychaete annelid with eggs containing several different cytoplasmic domains (Lillie, 1906; Eckberg, 1981). The mature oocyte contains three distinct regions: (1) the germinal vesicle (GV), (2) an endoplasmic domain containing yolk platelets, and (3) an ectoplasmic domain containing specific granules. We will be concerned primarily with the ectoplasmic domain because it contains localized mRNA. The ectoplasm is present in the animal-hemisphere cortex of oocytes. It is distinguished ultrastructurally by electron-dense granules, the ectoplasmic spherules (Eckberg, 1981; Jeffery, 1985), which are enmeshed in a cytoskeletal network (Fig. 1). The distribution of mRNA in sections of *Chaetopterus* oocytes and fertilized eggs has been examined by *in situ* hybridization with poly(U) and specific probes (Jeffery and Wilson, 1983). These studies show that poly(A)+ RNA, actin mRNA, and histone mRNA are localized in the cortical ectoplasm (Fig. 2A). When oocytes are released into seawater and fertilized, the GV breaks down and there is a concomitant rearrangement of cytoplasmic domains. During this rearrangement, material released from the GV enters the cytoplasm and becomes localized in the animal hemisphere. Simultaneously, the ectoplasm spreads vegetally to occupy the entire egg cortex, except for a small region around the animal pole. During this rearrangement, there is no mixing of mRNA between the various cytoplasmic domains; mRNA remains localized in the ectoplasm and can be visualized throughout the egg cortex, except for the animal pole region that lacks ectoplasm (Fig. 2B). It should be emphasized that the localization of mRNA in the ectoplasm does not occur simply because there is more cytoplasm in this region of the egg. It is true that the ectoplasm is depleted of yolk platelets, but it is filled with ectoplasmic spherules that are about the same size as these organelles (Eckberg, 1981; Jeffery, 1985). Moreover, the ooplasm derived from the GV also lacks yolk platelets, but does not contain localized mRNA.

Another cycle of cytoplasmic rearrangement accompanies the first cleavage. During prophase, the ectoplasm disintegrates into clusters of ectoplasmic spherules that migrate internally and surround the mitotic apparatus in the center of the cell. Subsequently, as the asters extend

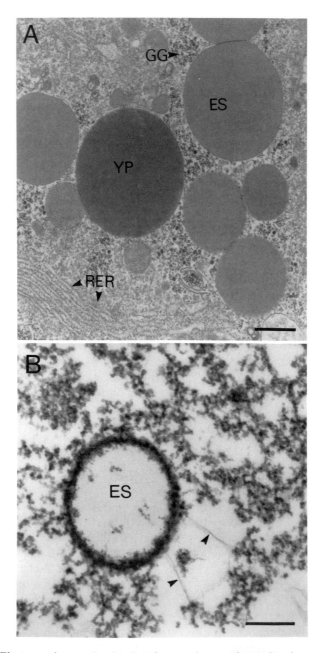

FIG. 1. Electron micrographs showing the ectoplasm and ectoplasmic cytoskeleton of *Chaetopterus* eggs. (A) Ectoplasm of an intact egg. ES, ectoplasmic spherules; YP, yolk platelet; RER, rough ER; GG, glycogen granules. Scale bar = 1 μm. (B) Ectoplasmic cytoskeleton of an NP-40-extracted egg. Arrowheads indicate filaments. Scale bar = 0.3 μm. Electron micrographs of sectioned eggs and NP-40-extracted cytoskeletons from Jeffery (1985).

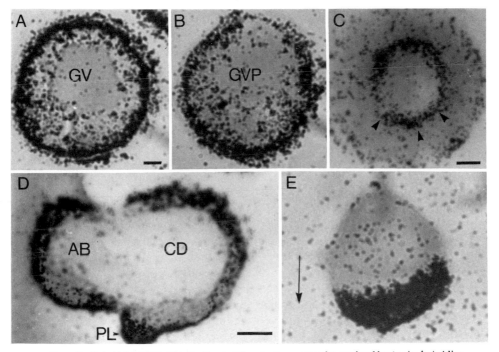

FIG. 2. Poly(A)+ RNA distribution in *Chaetopterus* eggs determined by *in situ* hybridization with poly(U). (A) Section through the animal hemisphere of a primary oocyte. GV, Germinal vesicle. (B) Section through the animal–vegetal axis of a fertilized egg that has completed maturation. The animal-pole region is at the top of the section. GVP, Ooplasm released from GV. In (A) and (B), poly(A)+ RNA is localized in the cortical ectoplasm. (C) Section through a fertilized egg immediately before first cleavage. Poly(A)+ RNA has accumulated around the periphery of the mitotic apparatus that has formed in the center of the egg. (D) Section through the animal–vegetal axis of an embryo at the trefoil stage. Poly(A)+ RNA is localized in the animal hemisphere cortex and the polar lobe (PL). AB, AB blastomere; CD, CD blastomere. (E) Section through a stratified egg. The arrow shows the direction of centrifugal force. Poly(A)+ RNA has been driven to the centrifugal end of the egg. All scale bars = 10 μm. Scale bar in (A) also refers to (B) and (E). *In situ* hybridization to sectioned oocytes and eggs from Jeffery and Wilson (1983) and Jeffery (1985).

toward the periphery of the egg, the ectoplasmic spherules move ahead of them until they reach the cortex again. Remarkably, mRNA is translocated with the ectoplasmic spherules during their inward and outward movements (Fig. 2C). When the ectoplasm is reestablished in the cortex it no longer exists as a single mass. Instead, it has separated into two domains, one in the animal hemisphere and another in the vegetal hemisphere (Fig. 2D). During first cleavage, the animal domain is split between the AB and CD blastomeres, whereas its vegetal counterpart enters the small polar lobe and is segregated to the CD blastomere. Likewise, the ectoplasmic mRNA localization is partitioned unevenly, so that more mRNA is distributed to the CD blastomere than to the AB blastomere. It is noteworthy that mRNA remains associated with the ectoplasmic spherules as they are translocated through the egg during activation and cleavage. This behavior suggests that mRNA is tenaciously bound to cytoplasmic structures in the ectoplasm.

The ontogeny of mRNA localization was studied by Jeffery et al. (1986). They showed that previtellogenic oocytes contain a large GV and a cytoplasm with uniformly distributed ectoplasmic granules. According to in situ-hybridization studies, poly(A)+ RNA is also distributed uniformly in the cytoplasm of previtellogenic oocytes. During subsequent vitellogenesis, however, yolk platelets accumulate in the perinuclear region and gradually form an exclusion zone between the ectoplasmic spherules and the GV. Thus, by the conclusion of vitellogenesis, most of the ectoplasmic spherules and associated poly(A)+ RNA are localized in the oocyte cortex. During the remainder of oogenesis, additional aggregates of ectoplasmic spherules and poly(A)+ RNA appear in the perinuclear cytoplasm, move peripherally between the yolk platelets, and join their counterparts in the cortical ectoplasm. This activity is greater in the animal hemisphere, where the GV is located, than in the vegetal hemisphere. Hence, the ectoplasm attains its typical animal cortical distribution in fully grown oocytes. These results suggest that the cortical localization of poly(A)+ RNA is formed by two mechanisms: (1) gradual exclusion of RNA-rich cytoplasm into the oocyte cortex during vitellogenesis and (2) directed movement of newly synthesized mRNA into the cortical region of the oocyte after vitellogenesis has been completed.

The role of ectoplasmic mRNA in development has been examined by two kinds of experiments (Swalla et al., 1985). These experiments depend on the displacement of the ectoplasm (Lillie, 1909) and its associated mRNA (Jeffery, 1985) from the cortex to the centrifugal region by low-speed centrifugation of Chaetopterus eggs. In the first experiment, strong centrifugation was applied to split the egg into two fragments. Because of their relative densities, the ectoplasmic spherules are driven to the centrifugal fragment, while the nucleus and most of the other egg cytoplasmic components (lipid bodies, ribosomes, mitochondria, ER, and some mRNA) are driven to the centripetal fragment. Ectoplasmic mRNA is also driven to the centrifugal fragment (Fig. 3). Thus, centrifugation is a means of producing portions of eggs (i.e., centripetal fragments) that contain a nucleus but lack ectoplasmic mRNA. These nucleate egg fragments do not cleave or develop any further, suggesting that they lack an essential com-

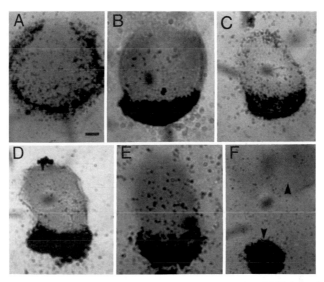

FIG. 3. Displacement of ectoplasm and associated poly(A)+ RNA into the centrifugal fragment during stratification of *Chaetopterus* eggs by centrifugation. (A) An uncentrifuged egg. (B–F) Eggs fixed during centrifugation.(B–E) Centrifugation drives the ectoplasm and poly(A)+ RNA to the centrifugal pole as an egg gradually elongates. (F) Centrifugation eventually splits an egg into nucleate centripetal (upward-facing arrowhead) and anucleate centrifugal (downward-facing arrowhead) fragments. Scale bar = 10 μm. *In situ* hybridization of sectioned eggs according to Swalla *et al.* (1985).

ponent for development—presumably the ectoplasmic mRNA. In the second experiment, weaker centrifugation was used to drive the ectoplasm and associated mRNA to the centrifugal pole of a zygote during first cleavage, without splitting the cell into two parts. After zygotes were removed from the centrifuge, the localized mRNA was seen to be enclosed in either one or two blastomeres in various embryos, depending on the position of the cleavage plane. When the cleavage plane was positioned so that both cells received localized mRNA, larval development was normal. In contrast, when the cleavage plane was positioned so that only one cell received the localized mRNA, the embryos developed into abnormal radialized larvae. These results suggest that localization of mRNA in the ectoplasm is required for normal development.

B. LOCALIZED mRNA IN *Drosophila* EGGS

Insect eggs are also characterized by specific cytoplasmic domains. The central region of the egg is a mass of yolk platelets, which contains the egg nucleus positioned within an island of cytoplasm. Most of the yolk-free cytoplasm is attenuated in the peripheral region of the egg, which is known as the periplasm. The periplasm is expanded at the posterior pole of the egg to form the pole plasm, a region involved in germ cell determination. As described later, the anterior-pole region also has specific properties that make it important in development.

After fertilization and ovulation, the zygotic nucleus begins to divide within the central yolk mass. These nuclear divisions occur in the absence of cytokinesis, so that cytoplasmic continuity is maintained throughout the embryo. After dividing several times, most of the nuclei, which are still located within cytoplasmic islands, migrate into the periplasm and continue to divide. A subset of the cleaving nuclei enters the pole plasm, a cytoplasmic region at the posterior pole of the egg which is thought to be responsible for germ cell formation, and becomes enclosed in the pole cells. This is called the syncytial blastoderm stage. Subsequently, membranes divide the periplasm into separate cells at the cellular blastoderm stage. A feature that distinguishes *Drosophila* and other dipteran eggs is that the oocyte nucleus is inactive in transcription. Consequently, egg RNA is derived from the nurse cells (Bier, 1963), which are connected to the anterior region of the oocyte by ring canals.

The spatial distribution of poly(A)+ RNA has been examined in *Drosophila* oocytes and developing embryos by *in situ* hybridization (Kobayashi *et al.,* 1988). During early oogenesis, poly(A)+ RNA was initially observed in the nurse cell cytoplasm. The nurse cell–poly(A)+ RNA is transported into the growing oocyte through the ring canals and becomes localized in the anterior region of the cell. Eventually, this poly(A)+ RNA must diffuse posteriorly, because after ovulation it is observed throughout the egg. When the nuclei migrate into the periplasm, the poly(A)+ RNA becomes concentrated in this region and enters the embryonic cells during cellularization. By the cellular blastoderm stage, some of the poly(A)+ RNA in the periplasm may be provided by zygotic transcription, which is activated at this stage of development.

Several cases of specific mRNA localization have been reported in *Drosophila* eggs. First, maternal mRNA, encoding the protein cyclin B, becomes localized in the posterior pole of the egg during the early cleavages (Whitfield *et al.,* 1989). Since cyclins are involved in controlling the cell cycle, and pole cells exhibit a cell cycle that is distinct from those of other embryonic cells, it is possible that cyclin B mRNA localization controls the timing of cell division in a specific region of the embryo. Second, caudal mRNA is localized in a shallow gradient that extends from the posterior to the anterior pole at the syncytial blastoderm stage (Mlodzik *et al.,* 1985; Macdonald and Struhl, 1986). Since caudal mRNA is evenly distributed at earlier stages of development, it seems likely that this gradient is produced by mRNA degradation in the more anterior regions of the egg. Although the function of the caudal gene is unknown, it contains a homeobox, and therefore could have a role in pattern formation. Third, bicoid mRNA is localized in the anterior region of oocytes, eggs, and early embryos. Bicoid (*bcd*) is a gene required for the development of the anterior portion of the embryo (Nüsslein-Volhard *et al.*, 1987). *bcd* mRNA is deposited in the anterior region of the oocyte by nurse cells during oogenesis (Fig. 4c, d). Unlike total poly(A)+ RNA, however, this anterior localization is maintained in the mature egg (Frigerio *et al.*, 1986; Berleth *et al.*, 1988). In very early embryos, *bcd* mRNA is localized in a cone-shaped region at the anterior tip of the egg (Fig. 4a). When the cleavage nuclei begin to migrate into the periplasm, *bcd* mRNA is also translocated to the egg periphery, and becomes restricted to a cap of periplasm at the anterior pole (Fig. 4b). Other genes have been identified with mutant alleles that affect normal development by disrupting *bcd* mRNA localization (Fröhnhofer and Nüsslein-Volhard, 1987; Berleth *et al.*, 1988). In swallow (*swa*) mutants, *bcd* mRNA is initially localized in the anterior region of the oocyte, but this localization must be disrupted later in oogenesis because *bcd* mRNA is dispersed throughout the mature egg and early embryo. In exuperantia (*exu*) mutants, *bcd* mRNA localization appears to be defective immediately and transcripts show a uniform distribution in

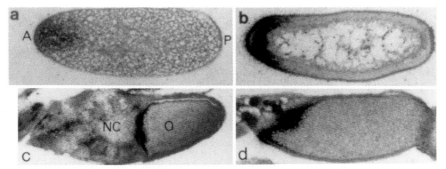

FIG. 4. Localization of *bcd* mRNA in *Drosophila* oocytes and eggs. (a) Section through a very early embryo showing *bcd* mRNA localized in a cone-shaped region at the anterior pole. A, anterior pole; P, posterior pole. (b) A section through a syncytial blastoderm stage embryo showing *bcd* mRNA localized in a cortical cap at the anterior pole. (c) An early follicle showing *bcd* mRNA in nurse cells (NC) and the anterior region of a young oocyte (O). (d) A later follicle showing *bcd* mRNA in the anterior region of a large oocyte. *In situ* hybridization to sectioned follicles and embryos from Berleth *et al.* (1988).

the oocyte and early embryo. Therefore, the *exu* and *swa* gene products appear to be involved in localizing *bcd* mRNA to the anterior region of the egg. Abnormal nuclear migration into the periplasm in *swa*⁻embryos raised at low temperature suggest that the *swa* gene product may be a component of the egg cytoskeleton (Fröhnhofer and Nüsslein-Volhard, 1987).

Thus, there is a specialized anterior domain in *Drosophila* eggs that is responsible for localizing *bcd* mRNA and perhaps other mRNAs. These studies are consistent with earlier work on eggs of other insects showing that ultraviolet irradiation or injection of RNase into the anterior-pole region inactivates localized RNA molecules and abolishes the ability of this region to develop into anterior parts of the embryo (Kalthoff, 1983; Kandler-Singer and Kalthoff, 1976).

C. LOCALIZED mRNA IN *Styela* EGGS

The eggs of the ascidian *Styela* are distinguished by cytoplasmic domains that are highlighted by different colors (Conklin, 1905). Two of these domains, the ectoplasm and myoplasm, are sites of mRNA localization (Jeffery and Capco, 1978; Jeffery *et al.*, 1983). The ectoplasm is a transparent cytoplasmic region, containing membranous vesicles, cytoskeletal filaments, and ribosomes (Section III, C), which is derived from the GV during oocyte maturation. *Styela* oocytes contain a prominent GV, wherein a large proportion of the total poly(A)+ RNA (Fig. 5A) and actin mRNA are localized, probably in the form of nuclear mRNA precursors. When the GV breaks down during maturation, the poly(A)+ RNA (Fig. 5B) and actin mRNA molecules (Fig. 6) do not disperse throughout the cytoplasm. Instead, they remain associated with the ectoplasm, which exhibits a poly(A)+ RNA concentration ~25- to 30-fold higher than the other cytoplasmic regions. The myoplasm is a yellow cytoplasmic domain that is initially restricted to the oocyte cortex. Poly(A)+ RNA levels are too low to be detected by *in situ* hybridization in the myoplasm, but this region is specifically enriched in actin mRNA (Fig. 6). The third cytoplasmic domain of *Styela* eggs is the endoplasm, a gray-colored region that is localized in the vegetal hemisphere and relatively depleted of poly(A)+ RNA and actin mRNA.

After fertilization, *Styela* eggs undergo ooplasmic segregation, an extensive rearrangement of cytoplasmic domains. During the first phase of ooplasmic segregation, the myoplasm and ectoplasm are translocated to the vegetal hemisphere and accumulate around the vegetal pole. Simultaneously, the endoplasm is displaced into the animal hemisphere. During the second phase of ooplasmic segregation, the ectoplasm and myoplasm move into the equatorial region of the egg. The myoplasm forms a yellow

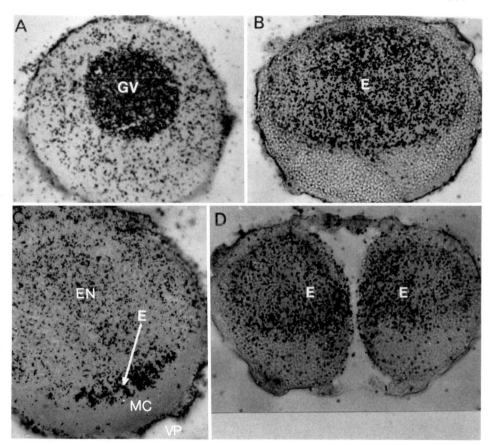

FIG. 5. Localization of poly(A)+ RNA in the GV and ectoplasm of *Styela* oocytes and eggs. (A) A primary oocyte showing poly(A)+ RNA concentrated in the GV. (B) A mature oocyte showing poly(A)+ RNA concentrated in the ectoplasm (E). (C) A fertilized zygote undergoing cytoplasmic rearrangement. Poly(A)+ RNA is still localized in the ectoplasm, which is compressed into a thin zone between the endoplasm (EN) and myoplasmic cap (MC). VP, Vegetal pole. (D) A two-cell embryo showing poly(A)+ RNA in the ectoplasm, which is localized in the animal region of the two blastomeres. *In situ* hybridization of sectioned oocytes and eggs from Jeffery *et al.* (1983).

crescent just below the equator and during subsequent cleavage is segregated to blastomeres that develop into muscle cells. The ectoplasm temporarily forms a transparent crescent just above the yellow crescent and then returns to the animal hemisphere, where it is segregated to the presumptive ectodermal cells during cleavage. In the meantime, the endoplasm returns to the vegetal hemisphere and enters vegetal cells during cleavage.

When zygotes undergoing ooplasmic segregation and cleavage were subjected to *in situ* hybridization, the same spatial distributions of mRNA
were observed as existed in the cytoplasmic domains before fertilization
(Fig. 5C, D; Fig. 6). These results suggest that localized mRNA molecules
are bound to a structural framework in each cytoplasmic domain and
therefore are not freely diffusible in the egg. All mRNAs do not behave like
ectoplasmic poly(A)+ RNA or actin mRNA. Histone mRNA, for instance,
is evenly distributed between the three cytoplasmic regions (Jeffery *et al.*,
1983).

These results show that mRNA molecules are localized in particular
cytoplasmic domains of *Styela* eggs and that these localizations are stable
during the extensive cytoplasmic movements and segregations that accompany early development. The localized mRNAs may be important in
determining cell fates during *Styela* development.

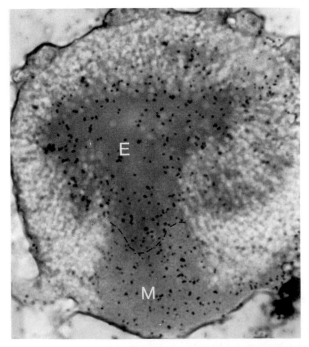

FIG. 6. Localization of actin mRNA in the myoplasm (M) and ectoplasm (E) of *Styela*
eggs. The border between the ectoplasm and myoplasm is indicated by the dashed black line.
In situ hybridization of a sectioned egg from Jeffery *et al.* (1983).

D. Localized mRNA in *Xenopus* Eggs

Different cytoplasmic domains are distributed along the animal–vegetal axis of amphibian eggs. In fully grown oocytes, the animal hemisphere contains dark cortical pigment granules and less yolk than the vegetal hemisphere, which is lightly pigmented and packed with yolk platelets. Animal–vegetal polarity is evident even in the smallest previtellogenic oocytes, in which the GV lies nearer to the animal pole than the vegetal pole (Coggins, 1973). In previtellogenic oocytes, the mitochondrial mass, a region enriched in mitochondria and electron-dense granules, is located immediately below the GV (Billet and Adam, 1976; Heasman *et al.,* 1984). The mitochondria disperse before the beginning of vitellogenesis and accumulate near the GV. In sections, the dispersed material appears as a halo surrounding the GV and contains mitochondria, stacks of membranes, and granular organelles (Balinsky and Devis, 1963). Presumably, the halo reflects the progressive movement of substances derived from the GV and/or mitochondrial mass to the oocyte cortex. Cortical pigment granules first appear during vitellogenesis. Initially both the animal and vegetal hemispheres contain pigment granules, but as large yolk platelets begin to fill the vegetal hemisphere, they become restricted to the animal hemisphere. Oocyte mRNA is transcribed and accumulates steadily during the previtellogenic period (Rosbash and Ford, 1974; Golden *et al.,* 1980). Afterward, RNA synthesis continues but transcription is balanced by degradation, and no net accumulation of mRNA occurs during the remainder of oogenesis (Dolecki and Smith, 1979).

In situ hybridization has shown that poly(A)+ RNA is evenly distributed in previtellogenic oocytes (Capco and Jeffery, 1982). Just before the beginning of vitellogenesis, however, poly(A)+ RNA becomes localized in a halo, which is positioned at various distances between the GV and the oocyte cortex (Fig. 7A). This localization may reflect the movement of mRNA to the cortex in concert with substances derived from the mitochondrial mass. During vitellogenesis, the halo of poly(A)+ RNA enters the cortex in both the animal and vegetal hemispheres. In fully grown oocytes, however, poly(A)+ RNA is observed only in the vegetal cortex (Fig. 7B). The fate of poly(A)+ RNA that was originally localized in the animal cortex is unknown; it could be dispersed, degraded, or deadenylated. After the beginning of maturation, there is another change in the distribution of poly(A)+ RNA, which moves from the vegetal cortex to more internal regions of the vegetal hemisphere (Capco and Jeffery, 1982; Larabell and Capco, 1988).

Some specific mRNAs also appear to be unevenly distributed in *Xenopus* oocytes and eggs. By separating oocytes and eggs into central and

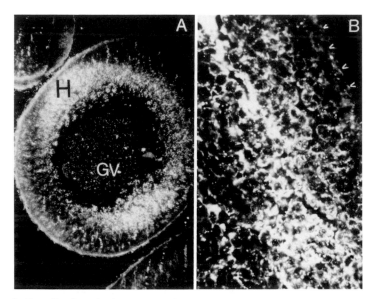

FIG. 7. Localization of poly(A)+ RNA in *Xenopus* oocytes. (A) A stage 3 oocyte showing a halo (H) of poly(A)+ RNA surrounding the GV. (B) A stage 6 oocyte showing the subcortical localization of poly(A)+ RNA. The arrowheads (upper right) indicate the position of the plasma membrane. *In situ* hybridization of sectioned oocytes from Capco and Jeffery (1982).

peripheral regions and measuring RNA titers in these regions by dot hybridization, it has been demonstrated that actin and tubulin mRNA are enriched in the periphery of fully grown oocytes (Perry and Capco, 1988). Actin mRNA remains enriched in the periphery of mature oocytes and unfertilized eggs, but the distribution of tubulin mRNA is changed during later development. After maturation, tubulin mRNA becomes concentrated in the central region of the oocyte, but after fertilization it is enriched in the periphery again. Since there is no new transcription between maturation and the midblastula transition, these results imply that tubulin mRNA is translocated during the course of development. The reason for this translocation is presently unknown, but it may be related to a differential requirement for tubulin synthesis in various regions of the cell.

Biochemical analysis of sectioned eggs provided the first evidence that specific mRNAs are localized in the vegetal hemisphere of *Xenopus* eggs. Poly(A)+ RNA was extracted from animal and vegetal sections of eggs and early embryos, and then compared by solution hybridization (Carpenter and Klein, 1982). Although the results showed that the vast majority of mRNA sequences are present in both the animal and vegetal hemispheres, the vegetal region appeared to contain a small number of specific mRNAs.

These studies were confirmed by *in vitro* translation of poly(A)+ RNA extracted from the animal- and vegetal-pole regions of fully grown oocytes (King and Barklis, 1985). Comparison of proteins synthesized by these RNAs showed that ~13 of 600 different mRNAs are localized in the most vegetal region of fully grown oocytes. Moreover, these results also showed that a few mRNAs were localized in the animal hemisphere (King *et al.*, 1987). Subsequent *in vitro*-translation studies (Smith, 1986) have confirmed the earlier studies and shown that the localized mRNAs remain in the vegetal hemisphere throughout early development (Fig. 8).

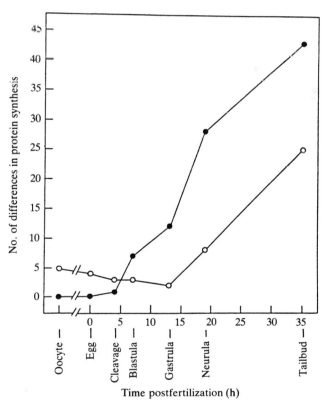

FIG. 8. Graph showing the relative number of specific mRNAs localized in the animal (●) and vegetal (○) hemisphere of *Xenopus* oocytes, eggs, and embryos. The number of localized mRNAs was determined by translating RNA isolated from these regions in an *in vitro* system and determining the number of protein products by two-dimensional gel electrophoresis. The number of mRNAs restricted to the vegetal hemisphere is small in oocytes, eggs, and early-cleaving embryos, but rises dramatically during later development. The dramatic rise is caused by the beginning of new transcription after the midblastula transition. From Smith (1986).

Localized mRNAs have also been identified in *Xenopus* oocytes and eggs by differential screening of an oocyte cDNA library with probes prepared from animal and vegetal poly(A)+ RNA (Rebagliati *et al.*, 1985). Four examples of localized mRNAs were identified in this fashion. Three of the mRNAs (An-1, An-2, and An-3) are localized in the animal hemisphere, whereas one mRNA (Vg-1) is restricted to the vegetal hemisphere (Fig. 9). Subsequently, three more vegetally localized mRNAs were identified by a similar screen (King *et al.*, 1987). Northern blots and RNase protection studies indicate that An-2 and An-3 mRNAs are present throughout embryogenesis and during later development, whereas An-1 and Vg-1 are strictly maternal; they are present in oocytes and early embryos but are degraded at gastrulation, and do not reappear during later development (Rebagliati *et al.*, 1985).

The ontogeny of Vg-1 mRNA localization has been studied by Melton (1987). *In situ* hybridization showed that Vg-1 mRNA is uniformly distributed in the cytoplasm of previtellogenic oocytes (Fig. 10a, b). During vitellogenesis, Vg-1 mRNA gradually becomes localized in the cortex of the vegetal hemisphere, after first accumulating in a halo around the GV (Fig. 10c, d; Fig. 11A). Vg-1 mRNA localization during oogenesis has also been examined by injecting *in vitro*-transcribed mRNA into young oocytes and allowing them to undergo vitellogenesis in culture (Yisraeli and Melton, 1988). Under these conditions, Vg-1 mRNA is distributed uniformly after injection, but gradually becomes localized to its normal po-

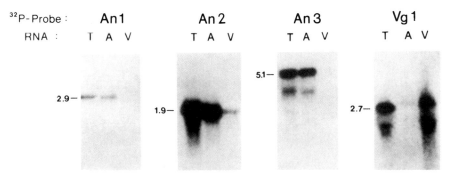

FIG. 9. Localization of An-1, An-2, An-3, and Vg-1 mRNAs in *Xenopus* eggs. Total RNA was extracted from animal (lane A) and vegetal (lane V) thirds of eggs. The RNA was then electrophoresed, blotted, and hybridized with An-1, An-2, and An-3 probes. (Lane T) RNA from whole eggs. An-1 and An-3 mRNAs appear as 2.9-kb and 5.1-kb transcripts, respectively, which are detected only in the animal third. An-2 mRNA appears as 1.9-kb transcript enriched in the animal third (but also detectable in the vegetal third). Vg-1 mRNA appears primarily as a 2.7-kb transcript which is detected only in the vegetal third. Northern blots from Rebagliati *et al.* (1985).

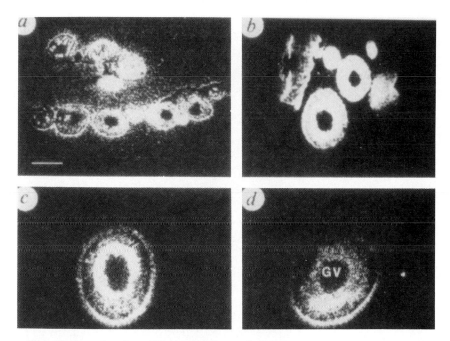

FIG. 10. Translocation of Vg-1 mRNA to the vegetal cortex during *Xenopus* oogenesis. (a–d) Sections through stage I–IV oocytes. GV, Germinal Vesicle. Scale bar = 200 μm. *In situ* hybridization of sectioned oocytes from Melton (1987). Reprinted by permission from *Nature (London)* Vol. 328, pp. 80–82. Copyright © 1987 Macmillan Magazines, Ltd.

sition in the vegetal cortex as the oocyte undergoes vitellogenesis. If injected oocytes are cultured in media lacking vitellogenin, however, they do not grow or localize Vg-1 mRNA. These studies suggest that Vg-1 mRNA is localized by an active mechanism that is coupled to oocyte growth during vitellogenesis. After the beginning of oocyte maturation, Vg-1 mRNA moves from the cortex, but maintains its localization in the vegetal hemisphere and is distributed to the vegetal blastomeres during cleavage (Fig. 11B, C). The distribution of Vg-1 mRNA differs from that of histone mRNA (Fig. 11D), which is uniformly distributed in oocytes and early embryos (Jamrich *et al.,* 1984; Melton, 1987; Perry and Capco, 1988).

Further information on the function of An-2 and Vg-1 mRNAs has been obtained from sequence data. The sequence of An-2 mRNA shows that it is homologous to the α subunit of a mitochondrial ATPase (Weeks and Melton, 1987a). Thus, localization of An-2 mRNA in the animal hemisphere might be required for translation and targeting of this ATPase to mitochondria, which also tend to be concentrated in this region. The

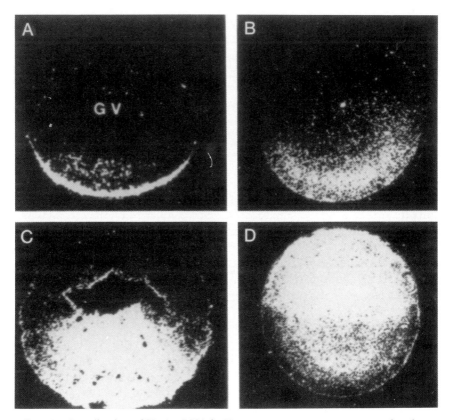

FIG. 11. Localization of Vg-1 mRNA in *Xenopus* oocytes, mature eggs, and early embryos. (A) Fully grown oocyte showing Vg-1 mRNA localized in the periphery of the vegetal hemisphere. GV, Germinal vesicle. (B) Mature egg showing Vg-1 mRNA localized in the vegetal hemisphere. (C) Blastula showing Vg-1 mRNA localized in the vegetal cells. (D) Histone mRNA distribution in an unfertilized egg. *In situ* hybridization of sectioned eggs and embryos from Weeks and Melton (1987b).

sequence of Vg-1 mRNA indicates a homology to transforming growth factor β (TGF-β) (Weeks and Melton, 1987b). It is known that vegetal cells of amphibian embryos have the capacity to induce equatorial cells to become mesoderm (Nieuwkoop, 1969). Several workers have shown that growth factors induce mesoderm formation (Slack *et al.*, 1987; Smith, 1987; Kimelman and Kirschner, 1987; Rosa *et al.*, 1988). Thus, the similarity between the sequences of the Vg-1 protein and TGF-β suggest that Vg-1 mRNA localization may be involved in mesoderm induction.

These studies indicate that there is a special cytoplasmic domain in the vegetal hemisphere of *Xenopus* oocytes and eggs in which poly(A)+ RNA and a small subset of specific mRNAs are localized. This domain appears to form during vitellogenesis and is reorganized after oocyte maturation. Surprisingly, the *in situ*-hybridization results indicate that Vg-1 mRNA and poly(A)+ RNA are distributed similarly during oogenesis (see Figs. 7 and 10). Both populations of molecules are distributed uniformly in the cytoplasm of early previtellogenic oocytes. Before the beginning of vitellogenesis, these molecules accumulate in the perinuclear zone and then become localized in the vegetal cortex. Finally, during maturation Vg-1 mRNA and poly(A)+ RNA are dispersed into the interior of the vegetal hemisphere. A possible explanation for the codistribution of Vg-1 mRNA and poly(A)+ RNA is that the poly(U) *in situ*-hybridization method may actually detect a small subset of vegetally localized mRNAs that are distinguished by especially long or unusually accessible poly(A) tails. Further experiments will be necessary to determine whether this interpretation is correct, but it is possible that the structure of the 3' noncoding regions including the poly(A) tail may be related to the localization of these messages in the vegetal cortex.

E. Summary

The examples cited above indicate that mRNA can be localized in egg cytoplasmic domains. Three patterns of mRNA localization have been identified. First, poly(A)+ RNA (presumably total mRNA) can be localized within cellular compartments (e.g., the GV of *Styela* oocytes) or in specialized cytoplasmic domains (e.g., the cytoplasms of *Chaetopterus* and *Styela* eggs and the vegetal cortex of *Xenopus* oocytes). Second, abundant mRNAs coding for egg structural proteins (e.g., actin mRNA in the ectoplasm of *Chaetopterus* eggs and the myoplasm of *Styela* eggs, and actin and tubulin mRNA in *Xenopus* oocytes) can be localized or enriched in particular cytoplasmic domains. Third, some mRNAs (e.g., *bcd* mRNA in the anterior-pole region of *Drosophila* eggs and Vg-1 mRNA in the vegetal hemisphere of *Xenopus* eggs) are tightly localized in positions where they are likely to have an important role in regulating development. One case of mRNA localization that has not been mentioned is that of early histone mRNA in sea urchin eggs (DeLeon *et al.*, 1983; Angerer *et al.*, 1984). Early histone messages are localized in the female pronucleus until just before first cleavage, when they are released into the cytoplasm and enter the polysomes. The sequestration of these messages within the egg nucleus is probably a mechanism for translational-level control of early histone synthesis.

III. The Egg Cytoskeleton

The egg cytoskeleton has been reviewed previously (Vacquier, 1981), and descriptions of its structure and function during early development are available for specific types of eggs (for examples see Wylie *et al.*, 1986; Fernandez *et al.*, 1987; Shimuzu, 1988). Here, we summarize information on the cytoskeleton in eggs exhibiting localized mRNA, as described in Section II. Initially, the egg cytoskeleton was thought to contain only Mt and actin filaments, with intermediate filaments appearing during later development (Lehtonen and Badley, 1980; Jackson *et al.*, 1980, 1981; Lehtonen *et al.*, 1983). Subsequent studies, however, have shown that certain types of intermediate filaments are components of the egg cytoskeleton. In many cases, the egg cytoskeleton is just beginning to be analyzed, and our information concerning this structure is limited.

A. The Cytoskeleton of *Chaetopterus* Eggs

The cytoskeleton of *Chaetopterus* eggs has been examined by transmission and scanning EM after extraction with the nonionic detergent NP-40 (Eckberg and Langford, 1983; Jeffery, 1985). These studies show that the ectoplasm contains a cytoskeleton consisting of filaments (Fig. 1B) connecting the surface of the ectoplasmic spherules to internal organelles and the plasma membrane. Biochemical analysis indicates that this cytoskeleton contains actin, tubulin, and intermediate filament proteins (J. Venuti and W. R. Jeffery, unpublished observations). The intermediate filament proteins range between 50K and 70K in molecular weight and probably include cytokeratinlike components. At present, little else is known about the ectoplasmic cytoskeleton.

B. The Cytoskeleton of *Drosophila* Eggs

Even though the early *Drosophila* embryo is a syncytium, each nucleus lies within an individual cytoskeletal domain. The egg cytoskeleton has been examined in whole mounts and sections stained with actin, tubulin, and intermediate filament protein antibodies (Warn and McGrath, 1983; Walter and Alberts, 1984; Karr and Alberts, 1986; Warn and Warn, 1986; Edgar *et al.*, 1987). Before the nuclei migrate into the periplasm, actin filaments, Mt, and intermediate filaments are distributed uniformly in a thin zone immediately beneath the plasma membrane. The cytoplasmic islands in the interior of the egg also contain a network of Mt. No specialized cytoskeletal domains have been detected at the anterior or posterior poles, although the search has yet to be extended to very early stages of development.

When the nuclei migrate into the periplasm, the individual cytoskeletal networks surrounding each nucleus become associated with the peripheral cytoskeleton. The cytoskeletal domains that are formed by this association are subsequently incorporated into the embryonic cells at the cellular blastoderm stage. There are three separate regions within each domain: (1) an apical region that lies adjacent to the plasma membrane and consists of actin filaments and intermediate filaments forming a cap above each nucleus, (2) an internal region that consists of a network of Mt and intermediate filaments surrounding each nucleus, and (3) a basal region bordering the central yolk mass. Further work is necessary to determine whether the apical and central regions contain distinct classes of intermediate filaments.

C. The Cytoskeleton of *Styela* Eggs

Cytoskeletal domains also appear to underlie the three cytoplasmic regions of *Styela* eggs (Jeffery and Meier, 1983). The structure of the ectoplasmic cytoskeletal domain has been examined by EM after extraction with Triton X-100. Before extraction, the ectoplasm contains membranous vesicles, ribosomes, glycogen particles, and fine filaments (Fig. 12A). Afterward, there is no trace of intracellular membranes, but filaments associated with ribosomes can be clearly discerned (Fig. 12B). During the second phase of ooplasmic segregation, sperm aster Mt appear in the ectoplasmic domain (Sawada and Schatten, 1988; Venuti and Jeffery, 1989).

Eggs extracted with Triton X-100 show that the myoplasmic cytoskeletal domain consists of two parts (Fig. 13). The first part is a submembrane network containing crosslinked actin filaments. The actin network is connected to a deeper lattice containing thicker filaments that radiate in three dimensions and connect the yellow pigment granules. These are probably intermediate filaments. Indeed, *Styela* eggs contain at least three polypeptides in the molecular weight range of 50–70K that are recognized by an intermediate filament protein antibody (Venuti and Jeffery, 1989). The endoplasm contains a more loosely arranged cytoskeletal domain than the ectoplasm or myoplasm, and its composition has not been investigated. Little is known about the behavior of these cytoskeletal domains during ooplasmic segregation and cleavage. However, the actin network segregates with the myoplasm during both phases of ooplasmic segregation and is still associated with the yellow crescent at first cleavage.

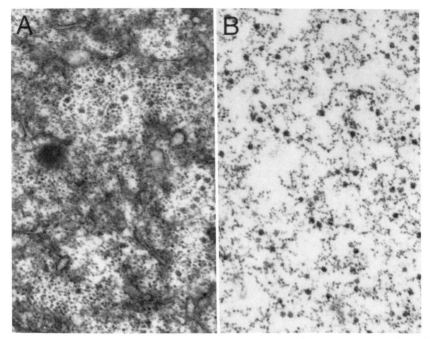

FIG. 12. Structure of the ectoplasm in (A) intact and (B) Triton X-100-extracted *Styela* eggs. Membranous regions have been extracted in the detergent-treated eggs, leaving a filamentous cytoskeleton containing ribosomes. (A) and (B) are transmission electron micrographs at the same magnification.

FIG. 13A.

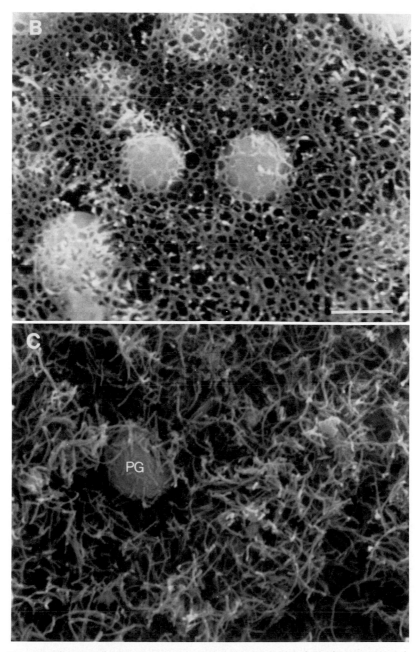

Fig. 13. Filaments in the myoplasmic cytoskeletal domain of *Styela* eggs. (A) A low-magnification photograph of the egg surface at the yellow-crescent stage showing the contrast between the myoplasmic (YC) and endoplasmic (EN) cytoskeletal domains. Scale bar = 5 μm. (B) The superficial network of actin filaments in the myoplasmic cytoskeletal domain and underlying pigment granules. Scale bar = 1 μm. (C) The deep filamentous network of the myoplasmic cytoskeletal domains showing associations with pigment granules (PG). Scale bar = 1 μm. Scanning electron micrographs of whole mounts from Jeffery and Meier (1983).

D. The Cytoskeleton of *Xenopus* Eggs

The cytoskeleton of *Xenopus* eggs also consists of actin filaments, Mt, and intermediate filaments. Some of these cytoskeletal networks are found in the cortex of oocytes and eggs, whereas others are located in the interior. These distributions have also been noted to change during development.

At various stages of development, the internal region of oocytes and eggs contains actin filaments, Mt, and probably intermediate filaments. The internal actin filaments have been reported to be present in a shell surrounding each yolk platelet (Columbo *et al.*, 1981). These actin shells may be derived from the cortical actin network (see later) by endocytosis during vitellogenesis. Antibody staining has shown that Mt are particularly abundant near the mitochondrial mass of previtellogenic oocytes and in the animal-hemisphere region above the GV of fully grown oocytes (Jessus *et al.*, 1986). After maturation, a transient localization of Mt also appears below the disintegrating GV, and in fertilized eggs, sperm aster Mt are present in the animal hemisphere. The intermediate filament protein vimentin has been reported to be present in the interior of *Xenopus* oocytes and early embryos. According to Godsave *et al.* (1984a), vimentin filaments radiate throughout the cytoplasm of previtellogenic oocytes, and there is a concentration of this protein in the mitochondrial mass. Before the beginning of vitellogenesis, vimentin appears in the perinuclear halo, but later is found throughout the cytoplasm. In fully grown oocytes, this protein is present in both the animal and vegetal hemispheres. After maturation and during early development, fine vimentin filaments are scattered throughout the egg. Recently these immunochemical studies have been substantiated by RNase protection assays, which indicate that vimentin is expressed throughout oogenesis and embryogenesis (Tang *et al.*, 1988). In contrast, other investigators have been unable to detect vimentin in *Xenopus* oocytes or early embryos by immunochemical, immunoblotting, or molecular methods (Franz *et al.*, 1983; Herrmann *et al.*, 1989; Dent *et al.*, 1989). A possible explanation for these conflicting results is that the component detected by Godsave *et al.* (1984a) may be a polypeptide that is related to, but distinct from, the vimentin expressed later during embryogenesis.

The cortical cytoskeleton contains actin and cytokeratin filaments (Franke *et al.*, 1976; Picheral *et al.*, 1982; Gall *et al.*, 1983; Godsave *et al.*, 1984b; Klymkowsky *et al.*, 1987). Both hemispheres of oocytes and early embryos contain actin filaments (Fig. 14). In contrast, there is a differential organization of cytokeratin filaments along the animal–vegetal axis (Klymkowsky *et al.*, 1987). In fully grown oocytes, these filaments appear

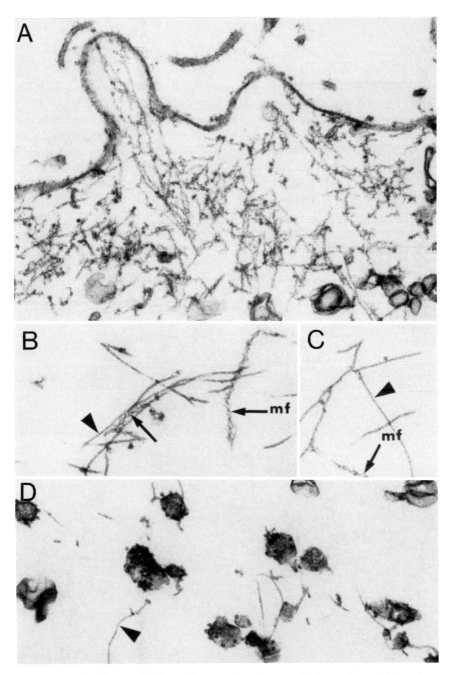

FIG. 14. Microfilaments (mf) and intermediate filaments in the cortex and internal cytoplasm of fully grown *Xenopus* oocytes. (A) The cortical region showing mf decorated with myosin S1 fragments. (B, C) Higher magnifications of (A) showing mf (arrows) decorated with myosin S1 fragments and intermediate filaments (arrowhead). (D) A more internal region showing intermediate filaments (arrowhead). Electron micrographs of myosin S1 fragment-decorated thin sections from Gall *et al.* (1983).

to be rather short and distributed irregularly in the animal hemisphere; however, they form a geodesic network in the vegetal hemisphere (Fig. 15). Temporal changes occur in the organization of cytokeratin filaments during egg maturation and early development (Fig. 16). The lattice of filaments that fills the vegetal hemisphere of fully grown oocytes disappears during maturation. Only a few short filaments are seen in mature eggs. About midway between fertilization and first cleavage, however, cytokeratin organization is reestablished. Thus, at the two-cell stage, each blastomere has an organized network of cytokeratin filaments in its vegetal

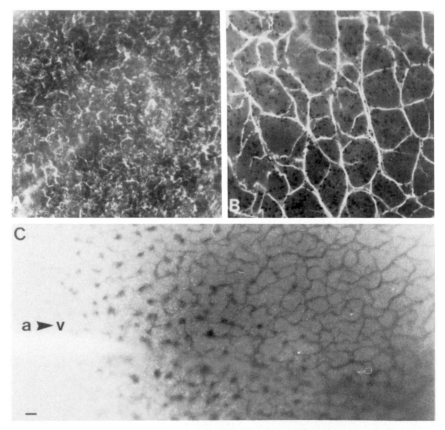

FIG. 15. Cytokeratin organization in the animal and vegetal hemisphere of early *Xenopus* embryos. (A) The animal hemisphere contains scattered short cytokeratin filaments. (B) The vegetal hemisphere contains a lattice of long cytokeratin filaments. (A, B) Scale bar = 10 μm. (C) The animal–vegetal gradient in cytokeratin organization. Scale bar = 10 μm. Fluorescent micrographs of antibody-stained whole mounts from Klymkowsky *et al.* (1987).

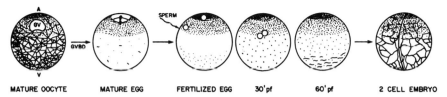

MATURE OOCYTE   MATURE EGG   FERTILIZED EGG   30' pf   60' pf   2 CELL EMBRYO

FIG. 16. Diagram showing changes in the pattern of cytokeratin filaments during maturation and early development of *Xenopus* eggs. A, animal; V, vegetal; GV, germinal vesicle;
GVBD, GV breakdown. 30' pf, 30 minutes postfertilization; 60' pf, 60 minutes postfertilization. From Klymkowsky *et al.*, (1987).

region (Fig. 15A, B), and there is a gradient of increasing organization from
the animal to the vegetal pole (Fig. 15C), which persists until the early-
blastula stage. Also, a transient array of parallel Mt appears in the vegetal
cortex about midway between fertilization and first cleavage (Elinson and
Rowning, 1988). The appearance of these Mt correlates with cortical
rotation, a rearrangement of the cortex and internal cytoplasm that is
thought to determine the dorsoventral axis (Vincent *et al.*, 1986).

IV. Association of mRNA with the Egg Cytoskeleton

In the preceding sections, we have compared the distribution of cytoplasmic domains, localized mRNA, and cytoskeletal elements in various
eggs. Two important correlations have been revealed by these compari-
sons. First, mRNA tends to be localized in regions that are rich in cytoskeletal architecture, such as the cortical ectoplasm of *Chaetopterus*
eggs and the vegetal cortex of *Xenopus* oocytes. Second, in several instances mRNA localizations are maintained in specific cytoplasmic domains while they are being rearranged, suggesting that mRNA is bound
tenaciously to structures within these regions.

In somatic cells, mRNA is known to be associated with two cytoplasmic
structures: the rough ER and the cytoskeleton. The association of mRNA
with the ER is indirect, mediated by the translation of a signal sequence
in the nascent polypeptide (Walter *et al.*, 1984). As described earlier,
the major evidence for cytoskeletal binding comes from experiments
showing that mRNA is enriched in the insoluble residue after cells were
extracted with nonionic detergents (Lenk *et al.*, 1977; Lenk and Penman,
1979; Cervera *et al.*, 1981; van Venrooij *et al.*, 1981; Jeffery, 1982; Pramanik *et al.*, 1986; Bagchi *et al.*, 1987; Brodeur and Jeffery, 1987; Davis *et
al.*, 1987). Independent evidence for mRNA–cytoskeleton interactions is

provided by the association of an mRNA cap-binding protein with the cytoskeleton (Zumbe *et al.,* 1982). Since ER and cytoskeletal elements are abundant in eggs, they are both candidates for mRNA-binding matrices. We now review experiments suggesting that mRNA is associated with the egg cytoskeleton.

A. RETENTION OF mRNA IN THE DETERGENT-INSOLUBLE FRACTION

Extraction with nonionic detergents has been used to examine mRNA–cytoskeleton interactions in the eggs of several different animals, including sea urchins (Moon *et al.,* 1983), *Chaetopterus* (Jeffery, 1985; Jeffery *et al.,* 1986), *Styela* (Jeffery and Meier, 1983; Jeffery, 1984), and *Xenopus* (King *et al.,* 1987; Pondel and King, 1988, 1989). In sea urchin, *Chaetopterus,* and *Styela* eggs, the results were similar. Although most of the lipids, proteins, and RNA were efficiently extracted from the eggs, a large proportion of the poly(A)+ RNA was retained in the cytoskeletal fraction, suggesting it is enriched in mRNA (Table I). The presence of mRNA in the detergent-insoluble fraction was confirmed by *in vitro* translation; however, few qualitative differences could be detected between mRNAs in the detergent-soluble and -insoluble fractions (Moon *et al.,* 1983; W. R. Jeffery, unpublished observations). These results suggest that many different mRNAs are associated with the cytoskeleton—at least in sea urchin, *Chaetopterus,* and *Styela* eggs. Similar results have been obtained for somatic cells (van Venrooij *et al.,* 1981; Cervera *et al.,* 1981).

Different results were obtained when *Xenopus* oocytes were extracted with Triton X-100 (Pondel and King, 1988). In this case, <1.5% of the total poly(A)+ RNA was recovered in the detergent-insoluble fraction, but this

TABLE I

PERCENTAGE OF LIPID, PROTEIN, AND RNA RETAINED IN THE
INSOLUBLE FRACTION AFTER EXTRACTION OF EGGS WITH
NONIONIC DETERGENTS[a]

Egg type	Percentage in insoluble fraction			
	Lipid	Protein	RNA	Poly(A)
Sea urchin	16	13	9	42
Chaetopterus	—	26	29	71
Styela	8	22	28	56

[a] Data from Moon *et al.* (1983), Jeffery and Meier (1983), and Jeffery (1985).

fraction was enriched 35- to 50-fold in Vg-1 mRNA (Fig. 17), suggesting that qualitative differences in mRNA distribution can occur between the detergent-soluble and -insoluble fractions. Histone mRNA, which is polyadenylated in *Xenopus* oocytes (Ruderman and Pardue, 1978), is also present in the detergent-insoluble fraction, although it is not enriched to the extent of Vg-1 mRNA. The conflicting results obtained for *Xenopus* oocytes and sea urchin, *Chaetopterus,* and *Styela* eggs with respect to the retention of poly(A)+ RNA in the detergent-insoluble fraction are probably caused by differences in extraction methods. The method used to extract sea urchin, *Chaetopterus,* and *Styela* eggs was more gentle than that used for *Xenopus* oocytes; hence, more poly(A)+ RNA may have been retained in the detergent-insoluble fraction. Perhaps general poly(A)+ RNA is less tightly bound to the cytoskeleton than Vg-1 mRNA.

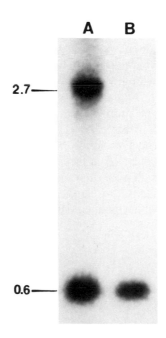

FIG. 17. Relative distribution of Vg-1 and histone mRNA in the (A) detergent-insoluble and (B) -soluble fractions of fully grown *Xenopus* oocytes. The 2.7-kb transcript is Vg-1 mRNA and the 0.6-kb transcript is histone mRNA. The blots were hybridized simultaneously with probes for each RNA. Northern blots from Pondel and King (1988).

These studies show that Vg-1 mRNA is specifically enriched in the cytoskeletal fraction of *Xenopus* oocytes.

The detergent extraction results just described do not rule out the possibility that localized mRNA is bound to intracellular membranes, because the latter also contain an associated cytoskeletal network (Ben-Ze'ev *et al.*, 1979). Moreover, detergent-resistant proteins are present in the intracellular membranes (Kreibich *et al.*, 1978a,b) and viral mRNAs translated on the ER are known to be associated with the detergent-insoluble fraction of somatic cells (Cervera *et al.*, 1981). Stronger evidence against the role of intracellular membranes in egg mRNA localization comes from studies in which *in vitro*-transcribed Vg-1 mRNA lacking its 5' signal sequence was injected into early *Xenopus* oocytes and still became localized in the vegetal cortex (Yisraeli and Melton, 1988). These results demonstrate that Vg-1 mRNA does not require the translation of a signal polypeptide to be properly localized.

In situ hybridization has also been used to investigate the relationship between mRNA localization and the egg cytoskeleton (Jeffery, 1984, 1985). The results demonstrate that poly(A)+ RNA and specific mRNAs are localized in the same regions of detergent-extracted and intact *Styela* and *Chaetopterus* eggs. For instance, when *Styela* oocytes or eggs are extracted with the nonionic detergent Triton X-100, the same patterns of poly(A)+ RNA localization in the GV and ectoplasm are observed (compare Figs. 5A–C and 18). Similar results have been obtained for localized actin mRNA in the myoplasm of *Styela* eggs (Jeffery, 1984). The results suggest that the spatial distribution of mRNA is determined by an association of mRNA with the egg cytoskeleton. However, they also suggest that mRNA localization is not caused entirely by binding to the cytoskeleton. This is shown by the fact that histone mRNA, an evenly distributed transcript, and actin mRNA, a localized transcript, are both associated with the detergent-insoluble residue of *Styela* eggs (Jeffery, 1984).

B. EXISTENCE OF mRNA–CYTOSKELETAL INTERACTIONS *in Vivo*

Although mRNA is retained in the egg cytoskeletal fraction, this is not conclusive evidence that mRNA–cytoskeleton associations exist *in vivo*. It is possible that mRNA binds to the cytoskeleton adventitiously in the presence of detergent. Even in somatic cells, in which mRNA–cytoskeleton associations have been investigated extensively, it has not been proven that these complexes exist *in vivo*. Adventitious interactions between nucleic acid and protein are difficult to rule out by experimental means, although some evidence against them is provided by the failure of high ionic strength buffers to disrupt mRNA–cytoskeleton associations

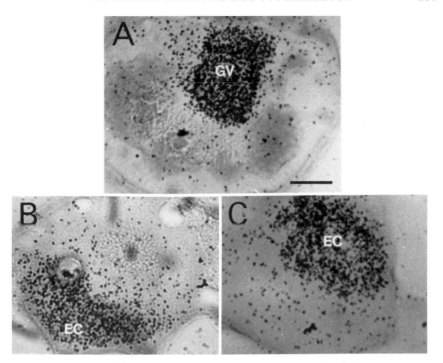

FIG. 18. Retention of poly(A)+ RNA in detergent-extracted *Styela* eggs. (A) Retention of poly(A)+ RNA in the GV area of detergent-extracted oocytes. (B, C) Retention of poly(A)+ RNA in the ectoplasm (EC) of detergent-extracted eggs undergoing cytoplasmic rearrangement. (A–C) Scale bar = 20 μm. *In situ* hybridization of sections of Triton X-100-extracted eggs from Jeffery (1984).

(Jeffery, 1984; Jeffery *et al.*, 1986; Pondel and King, 1988). The introduction of artifacts during the extraction procedure can also be monitored by adding an exogenous mRNA to the extract and following its distribution between the soluble and insoluble fractions. Thus, Pondel and King (1988) added globin mRNA during the extraction of *Xenopus* oocytes and found that it distributed with the soluble rather than the insoluble fraction, suggesting that no adventitious association of mRNA with the cytoskeleton occurred during the procedure.

Another approach has been developed to explore mRNA–cytoskeleton interactions in *Chaetopterus* eggs (Jeffery, 1985). Centrifugation stratifies the organelles of this egg into distinct zones based on their relative densities (Lillie, 1909). From the centripetal to the centrifugal pole of the egg, the stratified zones consist of (1) lipid droplets; (2) ER, nuclei, ribosomes,

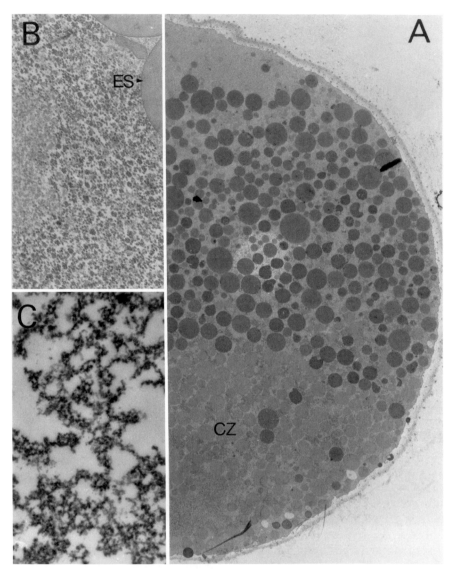

FIG. 19. Electron microscopy of centrifuged *Chaetopterus* eggs. (A) Section through the centripetal–centrifugal axis showing some of the zones to which specific organelles are stratified. The centrifugal zone is labeled CZ. (B) Higher magnification of the centrifugal zone. Edges of ectoplasmic spherules (ES) are seen at the upper right. (C) The structure of the centrifugal zone after extraction with NP-40. Compare with Fig. 1B.

and other membrane-bound organelles; (3) mitochondria; (4) yolk plate-lets; and (5) ectoplasmic spherules (Fig. 19A). In addition, EM showed that the ectoplasmic cytoskeletal network stratifies with the ectoplasmic spherules (Fig. 19B, C), presumably because of strong associations with the latter. Therefore, if mRNA is associated with the ectoplasmic cy-toskeletal network *in vivo,* it would also be expected to be driven to the centrifugal pole during centrifugation. If mRNA is not associated with the cytoskeleton *in vivo,* however, it would be expected to stratify in another zone during centrifugation, probably in the same zone as the ribosomes. When *in situ* hybridization was conducted on centrifuged eggs most of the poly(A)+ RNA (Fig. 2E), actin mRNA, and histone mRNA was lo-cated in the centrifugal zone. Furthermore, the stratified components of centrifuged eggs, including mRNA and ectoplasmic spherules, could re-turn to their former positions after removal from the centrifuge. Develop-ment of centrifuged eggs is normal; hence, the eggs are not irreversibly damaged during centrifugation. This experiment strongly suggests that mRNA is associated with one of the components of the centrifugal zone *in vivo.* However, the identity of these components remains to be deter-mined.

The stratification of ER and other membrane vesicles to a different zone during centrifugation suggests that these components are not binding sites for ectoplasmic mRNA. The ectoplasmic spherules are the only noncy-toskeletal structures detected in electron micrographs of the centrifugal region. When centrifuged eggs were treated with nonionic detergent, mRNA was retained in the residue of the centrifugal zone, but the matrix of the ectoplasmic spherules was extracted leaving a peripheral zone that appears to be attached directly to the surrounding cytoskeletal network (Jeffery, 1985). The detergent-resistant region surrounding the ectoplas-mic spherules may be an insoluble remnant of a membrane cytoskeleton or it could be an entangled mat of cytoskeletal filaments. *In situ* hybridization at the ultrastructural level will be necessary to resolve this issue and determine the precise binding site(s) of mRNA within the ectoplasmic cytoskeletal domain.

V. Molecular Basis of mRNA Localization

For an mRNA molecule to be localized, it should contain a localization site that interacts with a cytoplasmic receptor. The localization site is likely to be present in one of the untranslated regions of the mRNA molecule. Thus, mRNA binding would not interfere with its translational activity. The receptor must also be localized in the cell and firmly anchored to the cellular architecture. Indeed, it may be a part of the architecture itself. The binding of the localization site to the receptor may be direct or indirect. Since mRNA molecules exist as ribonucleoprotein particles (mRNP) (Preobrazhensky and Spirin, 1978), if the binding is indirect the interaction is likely to occur via an mRNA-associated protein. One of the protein-binding sites is the 3' poly(A) tail (Kwan and Brawerman, 1972; Baer and Kornberg, 1983), and other sites exist at the 5' terminus and in the 3' noncoding region of the mRNA molecule. More information about the organization of mRNP is necessary before we can understand how mRNA molecules are localized in cells. In the meantime, however, studies have been initiated to identify mRNA localization sites and receptors.

A. The mRNA Localization Site

The first evidence for the existence of an mRNA localization site was provided by microinjection experiments (Capco and Jeffery, 1981). In these experiments, poly(A)+ RNA was extracted from the animal- and vegetal-pole regions of unfertilized *Xenopus* eggs, end labeled *in vitro,* and then microinjected into the animal or vegetal hemisphere of a fertilized egg. When mRNA from the vegetal pole was injected into either hemisphere of a fertilized egg, it became distributed in a vegetal–animal gradient with about four to five times more mRNA at the vegetal pole than at the animal pole. In contrast, when mRNA from the animal pole was microinjected into either hemisphere, it became concentrated in the animal hemisphere, which normally contains more total RNA than the vegetal hemisphere (Capco, 1982; Phillips, 1982). The results suggest that the accumulation of vegetal-pole mRNA in a vegetal–animal pole gradient is a specific property of the mRNA sequences from the vegetal pole of the egg. Surprisingly, when poly(A) is microinjected into *Xenopus* eggs, it also accumulates in a vegetal–animal pole gradient (Froehlich *et al.,* 1977). A possible explanation for these results is that poly(A) has an affinity for specific binding sites in the vegetal hemisphere of the egg and that vegetal mRNA contains longer or more exposed poly(A) tails than animal mRNA.

Some evidence also suggests that poly(A) or a nearby sequence in the 3' noncoding region is responsible for the interaction of mRNA with the ectoplasmic cytoskeletal domain of *Chaetopterus* eggs (Jeffery *et al.,* 1986). After extraction of *Chaetopterus* eggs with NP-40, the detergent-insoluble residue can be recovered with associated mRNA molecules in the form of an mRNA–cytoskeleton complex. Information on the nature and position of mRNA-binding sites in this complex was obtained by treatment of the residues with pancreatic RNase A. This enzyme cleaves RNA molecules at the 3' side of pyrimidine nucleotides and thus completely hydrolyzes mRNA, except for its poly(A) tail and other regions of the mRNA that may be associated with protein in the cytoskeletal fraction. The strategy used to determine whether mRNA is associated with the cytoskeleton by its 5′ or 3′ region was to measure the amount of poly(A) in a limit digest after the cytoskeletal fraction was digested with RNase A. Since poly(A) is located at the 3′ end of the mRNA molecule, if mRNA binding occurs near the 5′ end, poly(A) should be released into the soluble fraction during hydrolysis. In contrast, if mRNA is associated by the poly(A) sequence or site in the 3′ noncoding region near the poly(A) tail, the latter should be retained in the cytoskeletal residue. After exhaustive RNase treatment, poly(A) is quantitatively retained in the detergent-insoluble residue. This result suggests that poly(A)+ RNA molecules are linked to the cytoskeleton by their 3′ ends. Whether the poly(A) sequence itself or a sequence in the 3′ noncoding region adjacent to poly(A) is the cytoskeletal binding site cannot be decided from this experiment. The possibility that the poly(A) tail of mRNA binds to the cytoskeleton is consistent with the observation that synthetic polyadenylated mRNAs diffuse through the cytoplasm more slowly than nonpolyadenylated mRNAs after microinjection into *Xenopus* oocytes (Drummond *et al.*, 1985).

A promising system for studying the localization site of Vg-1 mRNA has been developed by Yisraeli and Melton (1988). These investigators have microinjected Vg-1 mRNA into early oocytes and shown that it is localized normally to the vegetal hemisphere when the injected oocytes are cultured *in vitro*. In Section III, we described how this microinjection assay was used to demonstrate that Vg-1 mRNA does not require its 5′ signal sequence for localization. The same experiment eliminates the possibility that a direct localization site resides in the 5′ noncoding region of Vg-1 because *in vitro*-synthesized Vg-1 mRNA lacking this region still localizes to the vegetal hemisphere. Thus, it is likely that a localization site lies in the 3′ end of Vg-1 mRNA.

The localization site in *bcd* mRNA has been defined by experiments with transgenic *Drosophila* (Macdonald and Struhl, 1988). The strategy for identifying the localization site was to introduce altered *bcd* gene constructs into the germ line and determine whether mRNA transcribed from the introduced genes is localized in the anterior region of the egg. The basic gene construct used in these experiments was an 8-kb segment of genomic DNA containing a functional *bcd* gene. In previous studies, this DNA segment had been shown to rescue the development of *bcd*⁻ mutants. When this gene was introduced into the germ line by P-element-mediated transformation, the resulting *bcd* transcripts were localized as usual in the anterior portion of the egg (Fig. 20). When most of the coding region or the

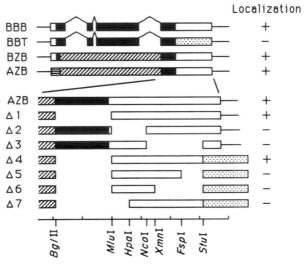

FIG. 20. A summary of the various *bcd* gene constructs that were introduced into the *Drosophila* genome and localization of the resulting mRNA in the anterior portion of the egg. The *bcd* gene is shown in the context of an 8-kb *Eco*RI restriction fragment (construct BBB), which rescues *bcd⁻* flies when inserted into their genome by P-element-mediated transformation. The exon sequences are boxed, with all but the 5′ and 3′ noncoding regions filled in, and the introns are indicated by bent lines. The transcripts derived from the construct BBB become localized in the anterior region of the egg, just as in wild-type flies. In construct BBT, the *bcd* 3′ noncoding region is replaced with the polyadenylation signal and flanking regions of a tubulin gene (shown as a stippled box). The transcripts derived from construct BBT do not become localized in the anterior region of the egg. In construct BZB, *lacZ* sequences (diagonal hatching) replace most of the *bcd* coding region. In construct AZB, all *bcd* 5′ regions are replaced by the alcohol dehydrogenase gene promoter and 5′ flanking sequences (horizontal hatching), leaving only the *bcd* 3′ regions. The transcripts from constructs BZB and AZB are localized in the anterior region of the egg. In constructs shown below the first AZB, the *bcd* 3′ noncoding region of AZB is expanded and presented to scale with respect to subsequent constructs Δ1–7 (deletions shown in empty spaces). Constructs Δ1–3 retain the *bcd* polyadenylation signal and downstream sequences, whereas in constructs Δ4–7 those sequences are replaced with a polyadenylation signal and flanking sequences from the tubulin gene. As shown, only those constructs retaining the entire 3′ noncoding region of the *bcd* gene localize their corresponding mRNAs in the anterior region of the egg. A partial restriction map of the *bcd* 3′ noncoding region is shown at the bottom (scale increments: 200 bp). From Macdonald and Struhl (1988). Reprinted by permission from *Nature (London)* Vol. 336, pp. 595–598. Copyright © 1988 Macmillan Magazines, Ltd.

5′ noncoding region was removed from the *bcd* gene and replaced with equivalent regions from other genes, *bcd* mRNA localization was not affected. These results show that the localization site is not in the coding or 5′ noncoding regions of *bcd* mRNA. In contrast, when the entire 3′ noncoding region was removed and replaced with the 3′ region of another gene, *bcd* mRNA localization was abolished, suggesting that the 3' noncoding region of the mRNA contains the localization site. When smaller sectors of the 3' noncoding region were excised and replaced with other sequences, *bcd* mRNA localization was prevented in each case, indicating that most or all of the 625-base, 3' noncoding region is required for localization. Computer analysis suggests that the *bcd* mRNA localization site can form an extended secondary structure (Fig. 21). This structure, perhaps by its association with an mRNP protein, may bind to a specific localization receptor at the anterior end of the egg.

In summary, current results indicate that the localization site of Vg-1 and *bcd* mRNA is located in the 3' rather than the 5' noncoding region of the message. Other results suggest that the cytoskeletal binding site also resides in the 3' noncoding region and/or the poly(A) terminus of mRNA. A possible model for mRNA localization and cytoskeletal binding is that the extended secondary structure of the 3' noncoding region is required to make the poly(A) tail more accessible to a factor responsible for attaching

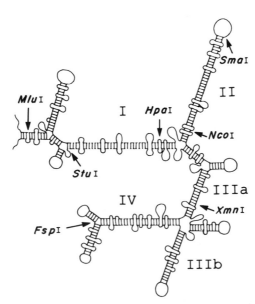

FIG. 21. Potential secondary structure formed by the localization site in the 3' noncoding region of *bcd* mRNA as predicted by computer analysis. From Macdonald and Struhl (1988). Reprinted by permission from *Nature (London)* Vol. 336, pp. 595–598. Copyright © 1988 Macmillan Magazines, Ltd.

the mRNA to the cytoskeleton. Further experiments will be necessary to determine the relationship between the localization and cytoskeletal binding sites in mRNA.

B. The mRNA Localization Receptor

According to the results discussed in Section III, the most probable mRNA localization receptor appears to be the egg cytoskeleton. In somatic cells, an association between mRNA (or nuclear pre-mRNA molecules) and actin filaments has been suggested by EM, cell fractionation, and cytochalasin release experiments (Reamakers *et al.*, 1983; Adams *et al.*, 1983; Ornelles *et al.*, 1986; Nakayasu and Ueda, 1985; Schröder *et al.*, 1987; see also Bagchi *et al.*, 1987, for a conflicting viewpoint). The possibility that localized mRNA is associated with actin filaments has also been tested in *Styela* and *Chaetopterus* eggs (Jeffery, 1984; Jeffery *et al.*, 1986) by treating detergent-insoluble residues with DNase I, which depolymerizes filamentous actin (Lazarides and Lindberg, 1974; Hitchcock *et al.*, 1976). The results showed that DNase I-treated residues retained as much mRNA as untreated controls, suggesting that actin filaments were not a binding site for localized mRNA. Likewise, treatment of these eggs with cytochalasin does not release mRNA into the detergent-soluble fraction. Similar experiments, in which mRNA binding was examined after cells were treated with low temperature or colchicine, also appeared to eliminate the possibility that Mt are required for mRNA–cytoskeleton interactions. These studies have narrowed the search for cytoskeletal elements that bind mRNA to the intermediate filaments.

Two kinds of indirect evidence support the possibility that mRNA may be bound to intermediate filaments or their associated components in the egg cytoskeleton. First, mRNA–cytoskeleton associations are stable in high ionic strength media (Jeffery, 1984; Pondel and King, 1988)—in the case of *Chaetopterus* eggs up to 1.5 M NaCl (Jeffery *et al.*, 1986). High ionic strength media are known to solubilize actin filaments and Mt, but intermediate filaments are not disrupted under these conditions (Renner *et al.*, 1981). Second, mRNA localizations often appear to be concentrated in cytoplasmic domains that are enriched in intermediate filaments. Notable examples are the ectoplasm of *Chaetopterus* eggs, the myoplasm of *Styela* eggs, and the vegetal cortex of *Xenopus* oocytes. Extensive localizations of Mt have not been reported in any of these domains, and some of them also lack an obvious actin filament network.

There is indirect evidence that Vg-1 mRNA is associated with the cytokeratin filament system in the vegetal hemisphere of *Xenopus* oocytes. The Vg-1 mRNA localization is in the same position as the organized array of cytokeratin filaments (see Figs. 11 and 15). Actin filaments are present in this region, but also exist in the animal hemisphere, where there is no localized Vg-1 mRNA. After the beginning of oocyte maturation, Vg-1 mRNA looses its tight cortical localization (Weeks and Melton, 1987b) during the same interval that the vegetal cytokeratin system becomes disorganized (Klymkowsky et al., 1987). Pondel and King (1989) have shown that progesterone-induced maturation causes Vg-1 mRNA to be released from the detergent-insoluble fraction. Interestingly, both poly(A)+ RNA and tubulin mRNA, which are localized in the same region as Vg-1 mRNA (Capco and Jeffery, 1982; Perry and Capco, 1988), are also dispersed after the induction of maturation by progesterone. This process appears to be mediated by a coupled influx of Ca^{2+} and efflux of $Cl-$ that occurs during oocyte maturation (Larabell and Capco, 1988). Thus, oocyte maturation may initiate a general disruption of mRNA localization in the vegetal cortex of Xenopus eggs that is correlated with the disappearance of a highly organized system of cytokeratin filaments. Although these biochemical and cytological correlations are intriguing, it should be pointed out that there is still no direct evidence for mRNA binding to intermediate filaments.

C. INDIRECT ROLE OF Mt IN mRNA LOCALIZATION

Although Mt do not have a direct role in mRNA localization, they appear to be indirectly involved in this process. This is shown by studies on the localization of fushi tarazu (ftz) mRNA during cellular blastoderm formation in Drosophila embryos (Edgar et al., 1987). The segmentation gene ftz is expressed in a single broad domain at the syncytial blastoderm stage and, as a result of continued transcription and local degradation, the expression pattern is reorganized into seven bands before cellularization (Fig. 22A, B). In each blastoderm cell where it is expressed, ftz mRNA is localized to the apical periplasm, a region known to contain actin and intermediate filaments. The regions below the apical periplasm, including the perinuclear region and the basal periplasm, are normally devoid of ftz mRNA. By inhibiting cellularization with cytochalasin, Edgar et al. (1987) showed that cell membrane formation is not required either for the development of seven ftz–mRNA expression bands or the localization of ftz mRNA in the apical periplasm (Fig. 22C, D). However, when these embryos were treated with colchicine to depolymerize the internal network of Mt, ftz mRNA spread from the apical periplasm into the region surrounding the nuclei and even into the basal periplasm (Fig. 22E, F). Thus, although the distribution of ftz mRNA in the apical periplasm may be due to a direct association with intermediate filaments (Edgar et al., 1987), the

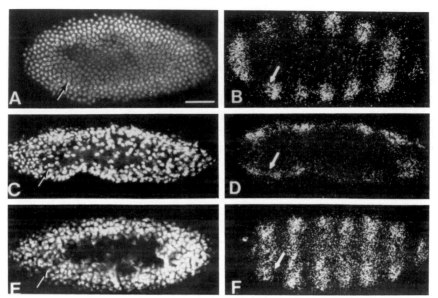

FIG. 22. The role of Mt in limiting *ftz* mRNA localization to the apical region of the *Drosophila* embryonic blastoderm. (A, C, E). Sections of embryos stained with Hoechst 33258 to visualize the position of nuclei. (B, D, F) Sections of embryos *in situ*-hybridized to show the location of *ftz* mRNA. (A–F) Each embryo is cut so that the central region of the section shows the basal periplasm and the peripheral part of the section shows the apical periplasm. (A, B) Normal embryos showing *ftz* mRNA localization in the apical periplasm. (C, D) Embryos injected with cytochalasin B showing normal *ftz* mRNA localization in the apical periplasm. (E, F) Embryos injected with cytochalasin B and colchicine showing *ftz* mRNA dispersed throughout the apical and basal periplasm. The boundary between the apical and central cytoplasmic domains is shown by the arrows. Scale bar = 40 μm. From Edgar *et al.* (1987).

underlying Mt network may also function in localizing *ftz* mRNA by restricting intermediate filaments to the apical periplasm. A similar situation may occur in cleaving *Chaetopterus* eggs, in which the astral Mt appear to exclude ectoplasmic mRNA from the internal regions of the cell (Jeffery and Wilson, 1983).

VI. Conclusions and Prospectus

New methods have been developed to determine the spatial distribution of mRNA in cells. Application of these methods during the last decade has shown that mRNA is not always distributed uniformly, but can be local-ized to particular cytoplasmic regions in both somatic cells and eggs. The eggs of diverse groups of animals, including annelids, insects, ascidians, and amphibians, show both general and specific mRNA localizations. Thus, mRNA localization is a widespread phenomenon and may have appeared early during the evolution of metazoans. Some specific localized mRNAs, such as *bcd* mRNA in *Drosophila* eggs and Vg-1 mRNA in *Xenopus* eggs, are likely to have an important role in development. The function of mRNA localization in other cases is still unclear. Part of the mRNA localization mechanism may involve the regional degradation of mRNA. Another possibility that remains to be explored is whether some egg mRNAs are localized because the proteins they encode are assembled into localized structures during translation. If this is so, one would expect that inhibiting protein synthesis would disrupt this type of mRNA local-ization.

Several lines of evidence suggest that the egg cytoskeleton may be involved in mRNA localization. First, regions of mRNA localization ap-pear to coincide with egg cytoskeletal domains. Second, mRNA is resis-tant to extraction with nonionic detergents and fractionates with the egg cytoskeleton. The cytoskeleton may not be the only factor involved in mRNA localization, however, because some evenly distributed messages also appear to be associated with the cytoskeleton. Further experiments are needed to demonstrate whether mRNA–cytoskeleton interactions oc-cur *in vivo,* and to characterize more effectively the cytoskeletal regions underlying the mRNA localizations. For example, it will be important to determine whether a cytoskeletal domain underlies the region of *bcd* mRNA localization in the anterior pole of *Drosophila* eggs.

In order to be localized, an mRNA molecule must contain a localization site that recognizes a receptor in the egg cytoplasm. Recent data suggest that both the localization site and a cytoskeletal binding site may be located in the 3′ noncoding region of the mRNA. The localization site of *bcd* mRNA has been sequenced and the sequence shown to be capable of generating an extended secondary structure. It will be important to deter-mine whether other localized mRNAs contain this or a related sequence. Presently, intermediate filaments are the most reasonable candidates for a localization receptor. However, the association between mRNA and inter-mediate filaments requires verification by controlled binding experiments, both *in vitro* and *in vivo*. Both mRNA localization sites and cytoplasmic receptors will ultimately need to be identified, and the nature of their interaction will have to be characterized to understand the mechanism of mRNA localization.

ACKNOWLEDGMENTS

I thank Drs. B. Edgar, M. L. King, M. W. Klymkowsky, and L. Gall for sending copies of preprints, photographs, or negatives. My work on localized mRNA and the egg cytoskeleton has been supported by NIH grant HD-13970 and NSF grant DCB-8812110.

REFERENCES

Adams, A., Fey, E. G., Pike, S. F., Taylorson, C. J., White, H. A., and Rabin, B. R. (1983). *Biochem. J.* **216**, 215–226.
Angerer, L. M., DeLeon, D. V., Angerer, R. C., Showman, R. M., Wells, D. E., and Raff, R. A. (1984). *Dev. Biol.* **101**, 477–484.
Baer, B. W., and Kornberg, R. D. (1983). *J. Biol. Chem.* **96**, 717–721.
Bagchi, T., Larson, D. E., and Sells, B. H. (1987). *Exp. Cell Res.* **168**, 160–172.
Balinsky, B. I., and Devis, R. J. (1963). *Acta Embryol. Morphol. Exp.* **6**, 55–108.
Ben-Ze'ev, A., Duerr, A., Solomon, F., and Penman, S. (1979). *Cell* **17**, 359–365.
Berg, W. E., and Humphreys, W. J. (1960). *Dev. Biol.* **2**, 42–60.
Berleth, T., Burri, M., Thoma, G., Bopp, D., Richstein, S., Frigerio, G., Noll, M., and Nüsslein-Volhard, C. (1988). *EMBO J.* **7**, 1749–1756.
Bier, K. (1963). *J. Cell Biol.* **16**, 436-440.
Billet, F. S., and Adam, E. (1976). *J. Embryol. Exp. Morphol.* **33**, 697–710.
Bonneau, A. M., Darveau, A., and Sonenberg, N. (1985). *J. Cell Biol.* **100**, 1209–1218.
Bravo, R., Small, J. V., Fey, S. J., Larsen, P. M., and Celis, J. E. (1982). *J. Mol. Biol.* **154**, 121–143.
Brodeur, R. D., and Jeffery, W. R. (1987). *Cell Motil. Cytoskel.* **7**, 129–137.
Brökelmann, J. (1977). *Cell Tissue Res.* **179**, 531–562.
Brown, S., Levinson, W., and Spudich, J. A. (1976). *J. Supramol. Struct.* **5**, 119–130.
Capco, D. G. (1982). *J. Exp. Zool.* **219**, 147–154.
Capco, D. G., and Jeffery, W. R. (1981). *Nature (London)* **294**, 255–257.
Capco, D. G., and Jeffery, W. R. (1982). *Dev. Biol.* **89**, 1–12.
Carpenter, C. D., and Klein, W. H. (1982). *Dev. Biol.* **91**, 43–49.
Cervera, M., Dreyfuss, G., and Penman, S. (1981). *Cell* **23**, 113–120.
Ciejek, E. M., Nordstrom, J. L., Tsai, M. -J., and O'Malley, B. W. (1982). *Biochemistry* **21**, 4945–4953.
Coggins, L. (1973). *J. Cell Sci.* **12**, 71–93.
Colombo, R., Benedusi, P., and Valle, G. (1981). *Differentiation* **20**, 45–51.
Conklin, E. G. (1905). *J. Acad. Nat. Sci. Phila.* **13**, 1–119.
Davidson, E. H. (1986). "Gene Activity in Early Development," 3rd Ed. Academic Press, New York.
Davis, L., Banker, G. A., and Steward, O. (1987). *Nature (London)* **330**, 477–479.
DeLeon, D. V., Cox, K. H., Angerer, L. M., and Angerer, R. C. (1983). *Dev. Biol.* **100**, 197–206.
Dent, J. A., Polson, A. G., and Klymkowsky, M. W. (1989). *Development* **105**, 61–74.
Dolecki, G. J., and Smith, L. D. (1979). *Dev. Biol.* **69**, 217–236.
Drummond, D., Armstrong, J., and Colman, A. (1985). *Nucleic Acids Res.* **13**, 7375–7394.
Eckberg, W. R. (1981). *Biol. Bull.* **160**, 228–239.
Eckberg, W. R., and Langford, G. M. (1983). *Biol. Bull.* **165**, 214.
Edgar, B. A., Odell, G. M., and Schubiger, G. (1987). *Genes Dev.* **1**, 1226–1237.
Elinson, R. P., and Rowning, B. (1988). *Dev. Biol.* **128**, 185–197.

Fernandez, J., Olea, N., and Matte, C. (1987). *Development* **100**, 211–225.

Fontaine, B., and Changeux, J. -P. (1989). *J. Cell Biol.* **108**, 1025–1037.

Fontaine, B., Sassoon, D., Buckingham, M., and Changeux, J. -P. (1988). *EMBO J.* **7**, 603–609.

Franke, W. W., Rathke, P. C., Seib, E., Trendelenberg, M. F., Osborn, M., and Weber, K. (1976). *Cytobiologie* **14**, 111–130.

Franz, J. K., Gall, L., Williams, M. A., Picheral, B., and Franke, W. W. (1983). *Proc. Nat. Acad. Sci. USA* **80**, 6254–6258.

Frigero, G., Burri, M., Bopp, D., Baumgartner, S., and Noll, M. (1986). *Cell* **47**, 1033–1040.

Froehlich, J. P., Browder, L. W., and Schultz, G. A. (1977). *Dev. Biol.* **56**, 356–371.

Fröhnhofer, H. G., and Nüsslein-Volhard, C. (1987). *Genes Dev.* **1**, 880–890.

Fulton, A. B., and Wan, K. M. (1983). *Cell* **32**, 619–625.

Fulton, A. B., Wan, K. M., and Penman, S. (1980). *Cell* **20**, 849–857.

Gall, L., Picheral, B., and Gounon, P. (1983). *Biol. Cell.* **47**, 331–342.

Garner, C. C., Tucker, R. P., and Matus, A. (1988). *Nature (London)* **336**, 674–677.

Godsave, S. F., Anderton, B., Heasman, J., and Wylie, C. C. (1984a). *J. Embryol. Exp. Morphol.* **83**, 169–187.

Godsave, S. F., Wylie, C. C., Lane, E. B., and Anderton, B. H. (1984b). *J. Embryol. Exp. Morphol.* **83**, 157–167.

Golden, L., Schafer, U., and Rosbash, M. (1980). *Cell* **22**, 835–844.

Heasman, J., Quarmby, J., and Wylie, C. C. (1984). *Dev. Biol.* **105**, 458–469.

Herman, R., Weymouth, L., and Penman, S. (1978). *J. Cell Biol.* **78**, 663–674.

Herrmann, H., Fouquet, B., and Franke, W. W. (1989). *Development* **105**, 279–298.

Heuijerjans, J. H., Pieper, F. R., Ramaekers, F. C. S., Timmermans, L. J. M., Kuijpers, H., Bloemendal, H., and van Venrooij, W. J. (1989). *Exp. Cell Res.* **181**, 317–330.

Hitchcock, S. E., Carlsson, L., and Lindberg, U. (1976). *Cell* **7**, 531–542.

Jackson, B., Grund, C., Schmid, E., Burki, K., Franke, W. W., and Illmensee, K. (1980). *Differentiation* **17**, 161–179.

Jackson, B., Grund, C., Winter, S., Franke, W. W., and Illmensee, K. (1981). *Differentiation* **20**, 203–216.

Jamrich, M., Mahon, M. A., Gavis, E. R., and Gall, J. C. (1984). *EMBO J.* **3**, 1939–1943.

Jeffery, W. R. (1982). *J. Cell Biol.* **95**, 1–7.

Jeffery, W. R. (1984). *Dev. Biol.* **103**, 482–492.

Jeffery, W. R. (1985). *Dev. Biol.* **1100**, 217–229.

Jeffery, W. R., and Capco, D. G. (1978). *Dev. Biol.* **67**, 151–167.

Jeffery, W. R., and Meier, S. (1983). *Dev. Biol.* **96**, 125–143.

Jeffery, W. R., and Wilson, L. (1983). *J. Embryol. Exp. Morphol.* **75**, 225–239.

Jeffery, W. R., Tomlinson, C. R., and Brodeur, R. D. (1983). *Dev. Biol.* **99**, 408–417.

Jeffery, W. R., Speksnijder, J. E., Swalla, B. J., and Venuti, J. M. (1986). *Adv. Invertebr. Reprod.* **4**, 229–240.

Jessus, C., Huchon, D., and Ozon, R. (1986). *Biol. Cell.* **56**, 113–120.

Kalthoff, K. (1983). *In* "Time Space and Pattern in Embryonic Development" (W. R. Jeffery and R. A. Raff, eds.), pp. 313–348. Alan R. Liss, New York.

Kandler-Singer, I., and Kalthoff, K. (1976). *Proc. Natl. Acad. Sci. USA* **73**, 3739–3743.

Karr, T. L., and Alberts, B. M. (1986). *J. Cell Biol.* **102**, 1494–1509.

Kimelman, D., and Kirschner, M. (1987). *Cell* **51**, 869–877.

King, M. L., and Barklis, E. (1985). *Dev. Biol.* **112**, 203–212.

King, M. L., Davis, R. E., Litvin, J., Klein, S. L., and Pondel, M. (1987). *In* "Molecular Approaches to Developmental Biology" (E. H. Davidson and R. A. Firtel, eds.), pp. 97–111. Alan R. Liss, New York.

Klymkowsky, M. W., Maynell, L. A., and Polson, A. G. (1987). *Development* **100**, 543.

Kobayashi, S., Mizuno, H., and Okada, M. (1988). *Dev. Growth Differ.* **30**, 251–260.

Kreibich, G., Ulrich, B. L., and Sabattini, D. D. (1978a). *J. Cell Biol.* **77**, 468–487.

Kreibich, G., Freienstein, B. N., Pereyra, B. L., Ulrich, B. L., and Sabattini, D. D. (1978b). *J. Cell Biol.* **77**, 488–506.

Kwan, S. -W., and Brawerman, G. (1972). *Proc. Natl. Acad. Sci. USA* **69**, 3247–3250.

Larabell, C. A., and Capco, D. G. (1988). *Wilhelm Roux's Arch. Dev. Biol.* **197**, 175–183.

Lawrence, J. B., and Singer, R. H. (1986). *Cell* **45**, 407–415.

Lawrence, J. B., Singer, R. H., Villnave, C. A., Stein, J. L., and Stein, G. S. (1988). *Proc. Natl. Acad. Sci. USA* **85**, 463–467.

Lawrence, J. B., Singer, R. H., and Marselle, L. M. (1989). *Cell* **57**, 493–502.

Lazarides, E., and Lindberg, U. (1974). *Proc. Natl. Acad. Sci. USA* **71**, 4742–4746.

Lehtonen, E., and Badley, R. A. (1980). *J. Embryol. Exp. Morphol.* **55**, 211–225.

Lehtonen, E., Lehto, V. -P., Vartio, T., Badley, R. A., and Virtanen, I. (1983). *Dev. Biol.* **100**, 158–165.

Lenk, R., and Penman, S. (1979). *Cell* **16**, 289–301.

Lenk, R., Ransom, L., Kaufman, Y., and Penman, S. (1977). *Cell* **10**, 67–78.

Lillie, F. R. (1906). *J. Exp. Zool.* **3**, 153–268.

Lillie, F. R. (1909). *Biol. Bull.* **16**, 54–79.

Macdonald, P. M., and Struhl, G. (1986). *Nature (London)* **324**, 537–545.

Macdonald, P. M., and Struhl, G. (1988). *Nature (London)* **336**, 595–598.

Mariman, E. C. M., van Eekelen, C. A. G., Reinders, R. J., Berns, A. J. M., and van Venrooij, W. J. (1982). *J. Mol. Biol.* **154**, 103–119.

Maundrell, K., Maxwell, E. S., Puvion, E., and Scherrer, K. (1981). *Exp. Cell Res.* **136**, 435–445.

Melton, D. A. (1987). *Nature (London)* **328**, 80–82.

Merlie, J. P., and Sanes, J. R. (1985). *Nature (London)* **317**, 66–68.

Milhausen, M., and Agabian, N. (1983). *Nature (London)* **302**, 630–632.

Miller, T. E., Huang, C., and Pogo, A. O. (1978). *J. Cell Biol.* **76**, 675–691.

Mlodzik, M., Fjose, A., and Gehring, W. J. (1985). *EMBO J.* **4**, 2161–2969.

Moon, R. T., Nicosia, R. F., Olsen, C., Hille, M. B., and Jeffery, W. R. (1983). *Dev. Biol.* **95**, 447–458.

Nakayasu, H., and Ueda, K. (1985). *Exp. Cell Res.* **160**, 319–330.

Nieuwkoop, P. (1969). *Wilhelm Roux's Arch. Dev. Biol.* **162**, 341–373.

Nüsslein-Volhard, C., Fröhnhofer, H. G., and Lehmann, R. (1987). *Science* **238**, 1675–1681.

Ornelles, D. A., Fey, E. G., and Penman, S. (1986). *Mol. Cell. Biol.* **6**, 1650–1662.

Perry, B. A., and Capco, D. G. (1988). *Cell Differ. Dev.* **25**, 99–108.

Phillips, C. R. (1982). *J. Exp. Zool.* **223**, 265–275.

Picheral, B., Gall, L., and Gounon, P. (1982). *Biol. Cell.* **45**, 208.

Pondel, M., and King, M. L. (1988). *Proc. Natl. Acad. Sci. USA* **85**, 7612–7616.

Pondel, M., and King, M. L. (1989). Submitted.

Pramanik, S. K., Walsh, R. W., and Bag, J. (1986). *Eur. J. Biochem.* **160**, 221–230.

Preobrazhensky, A., and Spirin, A. (1978). *Prog. Nucleic Acid Res. Mol. Biol.* **21**, 1–37.

Pudney, J., and Singer, R. H. (1980). *Tissue Cell* **12**, 595–612.

Ramaekers, F. C. S., Benedetti, E. L., Dunia, I., Vorstenbosch, P., and Bloemendal, H. (1983). *Biochim. Biophys. Acta* **740**, 441–448.

Rebagliati, M. R., Weeks, D. L., Harvey, R. P., and Melton, D. A. (1985). *Cell* **42**, 769–777.

Renner, W., Franke, W. W., Schmid, E., Geisler, N., Weber, K., and Mandelkow, E. (1981). *J. Mol. Biol.* **149**, 285–306.

Rosa, F., Roberts, A. B., Danielpour, D., Dart, L. L., Sporn, M. B., and Dawid, I. B. (1988). *Science* **239**, 783–785.

Rosbash, M., and Ford, P. J. (1974). *J. Mol. Biol.* **85,** 87–101.
Ruderman, J. V., and Pardue, M. L. (1978). *J. Biol. Chem.* **253,** 2018–2025.
Sawada, T., and Schatten, G. (1988). *Cell Motil. Cytoskeleton* **9,** 219–231.
Schliwa, M., van Blerkom, J., and Porter, K. R. (1981). *Proc. Natl. Acad. Sci. USA* **78,** 4329–4333.
Schröder, H. C., Trölltsch, D., Wenger, R., Bachmann, M., Diehl-Seifert, B., and Müller, W. E. G. (1987). *Eur. J. Biochem.* **167,** 239–245.
Shimuzu, T. (1988). *Dev. Biol.* **125,** 321–331.
Slack, J. M. W., Darlington, B. G., Heath, J. K., and Godsave, S. F. (1987). *Nature (London)* **326,** 197–200.
Small, J. V., and Celis, J. E. (1978). *J. Cell Sci.* **31,** 393–409.
Smith, J. C. (1987). *Development* **99,** 3–14.
Smith, R. C. (1986). *J. Embryol. Exp. Morphol.* **95,** 15–35.
Swalla, B. J., Moon, R. T., and Jeffery, W. R. (1985). *Dev. Biol.* **111,** 434–450.
Tang, P., Sharpe, C. R., Mohun, T. J., and Wylie, C. C. (1988). *Development* **103,** 279–287.
Vacquier, V. D. (1981). *Dev. Biol.* **84,** 1–26.
van Eekelen, C. A. G., and van Venrooij, W. J. (1981). *J. Cell Biol.* **88,** 554–563.
van Venrooij, W. J., Sillikens, P. T. G., van Eekelen, C. A. G., and Reinders, R. J. (1981). *Exp. Cell Res.* **135,** 79–91.
Venuti, J., and Jeffery, W. R. (1990). Submitted.
Vincent, J. -P., Oster, G. F., and Gerhart, J. C. (1986). *Dev. Biol.* **113,** 484–500.
Walter, M., and Alberts, B. M. (1984). In "Molecular Biology of Development" (R. Firtel and E. H. Davidson, eds.), pp. 263–272. Alan R. Liss, New York.
Walter, P., Gilmore, R., and Blobel, G. (1984). *Cell* **38,** 5–8.
Warn, R. M., and McGrath, R. (1983). *Exp. Cell Res.* **143,** 103–114.
Warn, R. M., and Warn, A. (1986). *Exp. Cell Res.* **163,** 201–210.
Weeks, D. L., and Melton, D. A. (1987a). *Proc. Natl. Acad. Sci. USA* **84,** 2798–2802.
Weeks, D. L., and Melton, D. A. (1987b). *Cell* **51,** 861–867.
Whitfield, W. G. F., González, C., Sánchez-Herrero, E., and Glover, D. M. (1989). *Nature (London)* **338,** 337–340.
Wilson, E. B. (1925). "The Cell in Development and Heredity," 3rd Ed. Macmillan, New York.
Wolosewick, J. J., and Porter, K. R. (1979). *J. Cell Biol.* **82,** 114–139.
Wylie, C. C., Brown, D., Godsave, S. F., Quarmby, J., and Heasman, J. (1986). *J. Embryol. Exp. Morphol.* **89,** Suppl., 1–15.
Yisraeli, J. K., and Melton, D. A. (1988). *Nature (London)* **336,** 592–595.
Zumbe, A., Stahli, C., and Trachsel, H. (1982). *Proc. Natl. Acad. Sci. USA* **79,** 2917–2931.

Regulation of Membrane Fusion during Exocytosis

HELMUT PLATTNER

Faculty of Biology, University of Konstanz, D-7750 Konstanz, Federal Republic of Germany

I. Introduction

Exocytosis involves the fusion of a secretory organelle with the cell membrane (DeRobertis and Vaz Ferreira, 1957; Palade, 1959, 1975). It is most important to discriminate between "regulated (triggered) exocytosis" (e.g., in gland and nerve cells, mast cells, protozoan extrusive organelles or "extrusomes") and "constitutive (untriggered) exocytosis" (release of intercellular matrix components, etc., and cell membrane biogenesis) (Kelly, 1985). Evidently, specific signals and signal transfer mechanisms would occur only with regulated exocytosis. Hence, mechanisms of membrane fusion regulation might be quite different with these two types of exocytosis. This basic difference may, however, also yield some clues pertinent to a basic concept of fusion regulation, albeit many more data—though bewildering and conflicting—are now available for regulated exocytosis. This type will therefore be discussed here in much more depth. A summary of hypotheses currently being considered is presented in Fig. 1.

Exocytosis is of the PF–PF fusion type (Stossel *et al.*, 1978); that is, plasmatic membrane faces fuse. It always follows the "focal [point] fusion scheme" (Plattner, 1981, 1989a,b) depicted in Fig. 2: fusion is restricted to a small area of ≤ 10 nm, before the exocytotic opening expands to release the secretory contents.

Regulated exocytosis is coupled in most cases with endocytosis (Palade, 1959; Douglas, 1968; Evered and Collins, 1982; Tartakoff, 1987; Meldolesi and Ceccarelli, 1988). This complicates some of the analyses dealing with exocytosis, in addition to the simultaneous occurrence of transcellular transport, which also is not a step of exocytosis per se.

Although reviews are continuously published on membrane fusion (Poste and Nicolson, 1978; Sowers, 1987; Ohki *et al.*, 1988) and exocytosis regulation (Meldolesi *et al.*, 1978; DeLisle and Williams, 1986; Knight, 1987; Baker, 1988; Meldolesi and Ceccarelli, 1988; Rink and Knight, 1988; Strittmatter, 1988; Winkler, 1988), most reviews are personally biased by the group's own work and by a set of preferred techniques. Yet a compari-

197

son of the different systems may increase the chances of unraveling a general mechanism of membrane fusion regulation during exocytosis, if indeed such a mechanism exists.

II. Development of Concepts

Since 1962, membrane fusion during stimulated exocytosis had been considered to be regulated by an increase of the intracellular free-calcium concentration ($[Ca^{2+}]_i$ or pCa_i),[1] according to the stimulus–secretion coupling hypothesis (Douglas and Poisner, 1962; Douglas, 1968, 1974; Rubin et al., 1985). In fact, when Ca^{2+}-sensitive fluorochromes (Campbell, 1983; Tsien, 1988) have been applied to various secretory cells, pCa_i values have been found to be in the range between 10^{-8} and 10^{-7} M in the resting state, whereas a pCa_i of 7–6 has been registered after stimulation (mast cells: Beaven et al., 1984a,b; White et al., 1984; platelets: Erne et al., 1987; Sage and Rink, 1987; basophilic leukemia cells: Fewtrell and Sherman, 1987; neutrophilic granulocytes: Lew et al., 1986; Barrowman et al., 1987; Chandler and Kazilek, 1987; pancreatic acinar cells: Dormer et al., 1987; parotid cells: Merritt and Rink, 1987a,b; neurons: Baker, 1972; Brethes et al., 1987; Thayer et al., 1988; adrenal medullary chromaffin cells: Kesteven and Knight, 1982; Kao and Schneider, 1986; egg cells: Busa and Nuccitelli, 1985; Swann and Whitaker, 1986a—to give just a few selected examples). So far only parathyroid hormone secretion has been found to operate with resting pCa_i values (Brown et al., 1984; Muff et al., 1988) and, of course, also constitutive (i.e., nontriggered) exocytosis. However, the stimulus–secretion coupling concept has now come into question as a general mechanism also for regulated exocytosis (see the following and Section XIII).

Interestingly, the high pCa_i levels reported often lag behind the onset of exocytosis (see, e.g., Cobbold et al., 1987). Under certain circumstances exocytosis can occur without a pCa_i increase above resting levels (see Sections VIII and XIII). Finally a cell disposes of pumps to regulate Ca^{2+}

---

[1] Abbreviations: AA, arachidonic acid; Ab, antibodies; AIDP, adenylyl-imidodi-phosphate; CaBP, calcium-binding proteins; Ca_i and Ca_o, intracellular and extracellular Ca^{2+}; $[Ca^{2+}]_i$ and $[Ca^{2+}]_o$, intracellular and extracellular free-Ca^{2+} concentration (pCa); CaM, calmodulin; CaN, calcineurin; DAG, diacylglycerol; γ-S-ATP, γ-thio-ATP; γ-S-GTP, γ-thio-GTP; GIDP, guanosylimidodiphosphate; G proteins, GTP-binding proteins; IMP, intramembranous particle, integrated membrane protein; IP_3, inositol 1,4,5-trisphosphate; LT, leukotriene; nd, "nondischarge" mutants of Paramecium; PAGE, polyacrylamide gel electrophoresis; PC, phosphatidylcholine; PE, phosphatidylethanolamine; PG, prostaglandins; PI, phosphatidylinositol; PKC, protein kinase C; PLA_2 and PLC, phospholipases A_2 and C; PP, phosphoproteins; PS, phosphatidylserine.

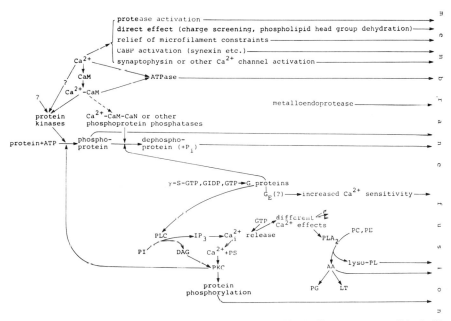

FIG. 1. Survey of hypothetical mechanisms proposed in the literature as possibly facilitating or inducing membrane fusion during exocytosis. As discussed throughout this review,
not all of these theories are shared by the author, but a multifactorial fusion control may be
assumed at this time. Some mechanisms included here will be applicable only to certain
systems. AA, Arachidonic acid; ATP, adenosine triphosphate; Ca^{2+}, free intracellular Ca^{2+};
CaBP, calcium-binding protein; CaM, calmodulin; CaN, calcineurin(-like protein); DAG,
diacylglycerol; G_E, exocytosis-relevant G protein; GTP, guanosine triphosphate; GIDP,
guanosyl-imidodiphosphate; γ-S-GTP, γ-thioguanosine triphosphate; IP_3, inositol trispho-
sphate; LT, leukotriene; PC, phosphatidylcholine; PE, phosphatidylethanolamine; PG, prostaglandins; PI, phosphatidylinositol; P_i inorganic phosphate; PKC, protein kinase C; PLA_2
and PLC, phospholipases A_2 and C; PS, phosphatidylserine.

homeostasis rather efficiently (Carafoli, 1987), so that different phenomena operate in an overlapping mode. Cells secreting selectively by different secretory organelles display different pCa_i threshold values (Lew *et
al.*, 1986). All this makes the original concept now appear too simplistic.

On the basis of the stimulus–secretion coupling concept, *in vitro* studies
with artificial membranes have been designed. Ca^{2+} has been seen to act as
a fusogen, though at considerably higher concentrations (pCa = 3–4)
than *in vivo*. Ca^{2+} has been assumed for some time to act directly on lipids
also during biological membrane fusion including exocytosis (Papahadjopoulos, 1978; Papahadjopoulos *et al.*, 1978). Considerable efforts were
made to determine the lipid compositions that prepare membranes for

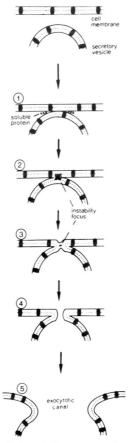

FIG. 2. Scheme of membrane fusion during exocytosis according to Plattner (1981). This scheme for the first time takes into account the presence of membrane-integrated (dark spots) and soluble proteins (-x-) at the fusogenic site, as well as the occurrence of "focal fusion." The exocytotic opening (canal) is very small (corresponding to the thickness of a membrane) at the beginning and then expands. Dotted lines indicate fracture planes in freeze–fracture replicas, exposing intramembranous particles (IMP)—particularly in the cell membrane— close to the fusion site.

fusion. In parallel, isolated secretory organelles or vesicle ghosts were used to analyze Ca^{2+}-mediated aggregation and fusion processes (see Section IV; see also Hopkins and Duncan, 1979; Sowers, 1987; Ohki *et al.*, 1988). Positive findings with these model experiments were all thought to support a fusion mechanism involving only lipids, but not proteins, in conjunction with Ca^{2+}. Only later on have proteins been included in liposome preparations and found to modulate fusion *in vitro* (Hong *et al.*,

1982, 1987). These experiments took into account the stringent ultrastructural evidence for the occurrence of proteins at sites of exocytotic membrane fusion (Plattner, 1981; see also Sections VI and XII).

Previous freeze–fracture studies, until the late 1970s, seemed to support the exclusive role of lipids, since exocytotic fusion sites showed removal of intramembranous particles (IMP) (Orci and Perrelet, 1978). In most cases, IMP represent (oligomeric) proteins, as derived mainly from reconstitution experiments (Zingsheim and Plattner, 1976). Membrane-associated proteins were also claimed, on the basis of the electron-microscopic (EM) analysis of ultrathin sections, to be removed from membrane–membrane contact sites before fusion can occur. Only in the present decade has it been realized that both these ultrastructural aspects are caused by insufficient preparation conditions (Plattner, 1981, 1989a,b; Burgoyne, 1984; Chandler, 1988).

Since then, advanced preparation schedules for EM analysis, including fast-freezing (cryofixation) techniques (see also Plattner and Bachmann, 1982), have been developed to a standard that allows verification of the presence of membrane-integrated and -associated proteins at sites of ongoing exocytotic membrane fusion. The "focal [point] fusion concept" (Figs. 2 and 3) has been advanced [Plattner (1981), based on own work and work of Chandler and Heuser (1979, 1980) and Ornberg and Reese (1981)]. This concept has now been verified by electrophysiological recordings of surface capacitance changes upon insertion of a secretory organelle mem-

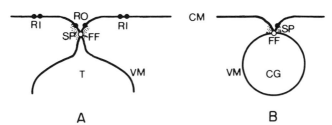

A   B

FIG. 3. Initial steps of membrane fusion during exocytosis in a *Paramecium* (A) and a chromaffin gland cell (B). FF indicates a fusion focus, where the cell membrane (CM) and the secretory vesicle membrane (VM) come into contact. The dotted area indicates the presence of soluble proteins (SP). Dark dots are membrane-integrated proteins, which only in *Paramecium* are assembled to a "fusion rosette" (RO); RI designates membrane-intercalated particles composing a ringlike structure that in *Paramecium* delineates the extrusion site of a trichocyst (T). No such regular particle aggregates occur at the extrusion site of a chromaffin granule (CG). Data are compiled from Momayezi *et al.* (1987a) and Plattner (1987, 1989a) for *Paramecium* (A), and from Schmidt *et al.* (1983) and Plattner (1989a) for the chromaffin cell of the adrenal medulla (B). The scheme in (B) appears to be applicable to other gland cells as well (see Section XVI).

brane into the cell membrane (Breckenridge and Almers, 1987; Zimmerberg *et al.*, 1987). Only advanced EM techniques and electrophysiological measurements can provide the temporal and spatial resolution to monitor the actual membrane fusion process. It is fascinating to see that both methods now come up with quite similar orders of magnitude: 1-msec events on a 10-nm scale. (Interestingly, the focal-fusion concept can now be extended also to a variety of biological membrane fusions other than exocytosis; see Knoll and Plattner, 1989.)

In the last decade, it has also been attempted to find out the molecular rearrangement of lipids, since it is, of course, the lipids that must be reoriented for fusion to occur. The debate on whether inverted micelles (Verkleij *et al.*, 1979; Verkleij, 1986; Siegel, 1984, 1986, 1987) or more stochastic arrangements (Hui *et al.*, 1981, 1988; Papahadjopoulos *et al.*, 1987) would account for fusion still goes on (Section XI). The direct interference of proteins in exocytotic fusion induction (see Section XII) is increasingly being substantiated (Almers and Breckenridge, 1988; Thomas *et al.*, 1988; Zimmerberg, 1988). This might entail new aspects for the rearrangement of lipids during fusion.

As mentioned, EM analyses also gave important hints to the involvement of both types of proteins (see Figs. 2 and 3). Some time ago their absence from "nondischarge" (*nd*) mutants of *Paramecium* (Beisson *et al.*, 1976; Plattner *et al.*, 1980; Pouphile *et al.*, 1986) was in striking contrast to the generally accepted hypothesis (which would have assumed the opposite). Another line of evidence came from the finding of synexin in chromaffin cells (Pollard *et al.*, 1981), one of the first Ca^{2+}-binding proteins (CaBP) described. Since then, numerous CaBP (proteins that bind to membrane phospholipids when Ca^{2+} is present, and thus were thought possibly to mediate aggregation and/or membrane fusion) have been found in different secretory systems (Geisow and Burgoyne, 1982; Creutz *et al.*, 1983, 1987a; Smith and Dedman, 1986; Geisow *et al.*, 1987; Martin and Creutz, 1987; Drust and Creutz, 1988; Peplinsky *et al.*, 1988). This line of thought is still being intensely pursued, although there is little evidence as yet that CaBP would be engaged in membrane fusion per se, since the pCa values they require to bind to membranes are for the most part too high (Klee, 1988).

The Ca^{2+}-regulatory protein calmodulin (CaM) has also been suggested to play a role, most frequently in connection with protein phosphorylation (DeLorenzo *et al.*, 1981; DeLorenzo, 1982; Knight *et al.*, 1988). Since then, protein dephosphorylation also has been envisaged (Plattner *et al.*, 1988; see also later).

Another line of research follows the GTP-mediated sensitization for

Ca^{2+} (Barrowman *et al.*, 1986), or Ca^{2+} independence of exocytosis mediated by GTP or its analogs (Neher and Almers, 1986; Neher, 1988). Specific GTP-binding proteins (G proteins), which would be directly relevant for membrane fusion during exocytosis, have not yet been identified (Section VIII). Some G proteins somehow act via phosphatidylinositol (PI) turnover (Section IX). The idea that PI turnover products would play a role has been launched by Michell (1975), Michell *et al.* (1981), and Berridge and Irvine (1984). These products can mediate a variety of modifications and metabolic steps with regard to proteins and lipids. Different systems display quite different reactions to PI breakdown products like diacylglycerol (DAG; the physiological activator of protein kinase C or PKC), or to phorbol esters (nonphysiological substitutes for DAG). There occurs either a small or a large increase of Ca^{2+} sensitivity, or none at all, even when the extent of secretion is increased—depending on the cell type investigated (Knight, 1987; Baker, 1988). Only some excitable cells may be operative directly by an increased pCa_i (Penner and Neher, 1988). A crucial aspect of this diversity may also be the activation of cells with or without receptor activation or by different kinds of receptors and the involvement of different ion channels, sometimes even in one and the same system (Meldolesi and Pozzan, 1987; Tartakoff, 1987; Partridge and Swandulla, 1988); see also Section IX. Therefore, second messengers, including PI turnover products, may play a diverse role from system to system, or none at all in others (see reviews in Miller, 1987; Meldolesi and Pozzan, 1987; Baker, 1988; Rink and Knight, 1988).

Lipids might interact with an (unknown) protein, which would have to be modified in order to impose a molecular rearrangement to lipids at the fusion spot. This concept is endorsed by the finding of fusogenic proteins (fusion proteins) in some viral systems (White *et al.*, 1983). A change of protein conformation, with a partial increase in hydrophobicity, might induce fusion (Gething *et al.*, 1986). Yet in contrast to viral systems (White *et al.*, 1983; Schlegel, 1987), acidic pH is unlikely a driving component; rather, covalent protein modification (as in other viral systems; see Schlegel, 1987) or a conformational change by Ca^{2+} binding, and so on, should be assumed relevant for exocytotic membrane fusion. Although phosphorylation is most commonly discussed in this context (see Section X), there is some evidence for protein dephosphorylation as a fusogenic step (Gilligan and Satir, 1982; Zieseniss and Plattner, 1985), possibly involving Ca^{2+}-dependent phosphatases (Momayezi *et al.*, 1987b). Other experiments point toward the involvement of metalloendoproteases (Mundy and Strittmatter, 1985; Strittmatter *et al.*, 1987).

A rearrangement of proteins and/or of lipids due to Ca^{2+} binding without covalent modifications are alternatives also envisaged in the literature (see Sections X–XII).

An interaction of actomyosin (microfilaments, Mf) with secretory granules also has been considered, after Ca^{2+} activation, to bring together the membranes to be fused (Poisner and Trifaró, 1967; Berl et al., 1973; Poste and Allison, 1973). A critical review by Trifaró and Kenigsberg (1987) denies this, however. Many pros and cons with regard to the involvement of cytoskeletal elements, particularly also of microtubules (Mt; see, e.g., references in Rindler et al., 1987; Soifer, 1986), were collected. Today it is generally believed that neither Mt nor Mf would play a role in the regulation of the actual membrane fusion process (chromaffin cells: Schneider et al., 1981; Baker, 1988; Paramecium: Plattner, 1987), although this judgment is frequently founded on results obtained with Mt- and Mf- disrupting agents that exert many side effects (Section V). Yet recent evidence with antibodies (Ab) against Mf components (Perrin et al., 1987) support some of the results previously obtained with drugs.

All these aspects have to be discussed in light of the small area size (involving perhaps only <0.01% of membrane lipids, as in the Paramecium cell) and duration (milliseconds) of membrane fusion during exocytosis (see Section XVI). Counterregulation phenomena (e.g., de/rephosphorylation, within seconds or fractions thereof in Paramecium: Zieseniss and Plattner, 1985) have frequently not been taken into account in the nonsynchronous systems mostly analyzed so far. Coincidence with other phenomena (not directly related to membrane fusion) also render the analysis of exocytotic membrane fusion extremely difficult. For instance, exocytosis-coupled endocytosis requires only seconds in chromaffin cells (Neher and Marty, 1982) or fractions thereof in nerve endings (Torri-Tarelli et al., 1985) and in Paramecium (Plattner, 1987).

We can now foresee the following aspects: A variety of molecular components have to be properly assembled before a secretory organelle can be attached to the cell membrane. An increased pCa_i is not the only messenger for membrane fusion. The reorientation of lipids during membrane fusion may be governed by a finely tuned, locally arranged "machinery," which might be Ca^{2+}-sensitive and comprise some modulatory and enzymatic components (probably different from one system to another). The effects of Ca^{2+} are so many (see Fig. 1) that it is difficult to pinpoint its crucial role in exocytosis. The situation is now even more complicated, since resting pCa_i levels are also considered by some authors to suffice for membrane fusion during exocytosis (Neher and Almers, 1986; Neher, 1988). This certainly holds true for constitutive exocytosis.

III. Systems and Techniques for the Analysis of
Exocytosis *in Vivo* and *in Vitro*

Secretion as a whole (from receptor stimulation until discharge of contents) has been thoroughly analyzed with many cell types. Most analyses dealt with regulated exocytosis. Only a limited number of systems appear appropriate, however, to analyze the crucial steps of membrane fusion without interference of other aspects (see, e.g., Section X).

Nerve terminals of motoneurons were a favorable system for a precise time-sequence analysis of ultrastructural features of synaptic-vesicle fusion (Heuser *et al.,* 1979). The rapid and synchronous response to electrical stimulation allowed workers for the first time to disprove the "fusion–fission model" (see Section XVI).

Mast cells are another frequently used system. Full exocytotic response occurs within ~20 seconds, in response to quite different stimuli, such as IgE receptor crosslinking, concanavalin A, or polycations (compound 48/80: Lagunoff and Chi, 1980; other polyamines: Foreman and Lichtenstein, 1980). "Compound exocytosis," or granule–granule fusion, is a feature rather unique for mast cells and only occurs after the outermost secretory granules have fused with the cell membrane. This should therefore not justify vesicle–vesicle interactions as model systems for exocytosis *in vitro* (see later).

Another extreme is represented by most gland cells in which secretory activity may continue for ≥1 hour. This frequently entails the problem of "masking" biochemical changes, when they are reversible or counterregulated (e.g., protein de/rephosphorylation within seconds; see Section X).

Attempts have frequently been made to determine parameters relevant for exocytosis regulation by inhibition studies. Inhibitory (or stimulative) drugs directed against certain pathways have frequently been used in this context; however, this is problematic, since many drugs exert multiple effects (see Sections IX and X). Antibodies to certain antigens were also applied to gland cells by vesicle-mediated injection or by using permeabilized cells (see later). Another approach is the application of nonmetabolizable nucleotide analogs (Sections VII and VIII) to such systems.

A limited number of cell types, including egg cells (sea urchins or vertebrates) and some protozoans, offer several important experimental advantages: (a) They may show a rather fast response (due to attachment of secretory organelles at the cell surface); (b) they are of a size that allows direct mechanical microinjection; and (c) they permit isolation of cell surface complexes (cell membrane with secretory organelles still attached, or cortices) for *in vitro* studies.

Among protozoans the ciliate *Paramecium tetraurelia* is of particular interest, because of the availability of secretory mutants (Cohen and Beisson, 1980; Pouphile *et al.,* 1986; Adoutte, 1988) and because potential secretory sites are found at predictable locations; these also display distinct ultrastructural features (Plattner *et al.,* 1973). In wild-type cells (or any normally secreting strain) these include (Plattner, 1987; Adoutte, 1988) the following: (a) the presence of a "plug" as a morphological receptor structure before a secretory organelle (trichocyst) becomes docked (see Section VI); (b) the formation of "connecting material" (proteins; see Sections VI and XII) between the cell membrane and the secretory organelle; and (c) the assembly of a group of about nine IMP (forming a "fusion rosette") in the plasmalemma at secretory sites. Features (b) and (c) are not expressed in trichocyst-free mutants ("trichless," *tl*), in *tam* mutants (with "free" trichocysts only—that is, without trichocyst docking), or in "nondischarge" (*nd*) mutants. Functional repair can, in some cases, be achieved by the transfer of wild-type cytosol, by pressure injection, or during cell conjugation (see Section VI). Microinjection or the use of isolated cortices allows testing of the effects of nucleotides, of certain proteins, or of Ab (as well as their fluorescent labeling). Synchronous exocytosis (<1 second) can be achieved with *Paramecium* cells for ≤90% of the secretory-vesicle population (Plattner *et al.,* 1985; Plattner, 1987).

The binding of secretory vesicles to the cell membrane ("docking") is now extensively analyzed also with yeast (*Saccharomyces*) cells. Mutant strains selectively deficient in this step have become available (Goud *et al.,* 1988). (See Note Added in Proof.)

Although not all molecular features may be transposable from lower to higher eukaryotes, a comparison shows extensive similarities in the occurrence of cation channels (Machemer, 1988; Saimi *et al.,* 1988), Ca^{2+}-activated proteins (see Sections VI and X), and second-messenger systems (Janssens, 1988). Extrapolation requires caution, though.

Important new methodological developments include (a) the use of permeabilized cells [Baker and Knight (1979, 1984b); see recent reviews by Knight and Scrutton (1986), Knight (1987), Baker (1988), and Rink and Knight (1988)]; (b) nonequilibrium-kinetic studies with such cells (Gomperts *et al.,* 1987); (c) electrophysiological recordings of exocytosis by surface capacitance changes occurring in discrete steps as seen with patch–clamp methods (Neher and Marty, 1982; Fernandez *et al.,* 1984; Neher, 1988); (d) the application of nonmetabolizable or γ-thio-nucleotides (ATP or GTP analogs; see Sections VII and VIII); (e) the application of calcium-sensitive fluorochromes *in vivo* (Campbell, 1983; Tsien, 1988); (f) the introduction of new, more specific chelators for Ca_i^{2+} buffering (Bers and Harrison, 1987); (g) the development of new stopped-flow procedures for precise time-sequence studies with synchronous systems in defined small intervals after exocytosis triggering (Knoll et al., 1989); and (h) mRNA injection into oocytes for expressing foreign signal transduction components (Jaffe et al., 1988).

Permeabilization can be accomplished in some (e.g., chromaffin) cells by electroporation (Baker and Knight, 1979, 1984b), or chemically with digitonin or saponin (for review see Knight and Scrutton, 1986). Recently streptolysin O or Staphylococcus α-toxin were also applied. Through the pores thus formed, Ab may gain access to the cell interior, while leakage of cytosolic components considered relevant for exocytosis may be negligible over short time periods (Baker and Knight, 1979; Ahnert-Hilger et al., 1985). Staphylococcus α-toxin yields smaller pores (Ahnert-Hilger et al., 1985) than streptolysin O (Bhakdi and Tranum-Jensen, 1987), so that one can produce access or leakage of components of different molecular size. Leaked components (<100 kDa) can be put back into such cell suspensions to restore their secretory activity (Sarafian et al., 1987).

Another approach was to recombine isolated cell membrane pieces and secretory organelles (see Section IV). These in vitro model systems are hampered by a variety of problems in most cases, however.

Much effort has been expended on the interaction between isolated secretory vesicles (particularly chromaffin granules from the adrenal medulla). A family of evolutionarily related proteins (Geisow et al., 1987; Crompton et al., 1988) have been identified by their capacity to bind to the lipids of some secretory vesicle in a Ca^{2+}-dependent fashion ("annexins" or "chromobindins," including "synexin"). The release of secretory contents by granule–granule interaction was assumed to mimic compound exocytosis (which occurs almost exclusively in mast cells). Yet it has not been shown unequivocally that this is equivalent to secretory vesicle–cell membrane interaction as it occurs during exocytosis in vivo (see Section IV).

A great number of investigations using artificial lipid membranes, planar and/or vesicular, have been published. Although this is undoubtedly an important approach, its relevance for membrane fusion during exocytosis is now increasingly questioned (see Section XI).

IV. What Is the Value of in Vitro Model Systems?

Results from liposome–liposome (Papahadjopoulos, 1978; Papahadjopoulos et al., 1978; Düzgünes et al., 1987) or liposome–planar membrane fusion (Cohen et al., 1980; Zimmerberg et al., 1980) were important for a general understanding of possible fusion mechanisms. They showed the relevance of Ca^{2+} for lipid phase transitions, charge screening, local dehydration (see Fig. 1), and the resulting close apposition and eventually fusion of membranes (Papahadjopoulos, 1978; Papahadjopoulos *et al.*, 1978; Parsegian *et al.*, 1984; Verkleij, 1984, 1986; Düzgünes, 1985; Rand and Parsegian, 1986; Blumenthal, 1987; Düzgünes *et al.*, 1987; Ohki, 1987). Yet a rather high pCa (3–4) is required, in contrast to membrane fusion during exocytosis. The latter also involves nucleotides, PI metabolites, and different proteins that might confer high Ca^{2+} sensitivity (Sections VII–IX and XII).

Another difference between *in vivo* and *in vitro* systems is the response to osmotic conditions. This is important for *in vitro* (Akabas *et al.*, 1984; Finkelstein *et al.*, 1986) but not for exocytotic membrane fusion (Zimmerberg, 1988). Differences between membrane fusion *in vitro* and during exocytosis are compiled in Table I. It is now understood that data obtained with artificial membranes are not directly applicable to fusion *in vivo* (Rand and Parsegian, 1986; Blumenthal, 1987; Papahadjopoulos *et al.*, 1987).

Artificial membrane systems may be useful to test the fusogenic effect of different lipid components and of integral membrane (Thomas *et al.*, 1988), or of cytosolic proteins isolated from gland cells (Hong *et al.*, 1987; Pollard and Rojas, 1988; see also section XII). Work along these lines has been stimulated by several observations: (a) Fusion can be induced *in vitro* (Burger *et al.*, 1988) via reconstituted viral fusion proteins (Kielian and

TABLE I

COMPARISON OF MEMBRANE FUSION *in Vitro* AND *in Vivo*[a]

	In vitro (pure lipids)	*In vivo* (exocytosis)
Ca^{2+} requirement	Millimolar	Submicromolar to micromolar
Fusion "anatomy"	Large contact areas	Focal (point) fusion
Leakiness	Yes; not with improved systems	No
Effect of proteins		
Nonenzymatic	Eventually stimulative	Positive evidence
Enzymatic	No effects known	Positive evidence
Specificity of interaction	None or little	Yes (high)
Reversibility of early stages	No	Yes
Osmotic driving force	Yes	No
Duration[b]	<0.1 msec	Milliseconds

[a] Compiled according to data cited in the text.
[b] *Paramecium;* 1 msec (Momayezi *et al.*, 1987a); nerve endings; see Section XIII. For a more comprehensive evaluation, see Zimmerberg (1987).

Helenius, 1985; Lapidot *et al.*, 1987) or amphipathic fragments thereof (Parente *et al.*, 1988). (b) Amphipathic proteins facilitate fusion of liposomes (Batenburg *et al.*, 1985; Hong *et al.*, 1987). (c) The electrophysiological analysis of early fusion events during exocytosis suggests some role of proteins (Almers and Breckenridge, 1988; Zimmerberg, 1988; see also Section XII). (d) Fusion can be achieved with synaptophysin-containing artificial membranes (Thomas *et al.*, 1988). Therefore, it now appears feasible to test any potentially fusogenic protein, isolated from secretory systems, by adding to or reconstitution with artificial membranes.

Isolated secretory vesicles (or membrane ghosts) have also frequently been used to mimic exocytosis *in vitro*, with Ca^{2+} as a "stimulant" (Morris and Bradley, 1984; Creutz *et al.*, 1987b; Stutzin *et al.*, 1987). Several aspects can be questioned: (a) *In vivo* fusion between secretory vesicles occurs almost exclusively in mast cells (compound exocytosis: Lagunoff and Chi, 1980) and (b) only when peripheral vesicles have fused previously with the cell membrane. (c) Fusion yield is usually small *in vitro;* vesicle aggregation might be more significant (Meers *et al.*, 1988b). (d) pCa values well above those occurring *in vivo* are required (Creutz and Pollard, 1983; Creutz *et al.*, 1983, 1987b; Morris and Bradley, 1984; Scott *et al.*, 1985). As shown with sea urchin egg components, interaction between cell membrane and secretory vesicles is in fact much more Ca^{2+}-sensitive than the interaction between vesicles alone (Whalley and Whitaker, 1988). (e) Some plasmalemmal contaminants could cause a release of secretory contents by an exocytosislike process (however, see later). (f) In some earlier experiments, leakage rather than fusion might have occurred, since one would not expect any content release with perfect vesicle–vesicle fusion.

Some of this work showed the aggregating or allegedly fusogenic effects of some cytosolic proteins, like chromobindins (Creutz *et al.*, 1987b; Stutzin *et al.*, 1987). In the case of calpactin, vesicle aggregation occurs in fact even with a physiologically low pCa (Drust and Creutz, 1988).

In another attempt to mimic exocytosis *in vitro,* secretory vesicles were recombined with cell membrane fragments. However, results obtained with the chromaffin system (Davis and Lazarus, 1976; Konings and De-Potter, 1982) have now been disproven (DeBlock and DePotter, 1987). Leakage may also explain the minute release of secretory contents observed by other groups with such systems (Lelkes *et al.*, 1980). Any significant membrane fusion cannot be expected, because most of the cell membrane vesicles isolated are right-side out and thus would not allow for a topologically correct interaction with secretory granules (Rosenheck and Plattner, 1986). The only system thus far that allows for cell membrane–secretory vesicle reconstitution *in vitro* is derived from sea urchin eggs

(Crabb and Jackson, 1985); exocytosis was monitored by fluorescent Ab against vesicle contents. The Ca^{2+} levels required are equivalent to or only slightly higher than those occurring *in vivo* (Jackson and Crabb, 1988; Whalley and Whitaker, 1988). These investigations also profoundly question the value of *in vitro* vesicle–vesicle interaction: True fusion occurs only when plasmalemmal components and cortical vesicles are recombined (Whalley and Whitaker, 1988).

Well-established *in vitro* systems are cell cortices prepared from egg cells (Detering *et al.*, 1977; Decker and Lennarz, 1979; Whitaker, 1987; Whitaker and Baker, 1983) and now also from *Paramecium* cells (Vilmart-Seuwen *et al.*, 1986). These systems are "frozen" in a stage between secretory organelle docking at the cell membrane and fusion (see Sections III and VI). The $[Ca^{2+}]_i$ required for maximal activation is 1–3 μm for egg (Whitaker, 1987) or 1 μM for *Paramecium* cortices (Lumpert *et al.*, 1989). In the latter case cortices from fusion-incompetent (*nd*) strains ascertained only a small (5%) rate of nonspecific leakage (Vilmart-Seuwen *et al.*, 1986). Another advantage is that exocytosis-coupled endocytosis does not occur with egg cells (Epel and Vacquier, 1978; Nishioka *et al.*, 1987) or with *Paramecium* cortices (in contrast to whole cells: Plattner *et al.*, 1988). One important finding with these systems is that they do not require ATP for membrane fusion (egg cells: Jackson and Crabb, 1988; *Paramecium:* Vilmart-Seuwen *et al.*, 1986; Lumpert *et al.*, 1989; for details see Section VII).

Cells permeabilized by electroporation or by chemical agents (see Section III) stay in between these *in vitro* and *in vivo* systems (see Sections VII–IX). Either method can be adjusted so as to retain soluble proteins, like lactate dehydrogenase (Baker and Knight, 1979; Ahnert-Hilger *et al.*, 1985), although leakage may occur with some permeabilizing agents (see earlier) and thus can cause aging of such preparations (Sarafian *et al.*, 1987). Permeabilized cells are valuable tools to elucidate ion and nucleotide requirements, and results obtained are discussed throughout this review.

V. Role of Microtubules and Microfilaments

Microtubules display a radial or polar arrangement in most cells (Kurihara and Uchida, 1987; Kreis *et al.*, 1988; *Paramecium:* Cohen and Beisson, 1988). They thus can mediate direct transport and docking of secretory or neurotransmitter vesicles at the cell surface by unidirectional, saltatory movements (Lacy *et al.*, 1968; Lacy, 1975; Allen *et al.*, 1985). Therefore, Mt would potentially represent pathways for targeting organelles to specific sites during regulated secretion (Lacy, 1975; Kern et al., 1979; Hall, 1982; Plattner et al., 1982; Gray, 1983; Westrum et al., 1983; Rogalski et al., 1984; Allen et al., 1985; Rindler et al., 1987; see also Soifer, 1986). In nerve terminals this aspect is particularly evident (Gray, 1983).

As far as constitutive secretion is concerned, the evidence is conflicting. Opposite results were obtained, when the polar delivery of influenza and vesicular stomatitis virus G protein was analyzed in response to Mt-affecting drugs (colchicine, nocodazole, taxol), (Rogalski et al., 1984; Salas et al., 1986; Rindler et al., 1987).

The role of Mt for the delivery of secretory organelles to the cell membrane and for their targeting to docking–fusion sites at the cell membrane can now be restricted in the sense that they might facilitate both these processes. However, they might not be obligatory for the following reasons: (a) Disruption of Mt by a variety of drugs [that shift the steady-state equilibrium from polymeric (microtubular) to monomeric tubulin] causes at the most some delay, but not an abolition of secretory activity (Virtanen and Vartio, 1986; Baker, 1988). (b) In Paramecium cells Mt originate from organizing centers associated with ciliary basal bodies, 1 μm remote from the actual docking site (Plattner et al., 1982), although they probably account for the saltatory transport of trichocysts (Aufderheide, 1977). In other words, Mt cannot be considered as precise targeting instruments, but perhaps as facilitating transport pathways and as long-range targeting signals. The interaction of a secretory organelle with the cell membrane probably requires a more specific narrow-range signal (see Section VI).

Microtubules can be excluded from any contribution to membrane fusion during actual exocytosis. In Paramecium cells, for instance, no permanent contact of Mt with trichocysts persists, once they are docked (Plattner et al., 1982). Many other cells also secrete by exocytosis, when antimicrotubular drugs had been applied (Schneider et al., 1981; Virtanen and Vartio, 1986; Baker, 1988). This might apply particularly to vesicles that already are in the vicinity of the cell membrane (Launay et al., 1980; Thuret-Carnahan et al., 1985).

Since so many of the data were obtained with inhibitory drugs, one should not overlook their considerable side effects, such as the anticholinergic effects of some actin- (cytochalasin B) and tubulin-binding drugs (Schneider et al., 1981; McKay et al., 1985), among many others. (For Mf-disrupting agents, see also Cooper, 1987.) Furthermore, they were frequently applied without controlling the actual effect on the cytoskeleton.

Aspects implying a role for Mf (actomyosin) in secretory activity have been particularly intriguing. A previous hypothesis on the binding of actomyosin to secretory organelles (Burridge and Phillips, 1975) and its possible involvement in their transcellular transport was replaced by assuming kinesin as a Mg^{2+}–ATP-driven motor (Kuznetzov and Gelfand, 1986; Schroer et al., 1988). A probable contribution of Mf is to control the access of secretory vesicles to the cell membrane (Orci et al., 1972; Perrin et al., 1987; Sontag et al., 1988). In fact, actin and myosin (Drenckhahn et al., 1977, 1983; Kurihara and Uchida, 1987) and some actin-binding proteins (α-fodrin: Perrin and Aunis, 1985; Langley et al., 1986; Sarafian et al., 1987) are displayed beneath the cell membrane, as shown by immunofluorescence. Other Ca^{2+}-sensitive proteins located in the cell cortex and affecting the crosslinking of cortical actin are gelsolin (Stossel et al., 1981; Sontag et al., 1988) and caldesmon (Burgoyne et al., 1986). Increased $[Ca^{2+}]_i$ would, by these different interactions, facilitate a gel $\rightarrow$ sol transition of the cortical cytoplasm and allow the acccess of secretory vesicles to the cell membrane. Only the population of vesicles already attached at the cell membrane can therefore be released without delay from gland cells, even when cortical actin filaments are crosslinked (Sontag et al., 1988). For sea urchin eggs and Paramecium, see Section VI and the following paragraphs.

This concept is substantiated by the localization of α-fodrin (Ca^{2+}-sensitive) not only by fluorescent-Ab labeling (Perrin and Aunis, 1985), but also by ultrastructural immunocytochemical localization (Langley et al., 1986) in the cortex of chromaffin cells and by blocking of their secretory activity with microinjected anti-α-fodrin Ab (Perrin et al., 1987). When actin severing thus had been inhibited, only ~10% of secretory granules (the percentage located close to the cell membrane, which would not need any further transport) could immediately be released. This concept is shared by others (Burgoyne and Cheek, 1987; Cheek and Burgoyne, 1987), who have manipulated cortical actin by varying the level of intracellular cAMP. A barrier function of cortical filamentous (F-) actin had been proposed a long time ago by Orci et al. (1972) on the basis of stimulatory effects of cytochalasin B treatment.

In Paramecium cells cortical actin (assayed by fluorescent-phalloidin microinjection) could be completely removed by depolymerization and reassembly to transcellular filament bundles (Kersken et al., 1986a,b). Yet such cells were perfectly able to secrete all their trichocysts attached at the cell surface. Moreover, F-actin is not present at the exocytosis sites proper in Paramecium, although it occurs in the cell cortex (Plattner et al., 1982; Kersken et al., 1986b; Cohen and Beisson, 1988). This holds true also for the "active zones" of nerve terminals (Smith and Augustine, 1988), where some conspicuous connecting material has also been found at exocytosis sites (see also Section VI).

All this evidence taken together indicates that neither Mt nor Mf directly participate in membrane fusion (Baker and Knight, 1981; Schneider *et al.*, 1981; Knight and Baker, 1982; Plattner, 1987; Baker, 1988). In particular, contractile-force generation by actomyosin, as previously proposed by Poisner and Trifaró (1967) and Berl *et al.* (1973), is no longer a tenable concept for membrane fusion regulation (Trifaró and Kenigsberg, 1987).

VI. Docking of Secretory Organelles and Assembly of Components Relevant for Exocytosis

Secretory organelles are only exceptionally firmly docked at the cell membrane for immediate exocytosis. This holds for egg cells and for some protozoans, notably for *Paramecium* cells. Only from these systems can one, therefore, isolate surface complexes (cortices) and use them for the analysis of membrane fusion during exocytosis *in vitro* (egg cells: Detering *et al.*, 1977; Decker and Lennarz, 1979; Schön and Decker, 1981; *Paramecium:* Vilmart-Seuwen *et al.*, 1986). In these two systems ultrathin sections, particularly after tannic acid staining, reveal the presence of electron-dense, amorphous to fibrillar connecting material between the two membranes (egg cells: Guraya, 1982; *Paramecium:* Plattner *et al.*, 1980; see also Plattner, 1987, 1989b). Similar structures are recognized, however, also on peripheral secretory granules of various gland cells, like chromaffin cells (Aunis *et al.*, 1979), pancreatic islet β- cells (Orci *et al.*, 1979), parathyroid cells (Setoguti *et al.*, 1981), and so on, as well as on "active zones" of nerve terminals (motor end plates: Couteaux, 1978; Rash *et al.*, 1988; noncholinergic fast-acting nerve terminals of the central nervous system: Smith and Augustine, 1988; see also Landis, 1988). This material is sensitive to proteolytic enzymes, as shown with *Paramecium* (Westphal and Plattner, 1981).

In *Paramecium* the docking process occurs at genetically predetermined sites in a regular arrangement and thus can be reliably analyzed by ultrastructural methods. This case will therefore be described in more detail: In wild-type cells docking of a secretory organelle (trichocyst) entails the assembly of characteristic IMP aggregates (fusion rosettes) and of connecting material, thereby conferring fusion capacity (Plattner 1987, 1989b). This correlation has been ascertained by (a) analyzing reinsertion of newly formed trichocysts after massive trichocyst discharge (Pape and Plattner, 1985), (b) reinsertion of trichocysts that had been experimentally detached from the cell surface *in vivo* (Pape and Plattner, 1989), (c) analysis of *nd* mutants (lacking connecting material and rosettes; see Plattner *et al.*, 1980; Pouphile *et al.*, 1986; Plattner, 1987, 1989b), and (d) analysis of the changes occurring during secretory function repair (Pape *et al.*, 1988).

In *Paramecium,* assembly of components relevant for exocytosis includes the formation of a "plug" (an electron-dense mass of proteins) as a kind of receptor structure for trichocyst docking beneath the cell membrane (Plattner, 1987, 1989b). Plug and connecting material (which, at least in part, is derived from the plug) contain CaM (Momayezi *et al.,* 1986; Plattner, 1987). Other proteins have not thus far been identified in detail. A plug is assembled within ~20 minutes following exocytosis and removal of a trichocyst ghost (Haacke and Plattner, 1984). Its occurrence is a prerequisite (among others) for establishing docking competence; yet docking of a trichocyst does not necessarily entail fusion competence (Beisson *et al.,* 1976; Cohen and Beisson, 1980; Plattner *et al.,* 1980; Pouphile *et al.,* 1986; Adoutte, 1988). (In *nd* mutants, trichocysts must be anchored separately to the surface outside the fusogenic zone proper.) In wild-type cells, docking of a trichocyst causes the plug materials to be smeared around the trichocyst tip. Docking entails, probably within several minutes, the assembly of rosette IMP (Pape and Plattner, 1985), which most likely are membrane-integrated proteins (Vilmart and Plattner, 1983). The causal connection or even the precise sequence of the rearrangement of membrane-integrated and -associated proteins during trichocyst docking, as well as their detailed molecular identity, remain unknown so far.

Rosettes and connecting material are also found in various other protozoans, where extrusomes are also docked at the cell membrane (Plattner, 1989a). For instance, the docking site of rhoptries (considered by some authors as mediators of host cell invasion) in parasitic sporozoans also displays this morphology and contains accumulated CaM as well (*Plasmodium:* Matsumoto *et al.,* 1987).

Since the docking process in *Paramecium* is reversible under certain conditions (Pape *et al.,* 1988), this suggests that some of the docking components would be reversibly bound. The occurence of CaM at exocytosis sites might imply a possible involvement of Ca^{2+}-CaM-stimulated kinases and/or phosphatases (see Fig. 4 and Section X). If so, these components, together with some target proteins and possibly some specific binding proteins (e.g., for CaM), would equally have to be assembled at preformed exocytosis sites in order to establish docking and/or fusion competence.

In sea urchin eggs CaM was also inferred to occur at docking sites (yet without assembly of rosettes) of cortical granules, as indicated by CaM depletion and reconstitution experiments as well as by Ab inhibition studies (Steinhardt and Alderton, 1982). Again CaM might contribute to the formation of connecting material between the cortical granules and the cell membrane, since fluorescent Ab yield a spotty pattern (Steinhardt and Alderton, 1982). However, according to Jackson and Crabb (1988), the

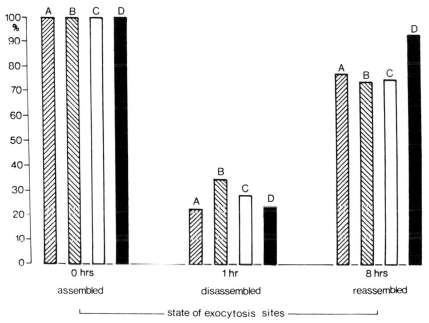

FIG. 4. Quantitative correlation between (A) the occurrence in *Paramecium* of trichocysts at a potential release site on the cell membrane (ultrathin-section evaluation),
(B) presence of a "rosette" IMP aggregate at such sites (freeze–fracture), (C) number of
trichocysts released in response to a maximal exocytosis trigger (light-microscope counting),
and (D) extent of phosphoprotein (PP65) dephosphorylation (^{32}P labeling and gel electrophoresis, combined with autoradiography or scintillation counting) under maximal trigger
conditions (as described in Section X). After a first trigger (0 hours), a second trigger was
applied 1 hour or 8 hours later, during which time structural components of exocytosis sites
are dispersed (1 hour) or widely reassembled (8 hours), respectively. Each column is referred
to the normalized values at 0 hours. Reassembly of structural components (also including
connecting materials; see text) during trichocyst reinsertion restores the capacity for exocytosis performance and for PP65 dephosphorylation. Data compiled from Pape and Plattner
(1985) and Zieseniss and Plattner (1985).

question of CaM involvement in the cortical granule reaction is not yet
settled.

The old postulate that CaM might confer high Ca^{2+} sensitivity to the
exocytosis mechanism (Means *et al.*, 1982) still deserves more detailed
investigation. It could imply phosphorylation or dephosphorylation of
fusogenic proteins (Sections X and XII), or of some of the "annexins" (see
the following), or of some enzymes related to second-messenger systems
(Section IX). Indirect effects could concern phosphorylation of Ca^{2+} channels (Section X) or of the widely distributed plasmalemmal Ca^{2+} pump
(Carafoli, 1987).

A docking protein has been described for chromaffin cells (Meyer and Burger, 1979), but these studies have been pursued only recently (Schäfer *et al.*, 1987). Synexin (Creutz and Pollard, 1983; Pollard *et al.*, 1987) and other chromobindins (Creutz *et al.*, 1987a; see also Section IV) were also proposed to mediate docking capacity. This concept has since been extended to nonchromaffin cells (annexins: Geisow *et al.*, 1987; Burgoyne and Geisow, 1989). Several aspects arise from this postulate.

a. The whole family of annexins (including synexin and other chromobindins, calelectrin, calpactin, endonexin, lipocortin, etc.) is not universally encountered in secretory cells (Geisow *et al.*, 1987; Burgoyne and Geisow, 1989), yet this could be the result of genetic diversification (Crompton *et al.*, 1988; Pollard *et al.*, 1988a) or the existence of analogous proteins in other systems.

b. Most of the chromobindin–annexin proteins, as far as analyzed, show maximal binding to secretory organelles only with pCa values exceeding those found in resting cells (see Section IV; see also Creutz and Pollard, 1983; Creutz *et al.*, 1983, 1987b; Pollard *et al.*, 1987). Since increased pCa does not occur during the docking process, the role of these proteins might be restricted to other phenomena. Only calpactin has a sufficiently high Ca^{2+} sensitivity and therefore might be considered as a possible docking protein (Drust and Creutz, 1988).

c. Synaptophysin (p38), a membrane-spanning Ca^{2+}-binding protein occurring in synaptic vesicles (Jahn *et al.*, 1985; Wiedenmann and Franke, 1985), has been proposed as a docking protein (Thomas *et al.*, 1988). It might interact with a putatative integral membrane protein called "mediaphore," found in *Torpedo* electroplax (Morel *et al.*, 1987), at the cell membrane (Section XII). Synaptophysin is identical to synaptin (Gaardsvoll *et al.*, 1988). Since it occurs only in neurons and some neuroendocrine cells (Wiedenmann *et al.*, 1986; Obendorf *et al.*, 1988a), but not in exocrine glands (Jahn *et al.*, 1985; Navone *et al.*, 1986), synaptophysin could be a docking protein only in some cases.

d. This might hold also for some other membrane-spanning proteins occurring in these systems, among them a 65-kDa protein (Matthew *et al.*, 1986; Lowe *et al.*, 1988). This p65 is again absent, though, from nonsecretory cells (Matthew *et al.*, 1986). A CaM-binding membrane-associated protein of the same size is assumed by Trifaró (Trifaró and Kenigsberg, 1987; Fournier and Trifaró, 1988a,b) to mediate docking in chromaffin cells. It interacts not only with chromaffin, synaptic, and neurohypophyseal vesicles (Fournier and Trifaró, 1988a), but also with the plasmalemma of chromaffin cells (Fournier and Trifaró, 1988b). This 65-kDa protein cross-reacts with Ab against p65 protein prepared from synaptic-vesicle membranes. It is unknown whether it is a phosphoprotein. In the adrenal medulla Gutierrez *et al.* (1988) have found a phosphoprotein of M_r 63,000, similar weight to a 63,000–65,000 M_r phosphoprotein (PP65) that we consider to be possibly relevant for exocytosis in *Paramecium* (Zieseniss and Plattner, 1985), also in conjunction with CaM binding (Momayezi *et al.*, 1987b; see also Section X). It is not known whether they represent related proteins in these two systems. Although Ab against PP65 from *Paramecium* cross-react with a band of similar size on Western blots from cell homogenates from chromaffin (Stecher *et al.*, 1987) and other mammalian cells (Murtaugh *et al.*, 1987), the *Paramecium* protein is not membrane-integrated in its phosphorylated form (Höhne *et al.*, 1989).

e. No possible role is as yet known for another membrane-integrated protein common to neuroendocrine and synaptic vesicles, the glycoprotein "SV2" of M_r 95,000–100,000 (Lowe *et al.*, 1988).

f. Another aspect concerns the possible mediation of synaptic-vesicle docking in nerve terminals by synapsin I. Since this is regulated by phosphorylation, see Section X for further discussion.

g. Nondocking yeast mutants in comparison to wild-type cells revealed the existence of a G protein that easily associates with secretory vesicles and the cell membrane (Goud *et al.*, 1988). This is to be seen in connection with the function of better known G proteins, as defined in Section VIII. One might speculate along the following lines: First, GTP confers high Ca^{2+} sensitivity to these membrane interactions in regulated systems (Section VIII). Second, at least constitutive (nonregulated) exocytosis evidently operates without any increase of pCa_i. Such G proteins are therefore candidates not only for docking proteins, but also in connection with the fusion process, although they have not yet been identified or localized in sufficient detail.

As mentioned in Section V, F-actin is absent from docking sites in *Paramecium* cells (Plattner *et al.*, 1982; Kersken *et al.*, 1986b) as in active zones of nerve endings (Smith and Augustine, 1988). It should be mentioned that this would be more difficult to ascertain with other systems with a less distinct morphology. This explains the long-lasting controversies on this topic in the literature.

Other soluble proteins have to be expected as components of docking sites, too. With *Paramecium,* the secretory function of *nd* mutants can be repaired by a transfer of soluble (undefined) cytoplasmic components, probably proteins, either by pressure injection (Aufderheide, 1978; Beisson *et al.*, 1980; Lefort-Tran *et al.*, 1981) or by cell conjugation (Lefort-Tran *et al.*, 1981; Pape *et al.*, 1988). A soluble protein component (different from CaM) restores the fusion capacity of aging sea urchin egg cortices (Sasaki, 1984) or of permeabilized chromaffin cells (Sarafian *et al.*, 1987). For all these reasons part of the components present at secretory organelle–cell membrane interaction sites can only loosely be bound to docking sites.

According to arguments presented in Section XI, lipid components are unlikely to be involved in specific membrane recognition.

In summary, it appears quite certain that docking requires or induces the assembly of membrane-bound and of integral proteins. The spectrum of proteins involved might be different from one system to another, and the actual trigger for assembly also remains unclear. At least in *Paramecium* and in nerve terminals, the occurrence of specific docking sites is quite evident. This might also apply to polar secretion, for which a targeting function of Mt evidently is not mandatory (see Section V).

VII. Is ATP Required for Membrane Fusion?

Until quite recently there has been widespread agreement that ATP is essential for exocytosis, tacitly or explicitly with the implication that it would be obligatory for membrane fusion. Now this view must be critically reevaluated.

Sustained secretory activity has been shown to require energy supply via oxidative phosphorylation or glycolysis (Kirshner and Smith, 1966; Matthews, 1975; Viveros, 1975). Also, ATP levels drop during secretory activity (Matt *et al.*, 1978; Lagunoff and Chi, 1980). Furthermore, ATP has been found to stimulate or to be obligatory for secretion from permeabilized cells (Baker and Knight, 1981, 1984b; Knight and Baker, 1982, and many papers since then), although evidence from isolated egg cortices was less consistent or contradictory (Baker and Whitaker, 1978; Schön and Decker, 1981; Whitaker and Baker, 1983).

A requirement for ATP was also expected from the "chemiosmotic hypothesis of secretion" proposed by Pollard's group (see Section XIV). Finally, this aspect was fundamental to an early hypothesis on biological membrane fusion by local ATP hydrolysis (Poste and Allison, 1973).

Clearly ATP is required for intracellular transport of secretory organelles (see Section V); it might also be required for a rearrangement of microfilamentous elements on the cell periphery during secretion (see Sections V, VI, and X). Most importantly, protein phosphorylation, another phenomenon generally occurring during secretory activity (see Section X), clearly requires ATP. However, protein phosphorylation is not yet unequivocally proven to be a final step in membrane fusion regulation (Vilmart-Seuwen *et al.*, 1986), as critically reviewed by Baker (1988) and

Knight *et al.* (1988). It may be, therefore, that ATP is required for some steps preceding or accompanying exocytosis proper. These might include (a) phosphorylation of a protein that is finally dephosphorylated to become fusogenic (Plattner *et al.*, 1988); (b) phosphorylation of receptors, of Ca^{2+} or other ion channels and pumps (Shenolikar, 1988); (c) phosphorylation of some vesicle-binding proteins (Michener *et al.*, 1986; Petrucci and Morrow, 1987) or of cytoskeletal components [binding of 7 of 22 chromobindins to secretory granules requires ATP in addition to Ca^{2+} (Creutz *et al.*, 1983)]; (d) autophosphorylation of kinases (Schworer *et al.*, 1988) or phosphatases (Hashimoto *et al.*, 1988) of possible relevance for step (a); (e) GTP levels required for other fusogenic mechanisms to be maintained via ATP (see later); (f) rephosphorylation of some metabolites from PI turnover. Problems related to PI turnover and protein phosphorylation will be analyzed in more detail in Sections IX and X, respectively.

One should also not overlook that exocytosis in leaky cells does not comprise solely membrane fusion, but in addition membrane internalization (Baker and Knight, 1981, 1984b), as it does in intact cells. In contrast, egg cells permanently integrate their cortical granule membranes into the cell membrane (Epel and Vacquier, 1978; Nishioka *et al.*, 1987). Also in the isolated cell cortex of *Paramecium* no exocytosis-coupled endocytosis occurs (Vilmart-Seuwen *et al.*, 1986; Plattner *et al.*, 1988). Interestingly, these two systems can perform exocytosis without any ATP supply (for egg cells, see Sasaki, 1984; Jackson and Crabb, 1988, and some data contained in the work of Whitaker and Baker, 1983; for *Paramecium* cortices, see Vilmart-Seuwen *et al.*, 1986). In addition, these systems also perform no transcellular transport of secretory organelles to the cell membrane. A recombined *in vitro* system consisting of isolated cortical granules and of cell membrane fragments from sea urchin eggs also carries out membrane fusion in the absence of ATP (Crabb and Jackson, 1985). Another example operating without ATP is the permeabilized PC12 cell (chromaffin cell-derived rat cell line), which also displays only docked but no free chromaffin granules in the cytoplasm (Ahnert-Hilger *et al.*, 1985; Ahnert-Hilger and Gratzl, 1987; Matthies *et al.*, 1988).

As far as membrane fusion per se is concerned, my group has proposed that ATP might keep fusogenic sites in a primed (reactive) state, while it inhibits fusion. This negative modulation of membrane fusion by ATP became evident with *Paramecium* cells, when ATP or nonhydrolyzable analogs, as well as γ-thio-ATP (γ-S-ATP) were either microinjected or applied to isolated cortices (Vilmart-Seuwen *et al.*, 1986). Thiophosphorylation by γ-S-ATP also inhibits exocytosis with chromaffin cells (Brooks *et al.*, 1984). With mast cells Howell and Gomperts (1987) discuss "the possibility that . . . the requirement for nucleotide triphosphates in the exocytotic process concerns a reaction other than a phosphorylation in which it might be expected that ATP would provide the best level of support," which is not the case.

Data obtained with *Paramecium* cells would be in line with the observation that a single phosphoprotein (M_r 63,000–65,000 in one-dimensional gel electrophoresis) is dephosphorylated during exocytosis (Gilligan and Satir, 1982; Zieseniss and Plattner, 1985). If maintenance of phosphorylation were the reason for the inhibitory effect of ATP on exocytosis (Vilmart-Seuwen *et al.*, 1986), this would imply a rapid turnover of phosphorylation sites—an aspect currently being analyzed in our laboratory.

What other role could ATP have in membrane fusion, and is there any at all? Baker (1988) leaves these questions open (as well as the significance of protein phosphorylation; see also Section X). In fact, data are accumulating from reanalysis of different systems that ATP is not mandatory for exocytotic membrane fusion (mast cells: Gomperts *et al.*, 1987; Howell *et al.*, 1987; Howell and Gomperts, 1987; neutrophilic granulocytes: Smolen *et al.*, 1986; platelets: Ruggiero *et al.*, 1985; neurosecretory nerve endings: Nordmann *et al.*, 1987; chromaffin cells: Ahnert-Hilger *et al.*, 1985; Matthies *et al.*, 1988). Gomperts and colleagues determined the secretory rate applying Ca^{2+} and GTP (or analogs), in conjunction with different levels of ATP, to permeabilized cells. In their kinetic analyses under nonequilibrium conditions, ATP sensitizes the system for Ca^{2+} and GTP lowers the pCa required for exocytosis in permeabilized cells (Barrowman *et al.*, 1986; Cockcroft *et al.*, 1987; Gomperts *et al.*, 1987; Howell *et al.*, 1987). Since ATP retards exocytosis, these analyses also entail the postulate of protein dephosphorylation (Tatham and Gomperts, 1989).

Hydrolysis of membrane-bound ATP has been proposed to be a fusogenic process (Poste and Allison, 1973). A Ca^{2+}-stimulated ATPase, as revealed by enzyme cytochemistry (Plattner *et al.*, 1977) at preformed exocytosis sites selectively in fusion-competent strains of *Paramecium* (Plattner *et al.*, 1980) could then be involved at the same time in controlling local Ca^{2+} concentration and/or fusion capacity. (However, it would be difficult to differentiate this reaction from de/rephosphorylation phenomena.) This assumption would be consistent with the finding that a competing phosphate substrate (unpublished observations) or a nonhydrolyzable analog also inhibits exocytosis (Vilmart-Seuwen *et al.*, 1986).

During synchronous (1 second) exocytosis of trichocyst in *Paramecium* cells, ATP consumption (frequently observed also in other systems and interpreted so far as an indication of an ATP requirement) was clearly seen to occur *after* membrane fusion (Vilmart-Seuwen *et al.*, 1986), so that any causal connection has to be denied. Alternatively, if ATP were required for the actual fusion process, it would have to be an immeasurably small pool. It has to be stressed how difficult it would be to obtain any correlation between ATP consumption and exocytosis in nonsynchronous systems (most secretory cells) or in systems with only a small percentage of vesicles released synchronously (nerve endings), since the cytoplasmic ATP pools are rapidly replenished.

Consumption of ATP was postulated also simply because exocytosis is a dynamic process. Yet any energy requirement for membrane fusion might be covered by non-energy-requiring modification of fusogenic proteins (e.g., via dephosphorylation, see Section X) or lipids (Section XI). The high curvature of fusing membranes (Section XV) might be equally important as another energy-conserving mechanism, so that a direct ATP requirement for membrane fusion cannot be postulated a priori.

It appears worth mentioning that the hitherto-unquestioned judgment on an ATP requirement for intracellular organelle fusions has also changed. Some discuss it also in terms of ATP → GTP regeneration (Melançon *et al.,* 1987; for a possible contribution of GTP to exocytosis regulation, see subsequent section). Others find ATP dependency only with *in vitro* systems when the components had been exposed to EGTA (Diaz *et al.,* 1988).

VIII. Interference of GTP and G-Proteins with Membrane Fusion

There are three possible levels on which GTP could interact with membrane fusion regulation (see Fig. 1): (a) by mobilizing Ca^{2+} from intracellular stores; (b) by activating adenylate cyclase (Gilman, 1987) or phospholipase C (PLC) on the cell membrane level, followed by the cascade of events (including PI turnover, as described in Section IX); and (c) possibly by facilitating membrane fusion in a rather direct way. Nonhydrolyzable GTP analogs [γ-thio-GTP (γ-S-GTP) or sometimes 5'-guanosylimidodiphosphate (GIDP)] can substitute for GTP in (b) but not in (a), whereas for activation by mode (c) contradictory data are reported. These different pathways are also expressed to a different extent depending on the system analyzed. Mode (a) operates without a known G protein (see Note Added in Proof), in contrast to mode (b); evidence for a G-protein involvement in mode (c) is only indirect so far.

A. Ca^{2+} Mobilization from Cellular Stores

Mode (a) has been summarized in depth by Gill *et al.* (1988). GTP regulates Ca^{2+} uptake into internal storage compartments and its subsequent release. The releasing capacity of GTP may surpass that of inositol 1,4,5-triphosphate (IP_3). Not only in regulated (neuroblastoma, parotids), but also in constitutive secretory systems (fibroblasts), GTP can release stored Ca^{2+}. Gill *et al.* (1988) also assume that GTP, by an unknown mechanism, would cause Ca^{2+} (which enters the cell from the outside medium) to be transferred to a subplasmalemmal compartment. Such a compartment has been postulated by Putney (1986), but it has not yet been localized by structural methods. Nonhydrolyzable analogs are without effect (Gill *et al.*, 1986; Henne and Söling, 1986; Nicchitta *et al.*, 1987). The pCa_i increase caused by GTP-mediated store leakage, like that due to IP_3, could account for sustained secretory activity.

B. Activation of Adenylate Cyclase or PLC

Mode (b) of GTP effects operates on the cell membrane level. G proteins couple an activated receptor to adenylate cyclase or to PLC (Stryer and Bourne, 1986; Gilman, 1987; Iyengar and Birnbaumer, 1987) and thus mediate second-messenger formation (cAMP, PI turnover products; see Section IX). Again the relevance for membrane fusion is probably only indirect, since none of the lipids or proteins formed or modified by second messengers are currently known to be fusogenic *in vivo*. (For lipids, see also Section XI.) As reviewed by Stryer and Bourne (1986), Iyengar and Birnbaumer (1987), and Allende (1988), G proteins are composed of α, β, and γ subunits. There exist stimulative and inhibitory (α_s, α_i) and α_0 subunits. G_s or G_i proteins are stimulated by ADP ribosylation using nicotinamide dinucleotide (NAD) as a substrate. Cholera toxin (CTX) and pertussis toxin (PTX) are experimentally used as catalysts for the activation of α_s or α_i subunits, respectively. When different receptors are activated on the cell membrane, G_s or G_i proteins (frequently of different M_r values in different cells) may be activated; concomitantly the toxins can be tested for their effects on G proteins and on the secretory response, when they are introduced into a cell (Allende, 1988).

Special types of G proteins have been assumed for some secretory cells, like G_n for neutrophils (sensitive to PTX, coupling to PLC), (Boroch *et al.*, 1988). One now differentiates among a variety of G proteins, like G_p, G_c, G_o, and G_i (Lo and Hughes, 1987; Lochrie and Simon, 1988; Neer and Clapham, 1988). They are all coupled to different receptors, they may be cell-specific (G_o occurring in nerve cells: Sternweis and Robishaw, 1984), and their toxin sensitivity is different: G_c responds to CTX, G_o and G_i to PTX, while G_p is insensitive.

G Protein activation can be fast: Within 30–100 msec they couple a muscarinic receptor to a potassium channel without any evident second messenger, at least in cardiac muscle cells (Pfaffinger *et al.*, 1985). In endocrine cells receptors are coupled via G_i-type proteins to voltage-dependent Ca^{2+} channels (Rosenthal et al., 1988), and γ-S-GTP or PTX affect channel inactivation by Ca^{2+} channel blockers (Scott and Dolphin, 1987). Inversely purified G proteins, microinjected during electrical recording, rapidly modulate Ca^{2+} channel activity (Hescheler et al., 1986). Such fast GTP-mediated responses could therefore also account for some regulation phenomena during exocytosis.

Any participation of the GTP-binding ras subunit (M_r of 21,000) of G proteins (Allende, 1988) in exocytosis regulation is still under discussion. This is denied by Adam-Vizi et al. (1987), but assumed by Burgoyne (1987) and supported by the observation of mast cell degranulation via ras protein microinjection, according to preliminary data discussed by Bar-Sagi et al. (1987). (See Note Added in Proof.)

There is also evidence that G proteins are at least in part associated with the cell membrane. This has been found by immunohistochemistry (with Ab against the α_o subunit) in the central nervous system, anterior pituitary lobe, adrenal medulla, and pancreatic islets, but not in pancreas acini (Asano et al., 1988). According to cell fractionation and biochemical analysis, some G-protein subunits showed association in part with the cytosol and only in part with the cell membrane, from which some components are easily detached (Boroch et al., 1988). Detachment of α_o, for example, occurs with nonhydrolyzable GTP analogs (McArdle et al., 1988). ^{32}P-GTP labeling of cell membranes also proved the binding of different molecular weight-type α subunits of G proteins, where they are accessible to ADP ribosylation (Schäfer et al., 1988).

G proteins are not bound to specific or azurophilic secretory granules in neutrophils, according to Boroch et al. (1988); in contrast to this, Rotrosen et al. (1988) found PTX-mediated ADP-ribosylation sites not only on the cell membrane (60%) but also on specific granules (35%)—but again none on azurophilic vesicles in the resting state. Stimulation reduced specific granule labeling in these experiments. A similar discrepancy exists between PTX-mediated ADP ribosylation of synaptic vesicles (negative: Toutant et al., 1987b) versus chromaffin granules (positive: Toutant et al., 1987a).

The impression prevails that at least some components of G proteins are loosely or reversibly attached to the cell membrane (Lo and Hughes, 1987; Lochrie and Simon, 1988; Neer and Clapham, 1988), where they could therefore interact in the docking process (see Section VI) and in the generation of second messengers (see also Section IX), or participate in fusogenic steps (see the following).

On the cell membrane level, adenylate cyclase activation represents the first signal transduction system detected for G proteins (Stryer and Bourne, 1986; Iyengar and Birnbaumer, 1987). It can affect exocytosis, for example, by cAMP-dependent protein kinases (relevant for ion channels and cytoskeletal elements). Since CaM is thought to interfere also with exocytosis (Sections VI and X), it should be mentioned that CaM stimulation of cyclic nucleotide phosphodiesterase is also thought to be controlled by G_i and G_o proteins (Asano *et al.*, 1986). Effects of PI turnover products generated on the cell membrane level by activation via *G* proteins are severalfold (Barnes, 1986; Merritt *et al.*, 1986a,b; Taylor *et al.*, 1986; Lo and Hughes, 1987; Altman, 1988; Casey and Gilman, 1988), as outlined in Section IX.

G-Protein activation by GTP binding to the α subunit is terminated via GTP hydrolysis; the α subunit displays GTPase activity (see, e.g., Casey and Gilman, 1988). This is the reason why application of γ-S-GTP or GIDP to permeabilized cells or by microinjection can cause long-lasting stimulation of secretory processes. Such effects have been obtained with various regulated exocytotic systems, excitable and nonexcitable, such as egg cells (Turner *et al.*, 1986; Cran *et al.*, 1988), mast cells (Fernandez *et al.*, 1987; Penner *et al.*, 1988), blood platelets (Baldassare and Fisher, 1986a,b), exocrine pancreas (Merritt *et al.*, 1986a,b), and pancreatic islet-derived cells (Vallar *et al.*, 1987). For this reason and because of results obtained with toxins (see later), a participation of G proteins in exocytosis can be reasonably assumed for a variety of systems, although the specific G proteins involved have not always been sufficiently identified.

With chromaffin cells different responses to GTP analogs were observed: Bittner *et al.* (1986) report induction of Ca^{2+}-independent secretion, which is opposed to previous findings by Knight and Baker (1985). Although this is one of the most thoroughly analyzed system (as reviewed in Winkler, 1988), it is currently difficult to reconcile such discrepancies. In cells derived from pancreatic islet β cells, GIDP also may remain inactive, whereas γ-S-GTP can fully stimulate PI turnover and simultaneously render the system independent of a pCa_i increase (Vallar *et al.*, 1987), as it does with the ACTH-producing AtT-20 cell line derived from a pituitary tumor (Luini and DeMatteis, 1988) and with HL60 cells (Stutchfield and Cockcroft, 1988). (See Note Added in Proof.)

Effects of toxins on different secretory systems are, of course, quite variable, since they all depend on the actual cell type and the kind of trigger used. With chromaffin cells PTX stimulates ADP ribosylation and delayed Ca_o^{2+}-dependent secretion; over longer time periods Ca_o^{2+} is no longer required (Brocklehurst and Pollard, 1988b). Microinjected CTX triggers the cortical granule reaction in sea urchin eggs (Turner *et al.*, 1987). The same procedure with hamster egg cells did not cause activation, however (Miyazaki, 1988). A toxin-insensitive PI turnover and Ca^{2+} release were assumed in this case, since a Ca^{2+} transient occurred not only with GTP (which could directly mobilize internal stores) but also with γ-S-GTP (unable to deplete intracellular stores; see earlier), even in much lower concentrations. With pancreatic acinar cells CTX (but not PTX) affects γ-S-GTP stimulation; both agents stimulate ADP ribosylation of G proteins of different molecular weights, which is reduced by cholecystokinin (Schnefel et al., 1988); carbachol effects were different. All this documents complicated interactions on the receptor–ligand and second-messenger-forming level.

Some G proteins are sensitive to botulinum toxin (BTX), which also causes ADP ribosylation (Matsuoka et al., 1987). Botulinum toxin inhibits exocytosis in adrenal chromaffin cells (Knight et al., 1985). Although this parallels the toxin effects on exocytosis, Adam-Vizi et al. (1988) deny any causal connection, because application of purified BTX components, together with their Ab, may stimulate or inhibit independently of each other. Curiously, BTX is reported to interfere with (sustained?) secretion by another mechanism not previously known, namely ADP ribosylation of actin (Matter et al., 1989).

Tetanus toxin (TTX) not only abolishes neurotransmission but also blocks exocytosis in chromaffin cell cultures following microinjection (Penner et al., 1986). With permeabilized chromaffin cells this effect has been shown to be due exclusively to the light chain of TTX (Ahnert-Hilger et al., 1989). It is not known whether TTX acts closely on the membrane fusion site; studies (e.g., with PC12 cells) would be interesting, because in normal gland cells TTX might also affect transcellular transport, a possibility reviewed by Bourne (1988).

C. Direct Facilitation of Membrane Fusion

A third line of GTP effects has also been recognized. In neutrophils and mast cells, GTP analog application to the cytosolic compartment induces exocytosis in a direct mode,—that is, without an increase of $pCa_i > 8$ by sensitizing the system for Ca^{2+} (Barrowman et al., 1986; Gomperts et al., 1987). Thereby, the effect achieved with GTP analogs increases, as ATP is added (which, however, is not mandatory); ATP analogs cannot be used (Cockcroft et al., 1987; Howell et al., 1987). The authors conclude that there exists a new species of G proteins (G_E), which would be directly relevant for exocytotic membrane fusion, perhaps by a preceding phosphorylation and subsequent dephosphorylation step. Thus, ATP would indirectly be a negative modulator of membrane fusion (Howell and Gomperts, 1987; see also Section VII). These data are corroborated by current analyses under conditions of nonequilibrium kinetics by the Gomperts group. These show the dependence of the rate of exocytosis on the ratio of nucleotides (ATP versus GTP) applied. Lohse *et al.* (1988) have argued against this concept, since the alleged ATP effect on mast cells occurs also with adenosine in their experiments, whereas the work cited before has obtained negative results with adenosine compounds other than ATP. [Yet the effects achieved by Lohse *et al.*, (1988) also appear much smaller.] Surprisingly, Rink and Knight (1988) indicate that exocytosis can be triggered, without agonist application, also by guanosine nucleotides other than GTP or its analogs, which is opposite to the findings by the other authors cited.

A stimulative effect of γ-S-GTP has been monitored also by patch–clamp analyses (Fernandez *et al.*, 1987). No pCa_i increase above resting levels was required in such experiments with mast cells. These effects of GTP or analogs on exocytosis and probably on membrane fusion proper are likely due to a "leftward shift" of the Ca^{2+} sensitivity curve by GTP—that is, sensitization to Ca^{2+} (Knight, 1987; see also Neher, 1988). This occurs also with *Paramecium* cells: Application of GTP or analogs to isolated cortices causes the system to respond to pCa_i levels close to values generally assumed for resting conditions, and microinjected γ-S-GTP also induces trichocyst release from intact cells (Lumpert *et al.*, 1989). In sum, GTP probably participates in the regulation of membrane fusion during exocytosis, possibly by interaction with G proteins, although G proteins specifically dedicated to this phenomenon still would have to be identified. ["Classical" G-protein types influence exocytosis as a whole by regulating different steps preceding membrane fusion. This function is also assumed to be superimposed to G_E-protein-mediated effects (Gomperts *et al.*, 1986).] (See Note Added in Proof.)

The possibility of membrane fusion with resting pCa_i levels in the presence of GTP might also have some importance for the regulation of constitutive exocytosis, although little is known about this at present. Cockcroft and Stutchfield (1988) include in their survey also a few data on γ-S-GTP stimulation of constitutively secreting systems.

The occurrence of G proteins in lower eukaryotes is established (e.g., for *Dictyostelium:* Lochrie and Simon, 1988). In ciliated and other protozoans their existence may be postulated from the response to guanine nucleotides although this has not yet been ascertained (Janssens, 1988).

G proteins postulated to be of potential relevance for secretory organelle docking at the cell membrane in yeast cells (Goud *et al.*, 1988) have already been mentioned in Sections III and VI.

IX. Are PI or Other Lipid Turnover Products Relevant for Membrane Fusion?

Michell (1975; Michell *et al.*, 1981; Michell and Putney, 1987) has generalized that hydrolysis of PI (polyphosphatidylinositol 4,5-bisphosphate, PtdIns4,5P$_2$) parallels elicited secretory activity. This has then been substantiated for a variety of cell types, including electrically nonexcitable and excitable cells: mast cells (Ishizuka and Nozawa, 1983; Dainaka *et al.*, 1986; Pribluda and Metzger, 1987), oocytes (Kamel *et al.*, 1985; Swilem *et al.*, 1986; Cobbold *et al.*, 1987), basophilic leukocytes (Beaven *et al.*, 1984a), thrombocytes (Lapetina *et al.*, 1985), pancreatic acinar cells (Powers *et al.*, 1985), adrenal chromaffin cells (Whitaker, 1985), as well as other excitable cells; see also detailed references later. Different stimuli may entail PI turnover in the same cell.

Metabolism of PI follows the scheme briefly outlined in Fig. 1. Hydrolysis of PI occurs on the cell membrane level by PLC (PtdIns4,5bisphosphatase) via activation of G proteins (see Section VIII). Diacylglycerol (DAG) and IP$_3$ are generated. This scenario, together with the subsequent steps, has been repeatedly summarized (Berridge, 1984, 1987a,b, 1988; Kikkawa and Nishizuka, 1986; Michell and Putney, 1987; Altman, 1988; Irvine, 1988). One mainly considers the release of Ca^{2+} from storage organelles as the function of IP$_3$ (Streb *et al.*, 1983; Gill *et al.*, 1986, 1988), whereas DAG activates PKC (phospholipid-dependent protein kinase) (Nishizuka, 1984; Kikkawa and Nishizuka, 1986). Protein kinase C then phosphorylates a variety of proteins in response to a secretory stimulus (see Section X). Inositol 1,4,5-trisphosphate is phosphorylated to inositol 1,3,4,5-tetrakisphosphate (IP$_4$), which is thought to affect the influx of extracellular Ca^{2+} (Morris *et al.*, 1987) and to fill up a subplasmalemmal store (Berridge, 1988). Gill *et al.* (1988) and Putney (1988) assume that IP$_4$ would also regulate Ca^{2+} uptake into IP$_3$-releasable pools (which are in addition under GTP control; see Section VIII). Cooperative effects between IP$_4$ and IP$_3$ are reported also for sea urchin eggs (Irvine and Moore, 1987).

A DAG-lipase splits off arachidonic acid (AA) from DAG. This is a Ca^{2+}- and CaM-independent plasmalemmal enzyme in chromaffin cells (Rindlisbacher *et al.*, 1987). Interestingly, both these lipids perturb artificial membranes and AA is fusogenic when applied to *in vitro* systems, although effects *in vivo* might be different (Section XI). A possible control of membrane fusion by PI turnover products now looks much more complex; they might essentially regulate different branches of Ca^{2+} signaling. This may be effectuated either by the mechanisms just mentioned for IP$_3$, or—as discussed later—DAG may sensitize the system to resting pCa$_i$

levels (thereby cooperating with a GTP-requiring mechanism; Section VIII). Not only PKC requires Ca^{2+} (unless activated by DAG), but also PLC (Kikkawa and Nishizuka, 1986).

Therefore, there is some difference in the relevance of some PI turnover products for (a) electrically "excitable" (neurons, chromaffin cells, pancreatic islet β cells) and (b) "nonexcitable" cells (blood platelets, neutrophilic leukocytes, mast cells), as emphasized by Penner and Neher (1988). In systems operating primarily by channel-mediated Ca^{2+} influx, rather than by Ca_i^{2+} release, IP_3 is less important. Diacylglycerol may still stimulate Ca^{2+} influx by PKC-mediated channel activation (see later), although PI turnover may not always be (as in neuromuscular junctions) a primary event. The same holds for some nonexcitable systems when they are triggered without receptor activation.

Mast cells are an interesting example to illustrate the widely varying and conflicting results on the relevance of PI turnover for exocytosis. Some assume that exocytosis induction by the polyamino compound 48/80 occurs by a diffuse increase of Ca^{2+} conductivity, as found with artificial membranes (Katsu et al., 1983). In the mast cell, Penner et al. (1988) observed leak currents possibly accounting for Ca^{2+} entry by nonspecific Ca^{2+} channels. This is supported by the finding that quite different polyamino compounds activate mast cells (Foreman and Lichtenstein, 1980). In contrast, IgE activates clearly defined receptors and clearly causes PI turnover (Ishizuka and Nozawa, 1983; Pribluda and Metzger, 1987), whereas some workers using 48/80 find little effect (Okano et al., 1985; Boam et al., 1988). Concomitantly, only IgE (but not 48/80) triggering is potentiated by DAG-like compounds (see later) (Boam et al., 1988). Intriguingly, when a tridecamer of 48/80, a more potent trigger agent than regular 48/80, was applied, precise time-sequence studies clearly showed PI turnover to precede histamine release (Dainaka et al., 1986). Penner et al. (1988) believe that IP_3 induces a Ca^{2+} influx during 48/80-induced mast cell degranulation; Gomperts et al. (1986), however, assume this process to be independent of Ca_o^{2+}.

With Paramecium cells we have found no evidence thus far for any extensive turnover of PI (although this lipid species is present in the usual concentrations: Kaneshiro, 1987) in response to polyaminoethyl dextran (AED) as a stimulant (Knoll et al., 1989). This supports our previous results obtained with different inhibitory or stimulatory compounds (Lumpert et al., 1987). Trichocyst exocytosis inhibition by exogenous neomycin (Zieseniss and Plattner, 1985) probably operates by a mechanism not related to PI turnover (see later). It would not penetrate into intact Paramecium cells, but it reduced the influx of Ca_o^{2+} through the cell membrane (Gustin and Hennessey, 1988). Paramecium is an excitable (and exciting) cell, but voltage-dependent Ca^{2+} channels are located in ciliary membranes (Kung and Saimi, 1982; Machemer, 1988) and are not relevant for the Ca^{2+} influx occurring during exocytosis (Plattner et al., 1985). Although the kinds of channels involved in AED triggering are not known, AED might act similarly to other polyamino compounds in mast cells.

In conclusion, all this would agree with the general assumption that PI turnover is generally initiated by receptor activation and thus serves for signal transduction in some cells (Nishizuka, 1984; Kikkawa and Nishizuka, 1986; Berridge, 1987a, b; Michell and Putney, 1987; Altman, 1988). Phosphatidylinositol turnover products may account for some of the exocytosis-regulating effects occurring not only with some nonexcitable but also with some excitable cells. In the first case the effects of DAG may prevail, in the second, those of IP_3. However, PI turnover may not be so significant with some nonexcitable cells in response to certain stimuli (see also later).

Neomycin, a nonpenetrable aminoglycoside antibiotic, is sometimes introduced into cells to inhibit PI turnover (Schacht, 1976). Its inhibitory effect is judged differently. As already mentioned, effects achieved with extracellular neomycin are due to depressing of Ca^{2+} entry from the outside medium into the cells (Gustin and Hennessey, 1988; Salzberg and Obaid, 1988). As far as intracellular effects of neomycin are concerned, Prentki et al. (1986) assumes it not to be specifically bound to inositol-containing phospholipids, but to bind weakly to a variety of anions such as IP_3 or ATP, and then to inhibit Ca^{2+} efflux from intracellular stores. Results from Swann and Whitaker (1986a,b) are in opposition: Microinjected neomycin, though reducing IP_3 formation, does not prevent Ca_i^{2+} mobilization when IP_3 is microinjected. In addition, neomycin can interact on the level of G proteins (Schäfer et al., 1988). Because of this uncertainty, neomycin use is limited in PI turnover studies. Furthermore, concentrations used in most work are orders of magnitude higher (millimolar) than required, for example, to stop trichocyst release from Paramecium cells (where PI turnover is not known to occur; see earlier), and neomycin is sometimes added to living cells where its effect is totally different from intracellular.

To analyze the relevance of PI turnover for exocytosis regulation, the usual approach is first to identify the PI hydrolysis products formed and then to mimic their effects by application to permeabilized systems or via microinjection.

As mentioned before, DAG activates PKC (Nishizuka, 1984; Kikkawa and Nishizuka, 1986). As discussed in Section X, there is no protein known to become fusogenic by phosphorylation, although many proteins become phosphorylated during elicited secretory activity. Any PKC effects might therefore have only some indirect bearing on membrane fusion regulation. Since PKC is stimulated by DAG and Ca^{2+}, it thus can activate, for example, the Ca^{2+} pumps (Kikkawa and Nishizuka, 1986) as well as Ca^{2+} and other ion channels by phosphorylation (Kaczmarek, 1987; Berridge, 1987a,b). Since DAG increases the sensitivity of PKC to Ca^{2+} and at the same time activates different ion channels following receptor activation, one might expect different PI activation patterns (e.g., with chromaffin cells, depending on whether they are stimulated by depolarization or via receptor-activated Ca^{2+} channels). In reality the situation with chromaffin as with nerve cells is not so clear-cut (see later). According to Oishi et al. (1988), the process of PKC activation is much more complex: It is activated not only by DAG and Ca^{2+} (eventually released by IP_3) and free fatty acids, but also by lysophospholipids [produced by Ca^{2+}-dependent phospholipase A_2 (PLA_2): Nakamura and Ui, 1985]. Considering, then, that L-type Ca^{2+} channels (for example), which account for an early Ca^{2+} influx in some excitable cells, are phosphorylated also by cAMP- or Ca^{2+}/ CaM-dependent kinases (Campell et al., 1988; Glossmann and Striessnig, 1988), which in turn depend on other regulatory cascades (see following section), one realizes the complexity of the problem. Annexins (Section VI) represent other examples of diverse phosphorylation phenomena that only in part depend on PKC activation by PI turnover (Burgoyne and Geisow, 1989). Such multiple feedback mechanisms may also explain some controversial results and deviations from system to system. Only some of them are considered in Fig. 1.

With neurons exocytosis is very fast (Section XIII), and regulation by PI turnover appears particularly intriguing. Among the voltage-operated Ca^{2+} channels, some are modulated by cAMP and PKC, but some do not respond to DAG (Meldolesi and Pozzan, 1987); see later. The response is fast in nerve terminals: Channels are open for several milliseconds (Salzberg and Obaid, 1988) and neurotransmitter release by exocytosis also lasts only 1–2 msec (Section XIII). This process might therefore be regulated directly by Ca^{2+} influx. But activation of PKC still causes some increase of miniature end plate potential frequency (Shapira et al., 1987; Light et al., 1988). Similarly, neurotransmitter release is enhanced by PKC stimulation also in brain neurons (Kaczmarek, 1987; Nichols et al., 1987) and in neurohypophyseal cells (Cazalis et al., 1987). Apart from channel activation by phosphorylation, phosphorylation of some CaBP is of importance (Section X). The interference of synapsin I at active zones of nerve terminals is another aspect pertinent to exocytotic transmitter release. Yet this operates by mechanisms independent of PI turnover (Section X).

The effects of DAG can be mimicked by the tumor-promoting phorbol ester TPA (12-*O*-tetradecanoylphorbol 13 acetate, also called phorbol 12-myristate-13-acetate, PMA; as a result of its structural analogy with DAG, TPA also stimulates PKC (Kaibuchi *et al.*, 1982) by increasing its Ca^{2+} sensitivity to resting levels (pancreatic acinar cells: Pandol *et al.*, 1985; neutrophils: DiVirgilio *et al.*, 1984; pancreatic islet cells: Jones *et al.*, 1985; chromaffin cells: Knight and Baker, 1983; Brocklehurst and Pollard, 1985; Pocotte *et al.*, 1985; see also Nishizuka, 1984, for an early proposal of this hypothesis). This effect is achieved with excitable and nonexcitable cells. Activation occurs even with pCa levels considerably below those actually required to activate Ca^{2+}-regulated proteins.

While TPA stimulates exocytosis in adrenal chromaffin cell cultures, increasing the influx of Ca^{2+} into cells via the ionophore ionomycin still enhances the secretory response in cultured adrenal medullary cells (Pocotte *et al.*, 1985). As also shown with chromaffin cells, TPA modulates secretion in response to carbachol stimulation (Morita *et al.*, 1985; Wakade *et al.*, 1986), in contrast to splanchnic nerve stimulation (Wakade *et al.*, 1986). Both groups obtained opposite results with stimulation via depolarization by high-K^+ media. All these data suggest some modulatory, but not mandatory effect of DAG on exocytosis regulation. As mentioned, it would be particularly interesting to know whether PI turnover is different during activation via membrane potential depolarization or via receptor activation.

There is evidence for a TPA-mediated rearrangement of PKC from an originally cytosolic to a particle-bound localization in pancreatic acinar cells (Wooten and Wrenn, 1984), blood platelets (Molski *et al.*, 1988), and chromaffin cells (Holz and Senter, 1988) upon exocytosis stimulation. Much like nicotinic-receptor activation or high-K^+ depolarization, TPA also causes, within seconds, a redistribution of PKC from an originally cytosolic to a membrane-bound form; the percentage of PKC rearrangement strictly parallels secretion stimulation (Terbush *et al.*, 1988). However, as mentioned, secretion can go on in these systems also without a TPA effect, as demonstrated by Holz and Senter (1988) as well as by Terbush *et al.* (1988). Whereas the TPA effect is generally assumed to be independent of Ca_i^{2+} (or of its increase), this is not so with AtT-20 cells (Luini and DeMatteis, 1988). Similarly, TPA cannot substitute for the γ-S-GTP effect with HL50 cells (Stutchfield and Cockcroft, 1988).

In permeabilized blood platelets DAG also lowers the threshold pCa required for activation (Knight and Scrutton, 1984), yet PI turnover (and protein phosphorylation) is not a sufficient or mandatory step for exocytosis to occur: Even though PI is hydrolyzed when permeabilized platelets are supplemented with Ca^{2+}, they are not activated (Lapetina *et al.*, 1985); serotonin is released only when thrombin is added to such samples. In the mast cell, TPA slows down secretion and PI turnover that otherwise may accompany (see earlier), to a certain extent, secretion induced by compound 48/80 (Okano et al., 1985).

Similarly to PKC (see earlier), PI-specific PLC becomes attached to chromaffin granule membranes with rising pCa_i levels (25% effect with $pCa = 6$), so that PLC has been classified as a chromobindin-type protein (Creutz et al., 1985; see also Section VI).

In neuromuscular junctions (e.g., of the frog), DAG or TPA effects are not so obvious: TPA has no or only little effect on stimulated acetylcholine release (Caratsch et al., 1988); the slight stimulatory effect observed was inhibitable with compound H7, which, however, does not inhibit PKC (Knight et al., 1988) as previously assumed (Shimomura et al., 1988). In opposition to these results, activation by microinjection of PKC or by phorbol ester treatment was achieved with *Aplysia* neurons (DeRiemer et al., 1985). In this case, the effect is due to synapsin I phosphorylation, which renders synaptic vesicles accessible to the presynaptic membrane (see Section X).

As far as IP_3 is concerned, it is thought to open functional Ca^{2+} channels in intracellular storage compartments. With permeabilized basophilic granulocytes three IP_3 molcules suffice; the lag time is <4 seconds, a tenth of the time required for fully depleting Ca^{2+} stores (Meyer et al., 1988). In the cells analyzed so far by analytical EM methods, this IP_3-releasable store is represented by calciosomes, single membrane-bound organelles, 50–250 nm in diameter, with a Ca^{2+}-binding calsequestrinlike protein in its core and a Ca^{2+}-pumping ATPase in its membrane (Meldolesi et al., 1988b; Pozzan et al., 1988). The authors have localized the calsequestrin-like protein by immunocytochemistry and Ca^{2+} sequestration by oxalate precipitation methods. Several CaBP occur, however, also in the rough endoplasmic reticulum (ER) (Macer and Koch, 1988). This has previously been considered, on the basis of x-ray microanalysis (though with restricted resolution), as the main store responsible for a fine-tuning of pCa_i under physiological conditions (Somlyo et al., 1985). Yet only calciosomes have been shown to release Ca^{2+} in response to IP_3 (Pozzan et al., 1988). A calsequestrinlike protein has also been identified with sea urchin eggs (Oberdorf et al., 1988). Other investigations on this subject postulate, on the basis of selective depletion and replenishment according to biochemical findings, the occurrence of different Ca^{2+} pools (Gill et al., 1988; see also later), yet without demonstrating them by methods that proved sufficiently useful to develop the calciosome concept.

The role of IP_3 has been thoroughly analyzed in egg cells, for instance. Phosphatidylinositol turnover with IP_3 formation takes place during the normal fertilization reaction (Kamel et al., 1985). Injection of IP_3 into

Xenopus eggs in the presence of a Ca^{2+} fluorochrome revealed a pCa_i increase close to the injection site, immediately followed by cortical granule exocytosis (Busa *et al.*, 1985). Injection of IP_3 into fish or hamster eggs also releases Ca^{2+} from internal stores, also without any appreciable lag phase (Iwamatsu *et al.*, 1988; Miyazaki, 1988). Injection of IP_3 into sea urchin eggs yielded similar results (Whitaker, 1984; Swann and Whitaker, 1986a,b); in contrast to some earlier reports, this also occurs in the absence of Ca_o^{2+} (Crossley *et al.*, 1988). The Ca^{2+} required for cortical granule reaction thus could be completely covered by the IP_3-mobilizable pool. A primary role of Ca^{2+} for fusion initiation can be derived from the inhibition of frog egg cell activation by depressing the pCa_i increase with injected Ca^{2+} buffer solution (Kline, 1988). Surprisingly, isolated cortices prepared from sea urchin eggs and exposed to a sufficient pCa not only secrete cortical granule contents, but at the same time also undergo PI hydrolysis (Whitaker and Aitchison, 1985); both processes are inhibited by neomycin. Given the uncertainty and complexity of neomycin's effects (see earlier), it might not yet be justified to derive from these findings the relevance of PI turnover products directly for membrane fusion.

In egg cells PI decays within 30 seconds and is restored within 2–3 minutes (Kamel *et al.*, 1985). In the exocrine pancreas IP_3 formation culminates within 5 seconds after cholinergic or cholecystokinin stimulation (Powers *et al.*, 1985; Doughney *et al.*, 1987). This would not be sufficient for fast-acting systems, but PI hydrolysis products could regulate sustained secretory activity in slowly secreting systems.

As mentioned, in electrically excitable cells (not only neurons, but under appropriate conditions also chromaffin cells and pancreatic islet β cells) the increased pCa_i levels accompanying elicited exocytosis can be accounted for mainly by a Ca^{2+} influx from the extracellular medium via voltage-dependent, receptor-activated, or other channels, rather than to GTP- (see Section VIII) or IP_3-mediated mobilization. (For a summary of these aspects, see Meldolesi and Pozzan, 1987; Penner and Neher, 1988.) The pCa_i increase required for fast neurotransmitter release from neurons is provided by voltage-dependent Ca^{2+} channels. Nevertheless, PI turnover products affect also the function of some excitable cells, thus indicating a multifactorial control of exocytosis also in these cells. For instance, IP_3 mobilizes Ca^{2+} from internal stores also, for example, in neuroblastoma (Gill *et al.*, 1986) and pancreatic islet cells (Dunlop and Larkins, 1988). It is possible that IP_4 formed from IP_3 provides a feedback control to Ca^{2+} entry via plasmalemmal channels (Gill *et al.*, 1988) or by activating a protein phosphatase, which might be relevant for Ca^{2+} release from particulate stores in brain cells (Zwiller *et al.*, 1988). Results obtained with chromaffin cells are quite contradictory, since Swilem *et al.* (1986) detected PI breakdown yet without Ca^{2+} mobilization, whereas Cobbold *et al.* (1987) report a slight increase of pCa_i detected by aequorin. [Species differences of chromaffin cells (Winkler, 1988) and Ca^{2+} chelation by some other fluorochromes in too high a concentration (Tsien, 1988) also require general attention.] With PC12 cells Vicentini *et al.* (1985) observed IP_3 formation in response to muscarinic receptor activation, yet without Ca_i^{2+} mobilization; they tentatively assign a second-messenger function to IP_3 independently from Ca^{2+}.

In permeabilized chromaffin cells the production of arachidonic acid (AA) by PLA_2 activity (Fig. 1) strictly parallels secretion activation (Frye and Holz, 1985). Similarly, PLA_2 blockage abolishes the cortical granule reaction (Ferguson and Shen, 1984). This is interesting because of the fusogenic properties of AA *in vitro* (see Section XI), although it inhibits the cortical granule reaction in sea urchin eggs *in vivo* (Elhai and Scandella, 1983). Lipocortin in dephosphorylated form is now considered as a natural PLA_2 inhibitor (Huang *et al.*, 1986), but its physiological role in membrane fusion regulation during exocytosis is not known.

Systems secreting independently via two different secretory granules, like azurophilic and specific granules in neutrophils, might be regulated by different branches of activation chains discussed so far and in Section XIII. Thus, TPA may stimulate or inhibit (Barrowman *et al.*, 1987), or exert no effect at all (Howell and Gomperts, 1987).

It would also be important to find out to what extent any of the mechanisms discussed in this as well as in the preceding and following section might contribute to constitutive exocytosis. But for methodological reasons, this will be a rather difficult task.

From all these data obtained with the different cell types, excitable or nonexcitable ones, under different trigger conditions, one might extract the following rule: PI turnover may play an important role in membrane fusion regulation with most receptor-activated excitable or nonexcitable systems. Diacylglycerol (or TPA as its nonphysiological equivalent) might sensitize for Ca^{2+} a mechanism preceding membrane fusion. (These preceding steps, however, could also take place otherwise.) Three types of secretory systems may now be differentiated, depending on their reaction to DAG or analogs (Baker, 1988; Rink and Knight, 1988): (1) a small leftward shift of the Ca^{2+} sensitivity curve (adrenal chromaffin cells); (2) this shift is much more pronounced, for example, for serotonin release from blood platelets; (3) other secretory events even in the same cell show no reaction at all (e.g., *N*-acetylglucosaminidase release from platelets). In addition, GTP (or analogs) reduces the pCa_i requirement for exocytosis to resting levels in some cells, particularly in system type (2) (Baker, 1988; see also Section VIII). The precise role of G proteins has yet to be settled.

Considerable variability is also observed with IP_3, which, particularly in nonexcitable cells, sustains continuous secretory activity by mobilizing Ca^{2+} from intracellular pools. None of the products generated by PLC or PLA_2 activity can be assumed to interfere directly on the level of membrane fusion, although lysophospholipids, DAG, and AA have been proposed as fusogens (see Section XI), and PKC-mediated protein phosphorylation is frequently inferred to be rather directly involved in the final steps of exocytosis (see Section X). This process, however, may only be engaged in a prelude before membranes actually undergo fusion.

X. Protein Phosphorylation and Dephosphorylation

When exocytosis is triggered, proteins of widely different molecular weights become phosphorylated in widely different cell types. To give just a few examples, this occurs with mast cells (Sieghart et al., 1978; Theoharides et al., 1980) lacrimal gland (Dartt et al., 1982), parotids (Baum et al., 1981), exocrine (Burnham and Williams, 1982; Roberts and Butcher, 1983) and endocrine pancreatic cells (Brocklehurst and Hutton, 1983, 1984), adrenal medullary chromaffin cells (Burgoyne and Geisow, 1982; Brocklehurst et al., 1985), and neurons (DeLorenzo et al., 1981; Nestler and Greengard, 1984; see also detailed recent references below). Only some of these phosphoproteins (PP) have been more clearly identified.

Occasionally a dephosphorylation of some proteins has been mentioned to accompany elicited secretory activities. This again holds for widely different cell types: parotid (Baum et al., 1981; Spearman et al., 1984), exocrine pancreas (Burnham and Williams, 1982; Burnham et al., 1988), pancreatic islet cells (Jones et al., 1988), chromaffin cells (Côté et al., 1986; Trifaró and Kenigsberg, 1987), and others.

Most data have been obtained by one-dimensional sodium dodecyl sulfate–polyacrylamide gel electrophoresis (SDS–PAGE). Two-dimensional gels reveal a more complex situation: In pancreatic acini (Burnham et al., 1988) and in adrenal medullary chromaffin cells (Gutierrez et al., 1988), more proteins become dephosphorylated than one would have previously expected from one-dimensional gel analyses.

The pattern of (de)phosphorylation depends on the stimulus applied, and one also notes some inconsistency of the data obtained with different triggers or by different authors.

Figure 5 shows the ambiguity of phosphorylation studies in vivo with nonsynchronous systems. The effects achieved may be contradictory, depending on the labeling conditions and the turnover of phosphorylation sites, particularly in the case of rapidly reversed dephosphorylation. An

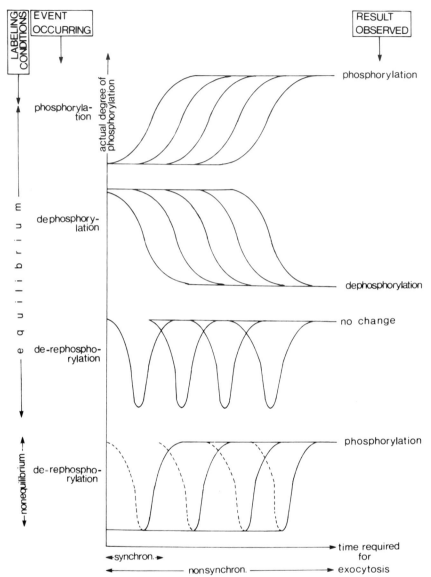

FIG. 5. Schematic time course of protein phosphorylation and dephosphorylation under equilibrium and nonequilibrium ³²P-labeling conditions (assuming a substrate being either fully phosphorylated or nonphosphorylated at the time exocytosis is triggered). The degree of phosphorylation during equilibrium labeling also depends on the turnover of phosphorylation sites. Whereas phosphorylation and dephosphorylation can easily be detected, when they are not reversed, reversible dephosphorylation would result in no net change under equilibrium-labeling conditions and in net phosphorylation under nonequilibrium conditions (when ³²P-ATP is present in a level sufficient to label selectively a protein undergoing a de-rephosphorylation cycle). The occurrence of this phenomenon has been demonstrated with *Paramecium* cells (Zieseniss and Plattner, 1985) and used for selective labeling of a phosphoprotein (PP65) that is originally dephosphorylated (<1 second) and rephosphorylated afterward (5–20 seconds) (Stecher *et al.*, 1987).

example of this is *Paramecium* cells (see later), where clearly one protein (one-dimensional SDS–PAGE) is rapidly dephosphorylated within <1 second (but only in exocytosis-competent strains: Fig. 6) and then re-phosphorylated within 5–20 seconds (depending on the strain: Zieseniss and Plattner, 1985). This phosphoprotein has a M_r of 65,000 (PP65) or 63,000 according to later determinations (Murtaugh *et al.*, 1987; Höhne *et al.*, 1989). Such a rapid de/rephosphorylation cycle would remain undetected in systems (like most secretory cells) that release their contents only

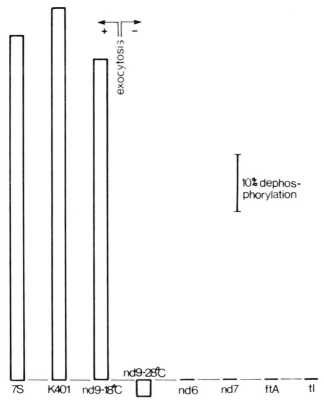

FIG. 6. Degree of dephosphorylation of the only phosphoprotein (PP65) dephosphorylated in *Paramecium* during synchronous exocytosis. Exocytosis-competent (left) and -incompetent (right) strains are widely different in PP65 dephosphorylation. Nonexocytosing strains include *nd* (trichocysts docked at the cell membrane, but not dischargeable), ftA (many free—that is, nondocked trichocysts), and *tl* (trichless, no trichocysts). The total amount of PP65 or its original degree of phosphorylation is not correlated with exocytosis capacity; this depends only on the extent of possible dephosphorylation. According to data from Zieseniss and Plattner (1985).

partially or during a time exceeding ~ 10 seconds. If in the presence of ^{32}P-labeled ATP, the turnover of phosphorylation sites at rest is slow, stimulation can result in net phosphorylation of a protein that was originally dephosphorylated. That this can occur has also been shown with *Paramecium* cells (Stecher *et al.*, 1987): When ^{32}P$_i$ uptake has been restricted to a time period that does not yet allow for any significant protein phosphorylation before synchronous exocytosis is triggered, PP65 is dephosphorylated and then it selectively incorporates ^{32}P (see Fig. 5, bottom).

Among the proteins phosphorylated in response to a secretory stimulus, there are several that have no bearing on membrane fusion, including some ribosomal proteins (Freedman and Jamieson, 1982), tubulin and Mt-associated proteins (Goldenring *et al.*, 1983), or enzymes engaged in secretory product formation, like tyrosine hydroxylase in chromaffin cells (Treiman *et al.*, 1983). Closer to the subject dealt with here is the phosphorylation of ion channels (Ewald *et al.*, 1985; Meldolesi and Pozzan, 1987) and pumps (Chan and Junger, 1983; Collins *et al.*, 1987), of enzymes engaged in second-messenger turnover, as well as of receptors in response to ligand binding (Hempstead *et al.*, 1983; White *et al.*, 1985; Shenolikar, 1988). For the wide diversity of kinases and phosphatases as well as for substrates for protein (de)phosphorylation in different cell types, see later and the preceding section, as well as the reviews by Nestler and Greengard (1984), Cohen (1988), and Shenolikar (1988).

Some of the proteins phosphorylated by exocytosis stimulation are associated with the secretory granule membrane, again in a variety of cell types (parotids: Spearman *et al.*, 1984; exocrine pancreas: Pfeiffer *et al.*, 1984; pancreatic islet β cells: Brocklehurst and Hutton, 1984; chromaffin cells: Burgoyne and Geisow, 1982; Summers and Creutz, 1985; Creutz *et al.*, 1987a; nerve: Pang *et al.*, 1988). Among them, there are also some substrates for protein kinases attached to chromaffin (Creutz *et al.*, 1987a) and neurotransmitter vesicles (see later). Some of the CaBP (see Section VI), like p34, 35, 36, 38, and p39, lipocortin- and calpactin-type proteins, as well as some of the chromobindins, are substrates for tyrosyl kinases (Hollenberg *et al.*, 1988); this is expected to hold also for calelectrin and calcimedin, all with partial sequential homologies. Soluble substrates may be rearranged within the cell, depending on the state of phosphorylation.

This holds for synapsin I. This is a protein of M_r 85,000 (type A) or 80,000 (type B) (Walker and Agoston, 1987); it is associated with all types of exocytotic vesicles containing classical neurotransmitters (Valtorta *et al.*, 1988). It is absent from adrenal medullary chromaffin tissue and PC12 cells, unless these were grown in the presence of nerve growth factor (Miller, 1985). Thus, synapsin I is a typical neuronal protein. It is also assumed to be a major constituent of active zones of nerve terminals (Smith and Augustine, 1988). Its binding depends on the state of phosphorylation regulated by two different protein kinases, PKA and PKII (see later); (Llinás et al., 1985). Protein kinase A acts on phosphorylation site 1, PKII on site 2 (Pang et al., 1988; Valtorta et al., 1988). Site 2 phosphorylation abolishes binding to the vesicles and thus allows for the access to the cell boundary and for membrane fusion. (This in turn is regulated by other components; see later.) Interestingly, PKII is mainly associated with vesicles (Walker and Agoston, 1987), but some binding to the cell membrane has also been reported (Schulman, 1984); EM immunocytochemistry has shown no restriction to fusion sites (McGuinness et al., 1985). The following experiments have demonstrated the relevance of synapsin I phosphorylation for exocytosis regulation: Microinjection of PKII or of dephosphosynapsin I into squid giant synapses caused an increase or a reduction, respectively, of amplitude and rate of rise of the postsynaptic potential as a consequence of stimulated or reduced transmitter release (Llinás et al., 1985). This might be one reason of neurotransmission inhibition by CaM-binding drugs (Imai and Onozuka, 1988), as shown before by DeLorenzo et al. (1981; DeLorenzo, 1982). The situation is complicated by the occurrence of phosphorylation of PKII (see later).

Even exocrine cells, like pancreatic acini, are assumed to contain synapsin I-like proteins that can be phosphorylated by a multifunctional Ca^{2+}/CaM-dependent kinase (Cohn et al., 1987). In adrenal medullary chromaffin cells, for instance, PKII substrates are proteins IIIa (M_r 74,000) and IIIb (M_r 55,000), (Haycock et al., 1988). Evolutionary diversification may be assumed.

Synaptophysin (p38, monomeric form: M_r 38,000) is an integral membrane protein of synaptic vesicles that can be phosphorylated intravesicularly by an intrinsic kinase (Pang et al., 1988). It occurs also with chromaffin and other endocrine cells. Phosphorylation of an epitope on the cytoplasmic side can be partly inhibited with microinjected Ab. It would be interesting to know whether this would also affect attachment at the cell membrane (as discussed in Section VI) and membrane fusion (as discussed in Section XII), since synapsin is now discussed as a fusogenic protein for which fusogenicity has been demonstrated with reconstituted artificial membranes by Thomas et al. (1988).

Chromaffin vesicles also contain synaptophysin (Obendorf et al., 1988a), ~20 molecules per organelle (Schilling and Gratzl, 1988). Among many others, a 38-kDa protein is phosphorylated upon stimulation (Trifaró and Kenigsberg, 1987; Gutierrez et al., 1988). Interestingly, PC12 cells that have their chromaffin granules attached at the cell membrane ready for fusion, secrete without any recognizable protein phosphorylation or in a protein kinase-deficient functional state (although not all types of kinases have been taken into account; see Matthies *et al.*, 1988).

All together, protein kinases depending on different second messengers, Ca^{2+}-activated proteins, or phospholipids (Schatzmann *et al.*, 1984; Cohen, 1988; Shenolikar, 1988) are important for secretion (Knight, 1987; Baker, 1988; Meldolesi and Ceccarelli, 1988; Winkler, 1988). This holds for PKA, PKII, and PKC [activated by DAG (or phorbol esters as analogs) from PI turnover; see Section IX]. Substrate specificities are variable with protein kinases of different types, although they are widely distributed in different tissues and phyla (Schatzman *et al.*, 1984; Cohen, 1988; Hashimoto *et al.*, 1988). Interlocking regulatory cascades are well known for kinases, but might exist also for phosphatases, like calcineurin (CaN; Cohen, 1988; see also later). Frequently different kinases regulate secretion in one and the same secretory cell type, as in pancreatic islet cells (Harrison and Ashcroft, 1982), in chromaffin adrenal medullary (Wise and Costa, 1985) or PC12 cells (Meldolesi *et al.*, 1988a), as well as in neurons (Pang *et al.*, 1988), among others. One substrate protein molecule may be phosphorylated at different sites by two different kinases, as demonstrated with synapsin I (see earlier). Autophosphorylation of PKII—as in nerve terminals—renders the enzyme Ca^{2+}-independent (Gorelick *et al.*, 1988) and capable of phosphorylating the Ca^{2+}-CaM-dependent phosphatase 2B (CaN) (Hashimoto *et al.*, 1988). Phosphorylated CaN can be dephosphorylated by either phosphatase 1 or 2A (for classification see Cohen, 1988) and then attains increased phosphatase activity (Hashimoto *et al.*, 1988). This illustrates the complex interactions between different phosphorylation–dephosphorylation steps (as summarized in Cohen, 1988) that might be relevant for the current topic.

Much work has been focused on the effects of PKC-mediated phosphorylation. Protein kinase C is initially cytosolic, but becomes membrane-bound—for example in exocrine pancreas cells upon carbamyl (but not isoproterenol) stimulation (Machado-DeDomenech and Söling, 1987; see also Section IX). For chromaffin cells there is circumstantial evidence against a participation of PKC in the final steps of exocytosis (Holz and Senter, 1988). In other experiments the inhibition of PKC activity achieved by downregulation via long-term stimulation of PKC by phorbol ester and/or by adding sphingosine as a PKC inhibitor to permeabilized cells went in parallel to the inhibition of secretory activity (Burgoyne *et al.*, 1988). However, in similar experiments by Knight *et al.* (1988) only a marginal exocytosis inhibition occurred with identical sphingosine concentrations, even when PKC activity was greatly reduced. Knight and colleagues reach the following conclusion: ''Although these data do not definitely establish an involvement of protein kinase C in exocytosis, none argues against it." In other words, there is no stringent evidence for a direct fusion control by PKC, and various investigators question its role in fusion regulation. For further details on PKC activation by PI turnover products and effects on other phenomena pertinent to secretion (not fusion) regulation, see Section IX.

A large number of investigations rely on drug effects to specify the involvement of certain kinases. However, considerable nonspecificity, lack of inhibition, or occurrence of side effects were noticed in several studies. Because of their hydrophobicity, the so-called CaM antagonists phenothiazines or calmidazolium affect membrane lipids because of a detergent effect in too high concentrations (Shenolikar, 1988); in particular trifluoperazine has been shown to interact also with membrane-integrated proteins (Wildenauer and Zeeb-Wälde, 1983) and to reduce inward ion currents in chromaffin cells (Clapham and Neher, 1984). Calmodulin antagonists also bind to other CaBP of the annexin type, which could account for some of the published effects (Burgoyne and Geisow, 1989). Criteria for a reasonable use of phenothiazines in exocytosis research have been summarized already by Roufogalis (1982). The same compounds, like some other drugs, also inhibit enzymes of opposite functions, like PKC (Schatzman et al., 1984) and phosphatase 2B (CaN) activity (Orgad et al., 1987). In retrospect the inhibitory effect achieved by anti-CaM Ab with sea urchin eggs (Steinhardt and Alderton, 1981) and chromaffin cells (Kenigsberg and Trifaró, 1985) could also be ambiguous, as they might have affected a kinase or a phosphatase.

The table presented by Knight et al. (1988) is also useful to consult with respect to inhibitory drugs; some drugs supplied as specific PKC inhibitors exert no effect at all (Knight et al., 1988). Staurosporine (Tamaoki et al., 1986; Vegesna et al., 1988), polymyxin B (Stutchfield et al., 1986; introduced in Wise and Kuo, 1983), and chlorpromazine (frequently used as a CaM antagonist), but not compound H7 (as used, e.g., in Shimomura et al., 1988), now appear to be the most useful PKC inhibitors (Knight et al., 1988), although many published data were obtained with other drugs.

If protein phosphorylation were the crucial step for inducing membrane fusion, irreversible thiophosphorylation would have to stimulate exocytosis. In fact, the opposite occurs with chromaffin cells (Brooks et al., 1984). From the time lag between secretion and protein phosphorylation in mast cells (in response to two different stimuli), it was also concluded that phosphorylation might terminate exocytosis rather than initiate it (Sieghart et al., 1978; Wells and Mann, 1983).

Some systems, like the *Paramecium* cell (to be discussed later in more detail) could be "frozen" in such a primed (reactive) state, before dephosphorylation might induce membrane fusion. With mast cells Tatham and Gomperts (1989) now reach a similar conclusion. As outlined in Section XII, one could imagine fusion introduction by a transiently increased hydrophobicity, although direct experimental proof for this—as for any other model—is still missing.

As mentioned earlier, synchronous exocytosis in *Paramecium* involves a dose–response-dependent dephosphorylation specifically of one phosphoprotein of 65 kDa (PP65; recently 63 kDa; see earlier). In this system there is some evidence that PP65 is dephosphorylated via a CaN-like phosphatase (Momayezi *et al.*, 1987b): (a) Anti-CaN IgG specifically inhibit exocytosis *in vivo* (microinjection) and *in vitro* (isolated cortices). (b) *In vivo* this inhibition is abolished by adding an excess of CaN. (c) The effect can be mimicked *in vivo* and *in vitro* by alkaline phosphatase. (d) Both types of phosphatases dephosphorylate PP65 (testable only *in vitro*). (e) Since CaN is stimulated by Ca^{2+}–CaM (Klee *et al.*, 1979; Pallen and Wang, 1985), we also tested the effects of anti-CaM IgG (Momayezi *et al.*, 1987b): The abundance of CaM in the cytosol means that IgG can inhibit exocytosis only with cortex fragments. (f) Calmodulin was shown to be bound to the preformed exocytosis sites (Momayezi *et al.*, 1986; see also Section VI). (g) PP65 and CaN-like protein components are also associated with the cortex (Stecher *et al.*, 1987; Momayezi *et al.*, 1987b), although a precise localization at exocytosis sites still has to be shown. Data available so far are from Western blots prepared from isolated cortices, using IgG, polyclonal or partly monoclonal Ab against PP65 bands extracted from gels (Stecher *et al.*, 1987), affinity-purified polyclonal anti-CaN IgG (reacting mainly with the Ca^{2+}-binding subunit B, M_r ~16,000; see Momayezi *et al.*, 1987b), and monoclonal anti-CaN Ab (reacting with a protein equivalent in size to subunit A, M_r ~60,000; see Kissmehl *et al.*, 1989).

To sum up, we assume protein dephosphorylation to parallel (or induce) membrane fusion during trichocyst exocytosis in *Paramecium* for the following reasons: CaM, a CaN-like protein, and the presumable target phosphoprotein PP65 all occur in the *Paramecium* cortex. Antibodies inhibit membrane fusion and PP65 dephosphorylation, which normally strictly parallels exocytosis performance (Zieseniss and Plattner, 1985). All of these effects can be mimicked by an exogenous phosphatase (Momayezi *et al.*, 1987b). According to enzyme cytochemistry on the EM level, a *p*-nitrophenylphosphate-splitting enzyme activity also occurs at the trichocyst–cell membrane interface (Plattner *et al.*, 1980). This is the substrate most commonly used for CaN phosphatase assays, whereas physiological substrates are mainly seryl-PP (Pallen and Wang, 1985); PP65 is a seryl-PP (Murtaugh *et al.*, 1987; Höhne *et al.*, 1989) Nevertheless, the precise biochemical and molecular identity of the CaN-like protein in this system has still to be clarified.

We are currently exploiting the possibility of selectively labeling PP65 in *Paramecium* by rephosphorylation (see earlier and Fig. 5, bottom) and extracting labeled PP65 (Höhne *et al.*, 1989). Several spots, labeled and unlabeled, then occur on isoelectric focusing gels. The protein isolated and characterized by Murtaugh *et al.* (1987) was not labeled, so that its relevance for exocytosis still would have to be shown. The type of kinase involved in PP65 phosphorylation also has to be elucidated in detail, although it might perhaps also be CaM-dependent (Lumpert *et al.*, 1987).

Both the CaM- and the Ca^{2+}-binding subunit of CaN (A, B, respectively) were detected in sperm from mammals and sea urchins (Tash *et al.*, 1988). At least subunit B has been detected also in PC12 pheochromocytoma cells (Farber *et al.*, 1987). In brain neurons CaN is associated with synaptic vesicles (Goto *et al.*, 1986). Any possible contribution of CaN or a CaN-like protein to exocytotic membrane fusion regulation is, nevertheless, still hypothetical. Even the involvement of CaN in nerve function and the occurrence of CaN or CaN-like proteins in other secretory cell types needs further analysis.

Until now little attention had been paid to a possible involvement of other protein phosphatases [as classified by Ingebritsen and Cohen (1983), Cohen (1988), and Shenolikar (1988)] in exocytosis regulation, since phosphatases besides CaN should equally be kept in mind. For instance, a polycation-stimulated form exists in *Xenopus* oocytes (Hermann *et al.*, 1988) that has striking enzymatic similarities to CaN. The interlocking network of different kinases and phosphatases, both with pleiotropic effects and even involving mutual (de)activation phenomena, has been summarized by Cohen (1988).

In conclusion, many proteins are phosphorylated by different kinases, when secretory activity is triggered; however, some proteins are dephosphorylated. Phosphorylation occurs on different levels of secretory activity. There is no stringent proof as yet for a direct involvement of a phosphorylation step in final fusion regulation. From the number of bands undergoing phosphorylation or dephosphorylation, one cannot derive any functional preponderance of one or the other step in exocytotic membrane fusion regulation. There is some evidence that a dephosphorylation step may be crucial. Previous phosphorylation might then be required to keep fusogenic sites in a reactive state.

XI. Fusogenic Properties of Lipids

Given that exocytosis involves the fusion of two lipid bilayers, it was logical to look for particular lipid species and for an asymmetric arrangement of lipids in the plasmalemma and in the secretory organelle membranes. Since the fluidity of lipids is also important for fusion, the "rigidi-fying" effect of cholesterol as well as of Ca^{2+} binding to acidic phospholipids (phosphatidylserine, PS) also received much attention (Papahadjopoulos, 1978; Papahadjopoulos *et al.*, 1978). In this context, model experiments with artificial membranes indicated a possible role of lipid phase transitions for fusion regulation (see later). The term phase here means arrangement of lipids as bilayers (fluid or solid), as micelles (hydro-phobic core of acyl chains in the center), or as "inverted micelles" (with polar head groups facing the center) as a focal form of otherwise tubular hexagonal phase HII (Fig. 7). When it was detected that some phospholi-pids undergo metabolic turnover during exocytosis stimulation, some of the resulting components were also analyzed *in vitro* for a possible fus-ogenic effect.

In a variety of secretory systems the overall lipid composition of the secretory vesicle membranes closely resembles that of the cell membrane: In contrast to the plasma membrane, vesicle membranes are rather poor in

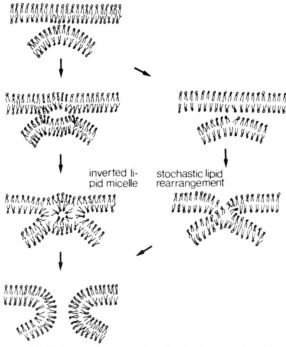

FIG. 7. Scheme of lipid fusion via inverted lipid micelle formation (hydrophilic core) or stochastic lipid rearrangement. For details, see text.

proteins (Cameron and Castle, 1984; Beaudoin et al., 1988). Both membranes have a high content of phosphatidylcholine (PC), phosphatidylethanolamine (PE), cholesterol, and sphingomyelin. Phospholipids contain predominantly saturated fatty acids; PS and PI are also present. This holds, for example, for chromaffin granules of the adrenal medulla (see data compiled in Westhead, 1987), synaptic vesicles (Breckenridge et al., 1973), exocrine pancreas zymogen (Meldolesi et al., 1971; Beaudoin et al., 1988), parotid and prolactin granules (Meldolesi et al., 1974). Amine-storing granules also have more PC, PE, and sphingomyelin than PS and PI (DaPrada et al., 1972). Cortical granules from egg cells contain even more cholesterol than the cell membrane (Decker and Kinsey, 1983). All of these characteristics would entail that secretory vesicles would be rather rigid structures (but see later for asymmetric arrangement of components).

Lipid asymmetry is known from the cell membrane, mainly of erythrocytes (Rothman and Lenard, 1977; OpDenKamp, 1979): Whereas the overall phospholipid content is equally distributed, cholesterol and sphingomyelin are localized predominantly in the outer leaflets, PE is enriched in the inner half, which contains all of the PS; PI is also on the inner side (OpDenKamp, 1979). As summarized by Westhead (1987), this might apply also to the cell membrane of secretory cells; he also assumes an analogous asymmetry for the secretory granule membrane, with PI and PS on the cytoplasmic side (as shown for synaptic vesicles in Michaelson et al., 1983) and a high content of saturated fatty acids and of cholesterol on the luminal side, although the number of data is restricted.

This would imply the following: (a) The lipid leaflets to fuse are more fluid than the other half of the bilayer. (b) Negatively charged phospholipid species could account for a high Ca^{2+}-binding capacity (Hauser et al., 1976) on the cytoplasmic surface area, where (c) PI turnover would also take place (see Section IX). (d) Rigidifying components like cholesterol and saturated fatty acid-containing phospholipids on the opposite side could stabilize the other leaflets when the cytoplasmic ones fuse. Interestingly, cis-unsaturated fatty acids can induce fusion of chromaffin granules after aggregation by synexin (Creutz, 1981). (e) The cytoplasmic leaflets would also have favorable conditions for inverted-micelle formation or other perturbations (see later). An alternative was proposed by Papahadjopoulos (1978; Papahadjopoulos et al., 1978; Portis et al., 1979): Ca^{2+} was thought to cause a liquid $\rightarrow$ solid phase transition when it is bound between two fusing membranes. The heat of melting released was proposed to drive fusion by local perturbation of adjacent lipid bilayers. Yet it is uncertain whether such effects play a role during exocytosis, since they require a pCa quite above physiological levels. In fact, the opposite—namely, increased membrane fluidity and deformation of isolated secretory organelles in response to Ca^{2+}—has been reported to occur (Miyamoto and Fujime, 1988; see also Section XV).

Tight junctions do not restrict diffusion of lipids within the inner leaflet of the cell membrane in polarized cells (Simons and VanMeer, 1988; VanMeer and Simons, 1988). Therefore, topological specificity of exocytosis (apical or basolateral) observed with such cells cannot be explained by a specific lipid composition.

An equivalent of the high Ca^{2+}-binding capacity of the secretory granule surface (Dean, 1974), which may reflect its high PS content, is the intense cationic ferritin binding, as visualized in the electron microscope (Bittiger and Heid, 1977; Howell and Tyhurst, 1977). A precedent of this are studies with synthetic lipid membranes (Hauser *et al.*, 1976). Yet, surprisingly, there was no equivalent cationic ferritin binding observed on the cytoplasmic side of the cell membrane, when quite different secretory systems were investigated (Bittiger and Heid, 1977; Howell and Tyhurst, 1977), so that any implications for membrane fusion properties due to PS partition appear doubtful.

Of particular interest was the finding of endogenous lysolecithin in the chromaffin granule membrane (Winkler, 1976). When negative results were obtained with other secretory organelles (amine storage organelles: DaPrada *et al.*, 1972; synaptic vesicles: Nagy *et al.*, 1976), a reinvestigation of chromaffin granules has ascertained that their lysolecithin content is not a postmortem artifact (Frischenschlager *et al.*, 1983). Its formation would depend on the widely distributed Ca^{2+}-dependent PLA_2 activity (VanDenBosch, 1980). Phospholipase A_2 is bound to the cell membrane (Frei and Zahler, 1979) also in cells secreting in a constitutive mode (Bar-Sagi *et al.*, 1988; see also Section IX). Lysolecithin was proposed to be a fusogen that, by virtue of its wedge shape, would cause formation of lipid micelles (Lucy, 1978); these were suggested to induce fusion. However, such micelles—in contrast to "inverted micelles" (see later—would cause leakiness of cells during exocytosis; this was not actually observed. Also, a comparative analysis of a variety of secretory systems did not reveal any significant amounts of lysophospholipids (Westhead, 1987). Furthermore, lysolecithin is located chiefly on the luminal side of the chromaffin granule membrane, where it is abundant (Westhead, 1987). For all these reasons the concept of lysolecithin-mediated fusion has been largely abandoned, although one micelle formed during fusion would suffice. This would be difficult to prove or disprove by lipid bulk analysis (see end of this section).

Other lipid species considered as fusogens are free fatty acids, particularly in a cis-unsaturated form (Creutz, 1981; Meers *et al.*, 1988b), as well as DAG and AA. These are formed by PI turnover or PLA_2 activity, respectively (see Fig. 1). Diacylglycerol causes distortions of the molecular packing of artificial phospholipid bilayers (Das and Rand, 1984). Arachidonic acid promotes the fusion of chromaffin vesicle ghosts (Creutz *et al.*, 1987b) and of liposomes (Hong *et al.*, 1987; Meers *et al.*, 1988b). It also favors formation of inverted micelles in artificial membranes, as also reported for DAG (Verkleij, 1986), so that both these components were considered as fusogenic.

As mentioned in Section IX, AA generation in chromaffin cells strictly parallels catecholamine release (Frye and Holz, 1984). Inhibitors of PLA_2 also abolish secretion in chromaffin cells (Frye and Holz, 1983) and cortical granule reaction in egg cells (Ferguson and Shen, 1984). In the chromaffin system, however, the effect could be caused on a step before fusion, because these inhibitors also curtail Ca^{2+} influx (Frye and Holz, 1983). (This again demonstrates the unreliability of responses to inhibitory drugs.) As pointed out by Winkler (1988), one should analyze permeabilized cells to discriminate between these two aspects. The effects of AA may be rather complex. Whereas it may be considered in some cases as a stimulating second messenger (Metz *et al.*, 1983; Luini and Axelrod, 1985), it inhibits exocytosis in sea urchin eggs (Elhai and Scandella, 1983). Its generation by PLA_2 is inhibited by lipocortin (Axelrod *et al.*, 1988), depending on its state of phosphorylation (Hollenberg *et al.*, 1988). Lipocortin as a CaBP of the annexin group (see Section VI) could thus provide some feedback mechanism. (Crossconnections with G proteins are discussed in Section VIII.)

It is also intriguing that the PLA_2 molecule per se causes liposome fusion (Jain and Vaz, 1987). Since other cellular enzymes totally unrelated to exocytosis, like glyceraldehyde-3-phosphate dehydrogenase, exert a similar effect (Lopez Vinals *et al.*, 1987), the effect observed *in vitro* with PLA_2 or some other alleged fusion-inducing proteins (see later) might be rather accidental.

For a more detailed account of DAG effects via PKC activation, see Section IX.

Formation of inverted lipid micelles was observed when liposomes, prepared from appropriate lipid constituents (thought to be fusogenic) were supplemented with Ca^{2+} (Verkleij *et al.*, 1979; Verkleij, 1984; Cullis *et al.*, 1986). There is a strong correlation between this HII phase induction and fusion promotion (Cullis *et al.*, 1986). Fusion was assayed by cryofixation and freeze-fracturing, as well as by light scattering. Inverted lipid micelle formation was monitored by nuclear magnetic resonance (NMR) or differential scanning calorimetry (DSC). An increased number of lipidic particles (or of equivalent pits), 10 nm in diameter, was considered as their ultrastructural equivalent that would also occur at the site of actual liposome fusion (which, however, has not been "caught" so far). The 10-nm large fusion intermediate that we saw in the cell membrane of fast-frozen exocytosing chromaffin cells (Schmidt *et al.*, 1983) was later on also interpreted as an inverted lipid micelle (Verkleij, 1986). There are, however, the following uncertainties with this correlation: (a) Fusion *in vitro* and lipid particle formation require millimolar Ca^{2+} (or somewhat lower Ca^{2+} levels when amphipathic proteins are added: Batenburg *et al.*, 1985). (b) Both NMR and DSC are slow methods that cannot directly monitor events during fusion. Lipidic particles persist for long time periods outside the fusion site, so that the relevance of lipidic particles for fusion is questioned by some authors (Hui *et al.*, 1988). (c) According to other investigators, lipids could undergo a less strict, but rather a stochastic short-lived rearrangement under these conditions (see Fig. 7 and Section XVI). (d) The fusion rate is maximal in an isotropic stage between fluid–lamellar and HII phase (Ellens *et al.*, 1986). (e) Lipid particles as fusion intermediates would not explain the flickering phenomena that were monitored by surface capacitance recordings during a reversible prefusion stage of exocytosis (Breckenridge and Almers, 1987; Zimmerberg *et al.*, 1987; see also Sections XII and XVI).

The latter investigations therefore entail the postulate of the involvement of pore-forming oligomeric proteins in exocytotic membrane fusion (Almers and Breckenridge, 1988; Cohen, 1988; see also Sections XII and XVI). These would impose a reorientation of surrounding lipids in a way probably different from any of the schemes just outlined. In addition, such proteins might attract a halo of appropriate lipid components. Lipid bulk analyses would be unable to reveal any clues to this aspect. One has to ask anyway how much of a certain lipid species one would have to expect if two membranes fuse during 1 msec on a 10-nm spot, as outlined in Section XVI.

To sum up, the relevance of specific lipid constituents for membrane fusion during exocytosis remains enigmatic, although local lipid composition might be important for fusion to occur. The problem could be solved perhaps, once a genuine fusogenic protein—if it exists—has been isolated, so that any effect of specific lipid constituents could be reevaluated.

XII. Fusogenic Proteins

Fusogenic proteins have been unequivocally identified so far only with viral systems (White *et al.*, 1983; Schlegel, 1987; Spear, 1987). Depending on the type of virus, fusogenicity is obtained by an acidic pH-induced conformational change or via proteolytic cleavage (Schlegel, 1987; Spear, 1987). An example is schematically shown in Fig. 8.

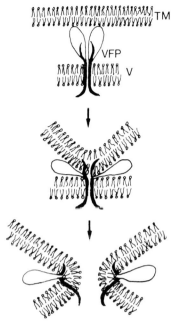

FIG. 8. Fusion of a virus (V) membrane with a cellular target membrane (TM). The viral fusion protein (VFP) has a hydrophilic (white) and two different hydrophobic regions (black), only one being inserted into the lipid bilayer of the viral membrane. A conformational change of the VFP depicted here allows the exposure of the upper hydrophobic part, which then inserts into the outer lipid leaflet of a target membrane. Both membranes thus come into intimate contact and then fuse. This scheme is drawn for a influenza virus according to Blumenthal *et al.* (1988). Viral fusion proteins are now considered as a paradigm of thus far unidentified fusogenic proteins that might be relevant for exocytosis.

The first alternative cannot be a model for membrane fusion during exocytosis, because this is not accompanied by a pH drop between membranes to be fused. However, some assume that a (Zn^{2+}-activated) metalloendoprotease participates in secretion in mast cells and adrenal medullary chromaffin cells (Mundy and Strittmatter, 1985), as well as in the acrosomal reaction of sperm cells (Farach *et al.*, 1987), but not in the cortical granule reaction of egg cells (Roe *et al.*, 1988). Other groups advocate against this hypothesis, which largely depends on the specificity of the inhibitory compounds used: In chromaffin cells, for example, their effect is caused by altering the Ca^{2+} influx through the cell membrane (Harris *et al.*, 1986; Lelkes and Pollard, 1987). Hence it is not yet settled whether proteolytic cleavage would entail fusogenicity of a certain protein, which, moreover, would still have to be identified. However, one should also bear in mind the occurrence of Ca^{2+}-dependent proteases in many cell types (Kawasaki *et al.*, 1986).

The evidence for protein involvement in fusion regulation during exocytosis is rather circumstantial. The importance of membrane-associated and -integrated proteins (IMP), as indicated in Fig. 2, became particularly evident from the ultrastructural analysis of *nd* mutations of *Paramecium* cells (Plattner, 1981, 1987). This aspect is discussed in more detail in Sections VI and X. Amphipathic proteins were found also to increase the fusion rate between liposomes (Batenburg *et al.*, 1985).

Essentially three potential candidates for fusogenic proteins have been proposed: synexin, phosphoproteins, and synaptophysin. Any candidate is expected to be or to become hydrophobic in response to some activator; possible mechanisms are the interaction with Ca^{2+} or covalent modification (e.g., by changing the degree of phosphorylation).

Binding of Ca^{2+} to CaBP might cause a conformational change. In fact, synexin (M_r of 47,000: Pollard *et al.*, 1980; Creutz *et al.*, 1983), one of the CaBP known as chromobindins or annexins (see Section VI), has been found in different secretory cells (Pollard *et al.*, 1980; Creutz *et al.*, 1982; Dowling and Creutz, 1985; Geisow *et al.*, 1987), including some secreting in the constitutive mode. It is assumed to expose a hydrophilic and a hydrophobic site upon Ca^{2+} binding (Pollard *et al.*, 1988a,b). Indeed, synexin also promotes fusion between liposomes in the presence of Ca^{2+}, provided arachidonic acid (a lipid considered to be fusogenic; see Section XI) is present (Papahadjopoulos *et al.*, 1987).

Pollard and co-workers now assume quite different steps for exocytosis regulation by synexin (Fig. 9): (a) Synexin is aggregated by Ca^{2+} (as seen with isolated synexin in the electron microscope: Pollard *et al.*, 1981). (b) Fusion would be caused by a conformational change exposing hydrophobic sites (Pollard *et al.*, 1988a,b). (c) This results in increased Ca^{2+} conductance (channel function: Pollard and Rojas, 1988). (d) Integration of synexin into aggregated membranes would cause their fusion (Rojas and Pollard, 1987; Pollard *et al.*, 1988a,b). However, even aggregation of isolated secretory organelles occurs only with pCa values well beyond physiological levels. The other assumptions are based on electrophysiological recordings with artificial systems, showing expansion of the surface area (interpreted as indicating protein insertion: Rojas and Pollard, 1987) and increased Ca^{2+} conductance. To transpose this to the situation *in vivo*, several questions would have to be answered: Why does synexin associate with and become integrated into membranes before pCa_i increases by synexin-mediated Ca^{2+} influx? How is synexin released from membranes after fusion has occurred? Further substantiation of the "hydrophobic-bridge hypothesis" proposed for synexin (Pollard *et al.*, 1988a,b) is required, as it is for any other model on protein-mediated fusion.

The second proposal of a potentially fusogenic protein comes from my

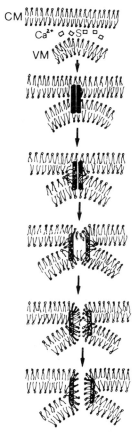

FIG. 9. "Hydrophobic-bridge hypothesis" of membrane fusion according to Rojas and Pollard (1987) and Pollard *et al.* (1988a,b). Synexin (S) is thought to aggregate when pCa$_i$ rises and then to become hydrophobic (black). These synexin aggregates are proposed to become integrated into the cytoplasmic lipid leaflets of the cell membrane (CM) and an adjacent secretory vesicle membrane (VM). Lipids are thought to redistribute around separating synexin oligomers, until an exocytotic opening is formed.

own group (Fig. 10). During synchronous (1 second) exocytosis one phosphoprotein band (PP65, ~63–65 kDa on one-dimensional SDS–PAGE) is selectively dephosphorylated (<1 second) and then rapidly (10 seconds) rephosphorylated (Zieseniss and Plattner, 1985); this would cause a transient increase in hydrophobicity. For more details, see Section X. Support comes from fusion induction by exogenous phosphatases (Momayezi *et al.*, 1987b) and from fusion inhibition by Ab against PP65 (Stecher *et al.*, 1987). Since dephosphorylation is reversible, such a protein might only

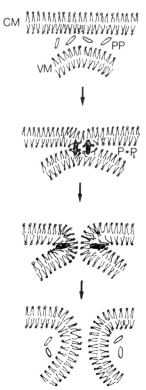

FIG. 10. Hypothetical fusogenic mechanism involving a (hydrophilic) phosphoprotein (PP, white) which turns hydrophobic (black by dephosphorylation (P + P_i) and then becomes integrated into the cytoplasmic lipid leaflets of the membranes to be fused (CM, cell membrane; VM, secretory vesicle membrane). The insertion of several protein molecules might cause a local perturbation of lipids and, thus, their fusion. Rephosphorylation (on sites that would have to be accessible to a kinase or by autophosphorylation) would force the protein to leave the fused membranes. This hypothetical scheme is based on results by the author's group with *Paramecium* cells (Zieseniss and Plattner, 1985; Stecher *et al.*, 1987; Plattner *et al.*, 1988).

transiently interact with membrane lipids. Remarkably, proteins immunologically cross-reacting with anti-PP65 Ab are widely distributed (Murtaugh *et al.*, 1987; Stecher *et al.*, 1987).

Along these lines, Gomperts and colleagues also infer protein dephosphorylation as a potentially fusogenic step (Howell and Gomperts, 1987; Tatham and Gomperts, 1989); see Sections VIII and X. A candidate would be a G protein specific for exocytosis (G_E), which, however, has not been identified so far—or another unknown protein. The opposite, protein phosphorylation, would not easily account for an increased hydrophobicity required for fusion induction, although this could be caused by increased Ca^{2+} binding and/or a conformational change. The controversies on this aspect are discussed in Section X.

One new proposal concerns the neuronal protein synaptophysin (Fig. 11), characterized in more detail in Sections VI and X. After reconstitution with liposomes, synaptophysin mediates fusion *in vitro*. Thomas *et al.* (1988) also demonstrated channel formation by synaptophysin hexamers. This small pore could then expand by dispersal of the monomers, which would allow the lipids to fuse and the exocytotic opening to expand. The rationale of these studies was to explain the conductance flickering observed with mast cells (from beige mouse mutants) immediately before fusion of a giant secretory granule occurs (Breckenridge and Almers, 1987; Zimmerberg *et al.*, 1987). This flickering involves discrete, reversible conductance changes that could hardly be explained without participation of proteins in the early fusion steps (Almers and Breckenridge, 1988; Zimmerberg, 1988). The authors assume matching hemichannels as in gap junctions. Again, there are some questions: Given the limited occurrence of synaptophysin (see Section VI), what would be the equivalent in cells where it does not occur, such as mast cells? How can these proteins in the secretory vesicle membrane match with a hemichannel in the plasma membrane that does not contain the same protein? In this regard, Thomas *et al.* (1988) refer to a "mediaphore" protein occurring in presynaptic membranes (Morel *et al.*, 1987) as a possible plasmalemmal counterpart to synaptophysin.

Given the size of a synaptophysin hexamer (230 kDa would be sufficient to yield a 10-nm IMP) and of the mediaphore protein (8 nm), one should be able to visualize their interaction on freeze–fracture replicas. As outlined in Section XVI, however, there are no matching IMP in the secretory organelle membrane—at least for example, in *Paramecium;* also, during exocytosis no such interaction is seen in rapid-frozen samples (although this is no proof against this possibility).

It would be interesting to know whether any of the other membrane-spanning proteins in common to synaptic and neuroendocrine vesicles (Section VI), particularly p65, would participate in the fusion process.

Studies with the black widow spider venom latrotoxin are exciting, since on the one hand it causes exocytosis of neurotransmitter vesicles (Meldolesi *et al.*, 1986), and on the other hand it is now known to fuse artificial membranes into which it becomes integrated (Sokolov *et al.*, 1987).

In conclusion, it remains open so far (a) whether there exists at all any fusogenic protein relevant for exocytosis (although this is likely),

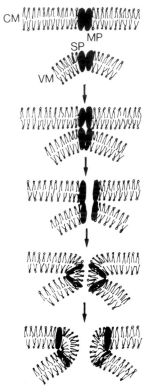

FIG. 11. Scheme of membrane fusion drawn on the basis of data combined from Almers and Breckenridge (1988), Zimmerberg (1988), and Thomas *et al.* (1988). Both the cell membrane (CM) and the vesicle membrane (VM) contain membrane-spanning proteins (black ovals). Thomas *et al.* (1988) propose neuronal synaptophysin (SP) and a "mediaphore" protein (MP) as candidates (see text) for oligomerically arranged matching membrane proteins. On this level, the scheme could account for flickering of electrical conductance observed with mast cells by Breckenridge and Almers (1987) and Zimmerberg *et al.* (1987) before fusion becomes irreversible. Fusion would allow monomers to shift centrifugally, thus forming an exocytotic opening. Its expansion would require the rearrangement of lipids around these fusogenic proteins. This could be more easily explained by globular proteins (as in the scheme drawn in Zimmerberg, 1988), which then, however, would contradict the actual molecular size of the monomers and the final particle (oligomer) size assumed by Thomas *et al.* (1988), since the protein oligomers are known to span each of the two membranes. In principle this scheme explains docking and fusion via matching integral proteins, although the combination of data from different systems and the absence of a corresponding freeze–fracture morphology entails some problems.

(b) whether such proteins would be universally distributed (this is less likely), (c) what kind of modification they would have to undergo to become fusogenic, and (d) how a fusogenic protein might perturb the membrane lipids to fuse ultimately with each other during exocytosis.

Enzymatic and nonenzymatic effects, possibly also involving Ca^{2+} and/or GTP, have to be taken into consideration. Despite all the uncertainties on the details, the participation of partly membrane-associated and partly integrated proteins in membrane fusion during exocytosis, as proposed before on the basis of ultrastructural evidence (Plattner, 1981), now appears quite certain. Figure 12 summarizes the possibilities currently being discussed.

XIII. Ca^{2+}: No Longer a Common Denominator?

Since the discovery of stimulus–secretion coupling in 1962, an increase of $[Ca^{2+}]_i$ had for a long time been considered as a common denominator for membrane fusion during exocytosis. This concept was founded on the dependency of exocytosis on a Ca_o^{2+} influx into the cells then analyzed. As outlined in Section I, this was supported later on by the increase of pCa_i observed with fluorochromes during elicited exocytosis and by the Ca^{2+} requirement for fusion *in vitro*. Additional support came from pressure (mast cells: Kanno *et al.*, 1973; neurons: Miledi, 1973) or ionophore injection (Rasmussen and Goodman, 1977) of Ca^{2+} into different cells that also provoked exocytosis. In *Paramecium,* a Ca^{2+} influx via the cell membrane mediated by A23187 or X537A also causes trichocyst release (Plattner, 1974); negative results with Ca^{2+} microinjection into the cytosol (Kersken *et al.*, 1986a) can be explained by rapid sequestration before Ca^{2+} can reach the cell boundary. Alternatively, complexation of Ca_i^{2+} by Ca^{2+} buffer injection into oocytes inhibits fertilization reaction (Kline, 1988).

Binding of Ca^{2+} to negatively charged lipids on the cytosolic side of interacting membranes, where it may cause local dehydration and charge

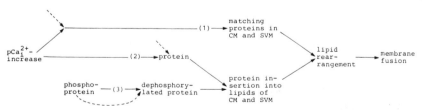

FIG. 12. Summary of membrane fusion mechanisms involving fusogenic proteins, as discussed in the text. CM, cell membrane; SVM, secretory vesicle membrane. The pathways indicated (1–3) correspond to the schemes presented in Figs. 11, 9, and 10, respectively; they apply to the hypotheses formulated for synaptophysin (1), synexin (2), and phosphoproteins (3). The current uncertainty on the relevance of a pCa^{2+}_i increase for individual fusogenic steps is also taken into account (dashed arrows).

screening (see Sections II–IV), is probably not sufficient to induce fusion (see Section XI and later).

In vitro systems such as cortices from sea urchin or *Paramecium* cells are triggered by $pCa^{2+} \geq 6$ (see Section IV). This corresponds to findings with other cells after permeabilization. Nevertheless these data provide no direct proof of a direct fusogenic effect of a pCa_i increase for different reasons: In these systems auxiliary components are still present, for example, for mediating Ca^{2+} sensitivity to a variety of Ca^{2+}-governed enzymatic or nonenzymatic processes (see Sections IX and X). A counterargument would be the evident Ca^{2+} dependency of many processes. Among them are receptor and ion channel activation as well as other phosphorylation–dephosphorylation phenomena (Section X), effects on cytoskeletal elements (Section V), activation of CaBP (Sections VI and XII), activation of PLA$_2$ (Frei and Zahler, 1979; VanDenBosch, 1980; see also Sections IX and XI), and activation of Ca^{2+}-dependent proteases (Kawasaki *et al.*, 1986), among others. Among them the neutral protease "calpain" is operating at micromolar Ca^{2+} levels (Inomata *et al.*, 1988). All these aspects are summarized in Fig. 1 and discussed throughout this review.

A quite novel aspect is that a transient increase of pCa_i may occur in some systems, yet without inducing exocytosis. This has been ascertained with patch-clamped mast cells (Neher and Almers, 1986; Neher, 1988), with collagen-activated blood platelets (Rink *et al.*, 1983), with muscarinic receptor-stimulated chromaffin cells (Yagisawa *et al.*, 1986), and with pancreatic β cells in response to certain stimuli (Pralong *et al.*, 1988). In cells with different secretory granules, their independent release requires different pCa_i levels (Lew *et al.*, 1986; Barrowman *et al.*, 1987).

This is closely related to another argument: Some factors have become known to sensitize some systems for resting pCa_i levels (leftward shift of the Ca^{2+} sensitivity curve). As described in Section VIII, GTP or analogs exert such an effect not only in neutrophils and mast cells, but also in *Paramecium* cells.

As discussed in various sections, the previous view of the role of Ca^{2+} for exocytosis regulation was therefore too simplistic. Its effects are multiple, with many crossconnections (Fig. 1). Yet if any ion has a primary effect on fusion control at all, Ca^{2+} is still the favorite. It still can be reasonably assumed to exert some control over membrane fusion in (only some?) exocytotic systems. It remains open how far pCa levels could be reduced below "resting levels" (7–8) in systems operative without a pCa_i increase.

A key function of Ca^{2+} may be executed particularly in systems exhibiting a rapid Ca^{2+} influx without any recognizable delay through the plasma membrane (e.g., nerves and some other cell types). The argument then arises that local subplasmalemmal [Ca^{2+}] might be higher than indicated by the cytosolic Ca^{2+} transient. This is observed with chromaffin cells, when aequorin is applied as a fluorochrome that has little effect on small pCa changes to be detected (Cobbold et al., 1987). For nerve terminals, during excitation, a local pCa ~4–5 was calculated (Simon and Llinás, 1986; Smith and Augustine, 1988) from the channel conductance (and assuming the number of channels to be equivalent to IMP seen at active zones: Llinás et al., 1981; Pumplin et al., 1981). From the short lag time between the trigger and the response (1–2 msec in neuromuscular junctions: Datyner and Gage, 1980; Llinás et al., 1981; Torri-Tarelli et al., 1985) and from the Ca^{2+} binding constant reported for CaM, Smith and Augustine (1988) derived the irrelevance of a Ca^{2+}–CaM activation process previously proposed by DeLorenzo et al. (1981) and DeLorenzo (1982). This is problematic for different reasons: (a) Some CaM is already membrane-bound before Ca^{2+} binding (Publicover, 1985). (b) Some of the possibly relevant processes change their Ca^{2+} requirement in the presence of CaM (Schworer et al., 1988; see also examples given in Section X). (c) Permeabilized neurohypophyseal cells operate with micromolar [Ca^{2+}], that is, ~100 times below levels occurring in triggered neurons in vivo (Cazalis et al., 1987). The Ca^{2+} spillover assumed might be preceded by a fusion reaction at lower pCa$_i$. Nonequilibrium kinetics, as initiated by Gomperts and colleagues, for mutual ATP–GTP–Ca^{2+} interactions (Sections VII, VIII, and X) look very promising to clarify some of these aspects.

All the different modes of Ca^{2+} involvement may or may not occur with one and the same system. I assume a multiple role of Ca^{2+} also for *Paramecium* for the following reasons: (a) Exocytosis with intact cells depends on Ca_o^{2+} (Plattner et al., 1985). (b) The isolated cortex requires a pCa ≥ 6 (depending on the pMg present) for triggering, yet with GTP (or analogs) present triggering occurs from a pCa ≥ 8 on and is maximal with pCa = 7 (Lumpert et al., 1989). (c) On the other hand, immuno- and affinity-labeling experiments simply show the presence of CaM at fusion sites (Momayezi et al., 1986), and Ab are inhibitory (Momayezi et al., 1987b). (d) A quite similar constellation occurs with chromaffin cells, since CaM binds to the vesicle and cell membrane (Fournier and Trifaró, 1988b), and Ab inhibit exocytosis (Kenigsberg and Trifaró, 1985), even though pCa increases selectively below the plasmalemma (Cobbold et al., 1987).

Oscillations of pCa$_i$ above resting levels accompany exocytosis stimulation, for example by cerulein or bombesin in pancreatic islet cells (Pralong et al., 1988). Since oscillations of pCa$_i$ were observed also in several other systems to accompany exocytosis (Berridge et al., 1988), a regulation by frequency modulation has been taken into account by some authors. This can be questioned, however, for various reasons: (a) Different periodicities occur in response to different stimuli, even when net exocytosis performance is equal (Grapengiesser et al., 1989). (b) Oscillations may simply reflect a feedback mechanism of pCa_i regulation, with alternating short phases of Ca^{2+} release and sequestration (Berridge and Galione, 1988; Berridge et al., 1988). (c) Oscillations vanish during sustained secretory activity (Penner and Neher, 1988).

Therefore, the possible interactions of Ca^{2+} in regulatory steps that finally may govern membrane fusion during exocytosis are so numerous (see Fig. 1) that its actual role is difficult to pinpoint. Given the local restriction of the fusion event (Section XVI), it also appears problematic to correlate directly the fusion event with a bulk change of pCa_i. Specific binding sites and/or sensitizing factors (like GTP) may explain the occurrence of membrane fusion, in nontriggered exocytosis and in some cases of triggered exocytosis, with resting levels of pCa_i.

XIV. Osmotic Effects

With different systems the release of secretory contents can be reduced by hyperosmotic media (sucrose, etc.) This has been observed with intact or permeabilized cells (for review see Holz, 1986) and with isolated sea urchin egg cortices (Zimmerberg et al., 1985). These results obtained by increasing the osmotic pressure in the medium, therefore, were interpreted by some authors as membrane fusion inhibition.

Others differentiated more clearly between different aspects: (a) Membrane fusion might still occur ("minifusions"), even though the contents are not released. (b) One should differentiate between effects achieved with intact cells and with "open systems" (cortex fragments, permeabilized cells), since in the latter case osmotic gradients would be sensed over the entire secretory vesicle membrane. (c) The ion composition of the media used might be important.

The first aspect (a) proved to be very important. Electrophysiological recordings with beige mast cells (containing giant secretory granules) ascertained the following sequence of events (Breckenridge and Almers, 1987; Zimmerberg et al., 1987): Membrane fusion (recorded by surface capacitance increase) clearly preceded, by several seconds, a subsequent granule swelling (monitored in the light microscope). Vesicle swelling, therefore, should be considered as a mechanism serving the extrusion of secretory contents, rather than promoting membrane fusion. This view is supported by our earlier findings with Paramecium cells: Under appropriate experimental conditions, membrane fusion can be induced without contents release (Matt and Plattner, 1983) by mobilizing Ca_i^{2+} by local anesthetics (Low *et al.*, 1979) in the absence of Ca_o^{2+} [Ca_o^{2+} would normally cause "decondensation" (i.e., swelling) of secretory contents (trichocysts) when it enters from the extracellular medium, once an exocytotic opening has been formed (Bilinski *et al.*, 1981)]. With $pCa_o < 6$ this decondensation no longer occurs, although EM analysis of ultrathin sections shows abundant, yet small exocytotic openings of unswollen trichocysts in the experiments by Matt and Plattner (1983). With sea urchin eggs a possible coincidence of Ca^{2+}-dependent granule swelling and membrane fusion still has to be settled (Whitaker, 1987); yet Zimmerberg and Liu (1988) have excluded with cortex fragments an influx of Ca^{2+} from the surroundings into cortical vesicles.

The second aspect (b) implies, for intact cells, that an osmotic gradient would have to be equilibrated through a very small area, where the cell membrane and the secretory vesicle get in contact when they fuse (focal-fusion concept; see Section XVI). Also, from this point of view it is much more likely that the influx of solutes occurs only after an exocytotic opening has been formed and then might induce the swelling or decondensation of contents to facilitate their release. A scrupulous analysis of permeabilized versus intact chromaffin cells revealed multiple effects on membranes and on proteins associated with them (Ladona *et al.*, 1987). With permeabilized chromaffin cells, even considerably shrunken vesicles may be discharged, and the effect achieved also depends more on the ionic strength than on the overall osmolality (Holz and Senter, 1986). The latter holds also for sea urchin eggs (Whitaker and Zimmerberg, 1987). Therefore, no straightforward interpretation of hyperosmotic effects is possible.

Finally, (c) a chemiosmotic hypothesis of secretion (membrane fusion) was proposed (Pollard *et al.*, 1979, 1980, 1981). Starting from the observation that chromaffin granules easily swell and lyse in media with ATP and Cl^-, they assume that Cl^- follows the H^+ gradient (maintained by the H^+-translocating ATPase present in most, if not all secretory granules; see, e.g., Winkler, 1988). This would cause water uptake by osmotic imbalance and, thus, swelling and "lysis" of secretory granules by formation of an exocytotic opening. The primary driving force for membrane fusion would be a ΔpH (Pollard *et al.*, 1979, 1981; Creutz and Pollard, 1983). In this work it was explicitly assumed (a) that a broad membrane contact area and a fusion diaphragm would be formed (see, however, Section XVI), and (b) that membrane fusion would require ATP (but see Section VII). In a careful reevaluation the authors themselves now consider the chemiosmotic hypothesis, at least in its original version, no longer valid (Brocklehurst and Pollard, 1988a; see also critical evaluation in Baker and Knight, 1984a; Holz, 1986).

A local osmotic effect as a driving force for membrane fusion is still assumed for polyethylene glycol-induced cell–cell fusion (Lucy and Ahkong, 1986). When erythrocyte membranes fuse, a diaphragm is formed indeed, as shown by fast-freezing methods (Hui *et al.*, 1985). Although EF–EF type fusions (see Section XVI), like cell–cell fusion, cannot be directly compared with exocytosis (PF–PF fusion) for a variety of reasons, Ahkong and Lucy (1988) have extended their concept to exocytosis. They suggest that secretory contents might swell only locally beneath the site of ongoing membrane fusion.

Fusion experiments with *in vitro* model systems (planar black lipid membrane and liposome fusion) were also analyzed by electrophysiological recordings. They unequivocally showed that an osmotic pressure gradient favors fusion (Zimmerberg *et al.*, 1980; Finkelstein *et al.*, 1986). This is therefore in contrast to what is now assumed for exocytosis and thus represents just another example of the ambiguity of any conclusions derived from artificial systems for membrane fusion during exocytosis (see also Table I).

XV. Is the Membrane Curvature Important?

In some instances a funnel- or pedestallike depression in the cell membrane (seen after fast freezing) accompanies exocytosis. This holds for amebocytes (Ornberg and Reese, 1981), for mast cells (Chandler, 1988), and for chromaffin cells (Schmidt *et al.*, 1983; R. L. Ornberg and T. S. Reese, unpublished observations, cited in Chandler, 1988). In *Paramecium* cells, the trichocyst membrane docked at the cell membrane bulges out to form a "nipple" before exocytosis is triggered (Beisson *et al.*, 1976), but only in secretion-competent strains, where some connecting materials (proteins) are found between cell and trichocyst membrane (Plattner *et al.*, 1980; Plattner, 1987). This material may help to maintain this shape of the trichocyst tip. During membrane fusion, the cell membrane also shows a funnellike depression in these cells (Momayezi *et al.*, 1987a; Plattner, 1989a). (See Fig. 3.)

These aspects might be relevant for membrane fusion in several respects: (a) A small curvature might provide increased fusion capacity by high free energy of the lipid bilayers involved (Blumenthal, 1987; Ohki, 1987). (b) Fusion sites might thus become locally restricted and allow for "focal fusion" to occur (see Fig. 3 and Section XVI). (c) Specificity of ultrastructural and biochemical properties can, by virtue of this backing material, more easily be maintained during and after fusion (resealing of "ghost" membranes for retrieval). In *Paramecium*, for instance, the failure rate (percentage intermixing of components) during exo-endocytosis is only 0.5% (Plattner *et al.*, 1985). For the occurrence of connecting material in other exocytotic systems, see Section VI.

Interestingly, chromaffin granule membranes become very flexible when exposed *in vitro* to a pCa similar to the Ca^{2+} transient during exocytosis (Miyamoto and Fujime, 1988). Neurotransmitter (acetylcholine) vesicles from motor end plates appear small enough to undergo fusion without further deformation (Heuser *et al.*, 1979; Torri-Tarelli *et al.*, 1985). It remains to be seen by further cryofixation analyses whether local membrane deformation during exocytotic membrane fusion is a general feature in common to all secretory cells, except for those with quite small secretory organelles.

XVI. A Unifying Ultrastructural Concept for Membrane Fusion during Exocytosis

Following the freeze–fracture terminology (Branton *et al.*, 1975), membrane fusion during exocytosis is designated as a PF-type fusion (Stossel *et al.*, 1978; Knoll and Plattner, 1989); that is, plasmatic faces fuse. This is opposite to endocytosis, virus–cell fusion, virus–lysosome fusion, or cell–cell fusion (EF type, involving exoplasmic faces). Exocytosis thus resembles fusion of intracellular compartments—for example, of different vesicles (endosomes and lysosomes, etc.) and/or of vesicles and stacks in the Golgi apparatus.

For a long time the Palade (1975) model, based on the EM analysis of ultrathin sections from mast cells (Palade and Bruns, 1968), has been accepted. It includes (a) close apposition of membranes, (b) elimination of the inner layers of apposed membranes, causing (c) the formation of a diaphragm [for (b) and (c) the term fusion was assumed], and (d) fission, that is, breakthrough of the diaphragm for formation of an exocytotic opening by an unexplained mechanism. (We now use the term fusion for the formation of a membrane continuum from two originally separated membranes.) This model has been supported by the finding of large contact zones in model membranes and in freeze–fracture replicas of different secretory cells (Orci and Perrelet, 1978). Dispersal of IMP (for definition, see Section VI) and of membrane-associated proteins (also observed in this work) supported the original Palade model.

A variation, proposed on the basis of EM work (also using conventional preparation techniques, including chemical pretreatments before freezing and freeze-fracturing) with *Phythophthora palmivora* (fungi) zoospores, has been propagated by PintoDaSilva and Nogueira (1977). It includes the formation of a micellar lipid "nucleus" from which the elimination of inner lipid layers proceeds in a zipperlike mode, thus forming a bilayer diaphragm devoid of IMP. Although no similar observations have been made so far with other cells, when prepared without artifact hazards (see later), this model was appreciated for a while because it implies thermodynamically favorable conditions for initiating membrane fusion (Shotton, 1978).

As far as analyzed ultrastructurally by reliable methods (including fast-freezing without chemical pretreatments) not only PF-type but also EF-type fusion processes all display "focal [point] fusion" characteristics (Plattner, 1981; Plattner, 1989a,b; Knoll and Plattner, 1989). The new scheme (Figs. 2 and 3) implies that no diaphragm is formed and, hence, that the fusion–fission scheme (Palade, 1975) is no longer valid. Also, no IMP rearrangement has been observed to precede any intravital fusion process (with the rare exception of goblet cells, in which secretory granules are very tightly packed: Specian and Neutra, 1980). These might, therefore, be features in common to all physiological fusion phenomena. Given the great number of exocytotic systems analyzed by cryofixation (i.e., fast-freezing without any chemical pretreatments) and freeze-fracturing, it is reasonable to assume that the "focal-fusion scheme" holds for practically all exocytotic events (Plattner, 1989a,b).

In retrospect, this discrepancy can be summarized and explained as follows. As stated before, previous investigations used chemical fixatives and revealed IMP clearing (Orci and Perrelet, 1978) as well as diaphragm formation by elimination of tightly apposed lipid layers from extended contact areas between the secretory vesicle and cell membrane (Palade and Bruns, 1968; PintoDaSilva and Nogueira, 1977). This scheme is currently maintained only exceptionally (PintoDaSilva, 1988). The older fusion schemes have lent support to concepts denying a direct participation of proteins (yet see Sections VI and XII) and favoring an involvement of osmotic forces (yet see Section XIV) in exocytosis. The reasons for the (reproducible) ultrastructural artifacts involved have been discussed by Plattner (1981) and Chandler (1984, 1988); they might be caused by a regional fluidization of membrane lipids (Chandler, 1984, 1988). Chandler and I also agree that IMP clearing and diaphragm formation are "an exaggeration of a physiological process" that normally requires only a 10-nm large focus.

In contrast to exocytosis, IMP clearing occurs during chemically induced fusion of cells (Roos *et al.*, 1983; Hui *et al.*, 1985), even when fast-freezing had been applied (Hui *et al.*, 1985; Huang and Hui, 1986). (Interestingly, myoblast fusion, when analyzed under physiological conditions by fast-freezing and freeze-fracturing, shows "focal fusion"; see Knoll and Plattner; 1989). Extended contact areas are observed with fusing liposomes (fast-frozen: Rand *et al.*, 1985), and formation of a diaphragm was also believed to occur (Papahadjopoulos *et al.*, 1978). Diaphragm formation was observed with fusing planar membranes (Neher, 1974). This is another lesson on the diversity of different fusion processes (see Table I), notably when various *in vitro* systems are compared with exocytosis.

At about the same time that point fusion was detected to occur during exocytotic membrane fusion (see earlier), inverted lipid micelles (HII hexagonal phase of lipids) were proposed as fusion intermediates, when artificial membranes fuse in the presence of a pCa $\geq$ 3 (Verkleij *et al.*, 1979; Siegel, 1984, 1986, 1987; Cullis *et al.*, 1986). Others assumed that a focal stochastic rearrangement of lipids occurs during fusion of artificial membranes (Hui *et al.*, 1981; Boni *et al.*, 1984; Düzgünes *et al.*, 1984, 1987), and that 10-nm large lipidic particles seen on freeze fracture replicas would not represent early fusion stages. This would be consistent with the fact that no lipidic particle has ever been seen precisely on a fusion site. Their lifetime in the millisecond range (Siegel, 1986) should allow their detection by fast-freezing methods (Knoll *et al.*, 1987; Plattner and Knoll, 1987), as one can identify 10-nm fusion stages during exocytosis (Plattner, 1989a; see also Section III). Thus, lipid particles might or might not represent equivalents of membrane fusion via inverted lipid micelles. The recent assumption of pore-forming fusion proteins (see Section XII) implies a third alternative for causing a rearrangement of lipids during exocytosis.

Regular IMP aggregates in the form of fusion rosettes have been found in the freeze-fractured cell membrane at preformed exocytosis sizes only in some protozoan cells (sporozoa, ciliates: Plattner, 1989a). Rosettes are, in most cases, composed of 6–10 particles, each ~10 nm in size, with a small area between particles where fusion can occur. The tightly apposed trichocyst membrane is also devoid of IMP (Plattner, 1987, 1989a), which hence do not display any structural equivalents of rosette particles. This is important with regard to the "matching-particle" hypothesis by others, as outlined in Section XII. There is as yet no evidence that rosette particles would represent Ca^{2+} channels (Gilligan and Satir, 1983), although this could be possible. As described in Section VI, *nd* mutants of *Paramecium* lack these fusion rosettes.

An analogous structure might be the double rows of IMP at active zones of nerve terminals (for references see Plattner, 1989a; Rash *et al.*, 1988, Smith and Augustine, 1988); see Section XIII. These as well as protozoan fusion rosettes are seen on ultrathin sections to be backed by electron-dense connecting material of a rather amorphous texture, probably representing proteins (see earlier and Section VI). Yet since this was found only rarely (perhaps due to the lack of appropriate fixation; see

Plattner, 1981) with other cell types, like gland or mast cells, this has for a long time not been considered as a general characteristic of exocytosis sites. Now the postulate that membrane-associated proteins occur at exocytosis sites, as proposed some time ago, has also been widely accepted (Plattner, 1981, 1989b).

Significant IMP rearrangements do not occur before fusion occurs in any of the exocytotic systems analyzed by fast-freezing methods (see also Section VI). This holds for chromaffin gland (Schmidt *et al.*, 1983) and mast cells (Chandler and Heuser, 1980), for *Limulus* amebocytes (Ornberg and Reese, 1981), for the cortical granule reaction in egg cells (Chandler and Heuser, 1979), as well as for trichocyst release in *Paramecium* (Olbricht *et al.*, 1984). For reasons mentioned before, goblet cells might be a rare exception (Specian and Neutra, 1980). The rearrangement of connecting materials between secretory organelles and the cell membrane during exocytosis has not yet been analyzed in detail.

With Golgi vesicles, a nonclathrin coat material has been identified (Orci *et al.*, 1986), which is thought somehow to mediate membrane fusion (Orci *et al.*, 1986). This is another example of PF fusion, which, therefore, might resemble exocytosis.

XVII. Conclusions and Outlook

Figure 1 presents a survey of hypotheses currently discussed for the regulation of membrane fusion during exocytosis. Many—if not all—of these aspects are still highly controversial, as discussed throughout this review, for various reasons: (a) Membrane fusion is very rapid (millisecond time range) and restricted to a small area (10-nm range; "focal [point] fusion"). Previous ultrastructural results (using cryofixation techniques) along these lines have since been corroborated by electrophysiological methods. (b) In most cases this process is superimposed by a variety of other phenomena that precede (intracellular organelle transport, receptor and ion channel activation, signal transfer mechanisms) or follow (discharge of secretory contents, exocytosis-coupled endocytosis, reestablishment of resting conditions—as of Ca^{2+} levels, etc.).

It is now assumed that an increase of $[Ca^{2+}]_i$ is not sufficient to produce exocytotic membrane fusion, although it might facilitate fusion by local charge screening and dehydration of phospholipid head groups. There is some evidence for the involvement of Ca^{2+}-regulated binding proteins and/or Ca^{2+}-dependent enzymatic processes. An increased $[Ca^{2+}]_i$ may come from an influx of Ca_o^{2+} or from intracellular storage compartments (ER or calciosomes), or from both sources. The increase of pCa_i^{2+} may be accompanied by formation of different second messengers (cAMP, DAG, IP_3 from PI turnover, etc.), and this might activate different functions, depending on the cell type analyzed and the trigger applied.

With some systems, protein phosphorylation via PKC (activated by DAG from PI turnover) is observed, although there is no stringent proof for its immediate participation in membrane fusion; several physiological substrates of PKC are known, but none that would be directly relevant for fusion induction. Another line of evidence suggests that protein dephosphorylation might be crucial. Calmodulin could also participate in exocytosis, via either protein phosphorylation or dephosphorylation. Possibly there occur quite different regulatory principles—for instance, either via PI turnover or via CaM, depending on the system. None of them might be universal.

The argument that exocytosis would require ATP (for protein phosphorylation, etc.) is no longer tenable; at least ATP is not required for the crucial step of membrane fusion. With *Paramecium* the dephosphorylation of a particular protein is strictly correlated with exocytosis, which can be inhibited by Ab against a phosphoprotein, CaM, or a phosphatase, or induced by cytosolic application of phosphatases.

Other CaBP, like some chromobindins (synexin) and, in addition, synaptophysin, are also taken into consideration for establishing contact between membranes and eventually to serve as fusogens. Stringent proof from *in vivo* experiments (with Ab inhibition, etc.) is still missing.

Lysophospholipids or arachidonic acid, produced by phospholipases, were also assumed to contribute to fusion, since they are known to be fusogenic with *in vitro* model experiments; yet their role *in vivo* remains enigmatic.

Most importantly, GTP analogs have been found to stimulate membrane fusion in some systems; they might act by various mechanisms. (a) During PI turnover PLC is modulated via different G proteins. (b) Since GTP analogs can produce exocytotic membrane fusion without a $[Ca^{2+}]_i$ increase, there might also occur an independent fusogenic mechanism operating at resting $[Ca^{2+}]_i$ levels. A special type of G protein (G_E) has been proposed to be more directly engaged in exocytosis, although its molecular identity remains unknown. This line of work also revealed an inhibitory effect of ATP and thus underscores the possible relevance of protein dephosphorylation for fusion induction.

The following parameters are no longer considered as driving forces for membrane fusion: membrane potential (Zucker and Haydon, 1988), acidic pH in secretory granules (see Section VII), or osmotic effects (see Section XIV), as well as any direct effect of Mt or Mf (Section V). The irrelevance of ΔpH or ΔV was stressed some time ago by Baker and Knight (1984a,b).

Therefore, one currently has to pursue the possibility that exocytotic membrane fusion is subjected to a multifactorial control (possibly with different lines expressed in different systems). If any ion would be directly involved at all, it would be Ca^{2+}, although an increased pCa_i is no longer generally accepted to be either mandatory or sufficient for controlling membrane fusion during exocytosis.

Of course, some rearrangement of lipids is necessary for fusion to occur. The precise molecular structure of fusing lipids (micelles or rather random) has not yet been resolved. Another new point is the likely involvement of proteins (different candidates being currently discussed) that might transiently be integrated into the membranes to be fused and thus could impose some rearrangement of the lipids. However, no fusogenic protein has yet been unequivocally identified with any exocytotic system. Future work will certainly be focused on this aspect.

The funnel- or nipplelike deformation of the membrane contact areas before or during exocytosis could reflect some importance of the local curvature for the fusion process.

In conclusion, further progress in this field has to overcome a variety of difficulties: (a) The process to be analyzed might be subjected to a multifactorial control. (b) Diverse mechanisms might be executed in different systems. (c) Some mechanisms are quickly counterregulated and thus might be obscured (like de/rephosphorylation cycles), particularly in nonsynchronous systems. (d) Some messengers or effectors (e.g., Ca^{2+}) probably exert multiple effects. (e) Drugs applied to specify certain mechanisms have frequently multiple effects; Ab should be used as far as possible. (f) Exocytotic membrane fusion takes only extremely little time, sometimes without any appreciable lag period after trigger application. (g) It is also considerably restricted in space. It follows from points (f) and (g) that relevant changes might comprise time periods far beyond the resolution of most methods (except for freeze–fracture EM analysis and electrophysiology) and that only as little as thousandths of a percent of the total membrane area participate in membrane fusion during secretion. (h) Finally, it still appears particularly difficult to get access to the mechanisms controlling constitutive exocytosis.

All this might explain why the elucidation of the regulation of membrane fusion during exocytosis is so difficult and why the progress we achieve is so sluggish, despite the considerable efforts in this field.

Acknowledgments

I thank Mrs. A. Lippus-Broll for her patience and skill with text processing of the different versions of this manuscript. Some of the author's original work cited in this review has been supported by SFB 156.

REFERENCES

Adam-Vizi, V., Knight, D., and Hall, A. (1987). *Nature* (*London*) **328**, 581.
Adam-Vizi, V., Rosener, S., Aktories, K., and Knight, D. E. (1988). *FEBS Lett.* **238**, 277–280.
Adoutte, A. (1988). *In* "Paramecium" (H. D. Görtz, ed.), pp. 325–362. Springer-Verlag, Berlin.
Ahkong, Q. F., and Lucy, J. A. (1988). *J. Cell Sci.* **91**, 597–601.
Ahnert-Hilger, G., and Gratzl, M. (1987). *J. Neurochem.* **49**, 764–770.
Ahnert-Hilger, G., Bhakdi, S., and Gratzl, M. (1985). *J. Biol. Chem.* **260**, 12730–12734.
Ahnert-Hilger, G., Weller, U., Dauzenroth, M. E., Habermann, E., and Gratzl, M. (1989). *FEBS Lett.* **242**, 245–248.
Akabas, M. H., Cohen, F. S., and Finkelstein, A. (1984). *J. Cell Biol.* **98**, 1063–1071.
Allen, R. D., Weiss, D. G., Hayden, J. H., Brown, D. T., Fujiwake, H., and Simpson, M. (1985). *J. Cell Biol.* **100**, 1736–1752.
Allende, J. E. (1988). *FASEB J.* **2**, 2356–2367.
Almers, W., and Breckenridge, L. J. (1988). *In* "Molecular Mechanisms of Membrane Fusion" (S. Ohki, D. Doyle, T. D. Flanagan, S. W. Hui, and E. Mayhew, eds.), pp. 197–208. Plenum, New York.
Altman, J. (1988). *Nature* (*London*) **311**, 119–120.
Asano, T., Ogasawara, N., and Sano, M. (1986). *FEBS Lett.* **203**, 135–138.
Asano, T., Semba, R., Kamiya, N., Ogasawara, N., and Kato, K. (1988). *J. Neurochem.* **50**, 1164–1169.
Aufderheide, K. J. (1977). *Science* **198**, 299–300.
Aufderheide, K. J. (1978). *Mol. Gen. Genet.* **165**, 199–205.
Aunis, D., Hesketh, J. E., and Devilliers, G. (1979). *Cell Tissue Res.* **197**, 433–441.
Axelrod, J., Burch, R. M., and Jelsema, C. L. (1988). *Trends Neurosci.* **11**, 117–123.
Baker, P. F. (1972). *Prog. Biophys. Mol. Biol.* **24**, 177–223.
Baker, P. F. (1988). *Curr. Top. Membr. Transp.* **32**, 115–138.
Baker, P. F., and Knight, D. E. (1979). *Trends Neurosci.* **2**, 288–291.
Baker, P. F., and Knight, D. E. (1981). *Philos. Trans. R. Soc. London, Ser. B* **296**, 83–103.
Baker, P. F., and Knight, D. E. (1984a). *Biosci. Rep.* **4**, 285–298.
Baker, P. F., and Knight, D. E. (1984b). *Trends Neurosci.* **7**, 120–126.
Baker, P. F., and Whitaker, M. J. (1978). *Nature* (*London*) **276**, 513–515.
Baldassare, J. J., and Fisher, G. J. (1986a). *Biochem. Biophys. Res. Commun.* **137**, 801–805.
Baldassare, J. J., and Fisher, G. J. (1986b). *J. Biol. Chem.* **261**, 11942–11944.
Barnes, D. M. (1986). *Science* **234**, 286–288.
Barrowman, M. M., Cockcroft, S., and Gomperts, B. D. (1986). *Nature* (*London*) **319**, 504–507.
Barrowman, M. M., Cockcroft, S., and Gomperts, B. D. (1987). *J. Physiol.* (*London*) **383**, 115–124.
Bar-Sagi, D., Fernandez, A., and Feramisco, J. R. (1987). *Biosci. Rep.* **7**, 427–434.
Bar-Sagi, D., Suhan, J. P., McCormick, F., and Feramisco, J. R. (1988). *J. Cell Biol.* **106**, 1649–1658.
Batenburg, A. M., Bougis, P. E., Rochat, H., Verkleij, A. J., and DeKruijff, B. (1985). *Biochemistry* **24**, 7101–7110.
Baum, B. J., Freiberg, J. M., Ito, H., Roth, G. S., and Filburn, C. R. (1981). *J. Biol. Chem.* **256**, 9731–9736.
Beaudoin, A. R., Gilbert, L., St.-Jean, P., Grondin, G., and Cabana, C. (1988). *Eur. J. Cell Biol.* **47**, 233–240.

Beaven, M. A., Moore, J. P., Smith, G. A., Hesketh, T. R., and Metcalfe, J. C. (1984a). *J. Biol. Chem.* **259,** 7137–7142.

Beaven, M. A., Rogers, J., Moore, J. P., Hesketh, T. R., Smith, G. A., and Metcalfe, J. C. (1984b). *J. Biol. Chem.* **259,** 7129–7136.

Beisson, J., Lefort-Tran, M., Pouphile, M., Rossignol, M., and Satir, B. (1976). *J. Cell Biol.* **69,** 126–143.

Beisson, J., Cohen, J., Lefort-Tran, M., Pouphile, M., and Rossignol, M. (1980). *J. Cell Biol.* **85,** 213–227.

Berl, S., Puszkin, S., and Niklas, W. J. (1973). *Science* **179,** 441–446.

Berridge, M. J. (1984). *Biochem. J.* **220,** 345–360.

Berridge, M. J. (1987a). *ISI Atlas Sci. Pharmacol.* **1,** 91–97.

Berridge, M. J. (1987b). *Annu. Rev. Biochem.* **56,** 159–193.

Berridge, M. J. (1988). *Proc. R. Soc. London, Ser. B* **234,** 359–378.

Berridge, M. J., and Galione, A. (1988). *FASEB J.* **2,** 3074–3082.

Berridge, M. J., and Irvine, R. F. (1984). *Nature (London)* **312,** 315–320.

Berridge, M. J., Cobbold, P. H., and Cuthbertson, K. S. R. (1988). *Philos. Trans. R. Soc. London Ser. B* **320,** 325–343.

Bers, D. M., and Harrison, S. M. (1987). *J. Physiol. (London)* **386,** 58P.

Bhakdi, S., and Tranum-Jensen, J. (1987). *Rev. Physiol. Biochem. Pharmacol.* **107,** 147–223.

Bilinski, M., Plattner, H., and Matt H. (1981). *J. Cell Biol.* **88,** 179–188.

Bittiger, H., and Heid, J. (1977). *J. Neurochem.* **28,** 917–922.

Bittner, M. A., Holz, R. W., and Neubig, R. R. (1986). *J. Biol. Chem.* **261,** 10182–10188.

Blumenthal, R. (1987). *Curr. Top. Membr. Transp.* **29,** 203–254.

Blumenthal, R., Puri, A., Walter, A., and Eidelman, O. (1988). *In* "Molecular Mechanisms of Membrane Fusion" (S. Ohki, D. Doyle, T. D. Flanagan, S. W. Hui, and E. Mayhew, eds.), pp. 367–383. Plenum, New York.

Boam, D. S. W., Spanner, S. G., Ansell, G. B., and Stanworth, D. R. (1988). *Biochim. Biophys. Acta* **971,** 215–222.

Boni, L. T., Hah, J. S., Hui, S. W., Mukherjee, P., Ho, J. T., and Jung, C. Y. (1984). *Biochim. Biophys. Acta* **775,** 409–418.

Boroch, G. M., Bickford, K., and Bohl, B. P. (1988). *J. Cell Biol.* **106,** 1927–1936.

Bourne, H. R. (1988). *Cell* **53,** 669–671.

Branton, D., Bullivant, S., Gilula, N. B., Karnowsky, M. J., Moor, H., Mühlethaler, K., Northcote, D. H., Packer, L., Satir, B., Speth, V., Staehelin, L. A., Steere, R., and Weinstein, R. S. (1975). *Science* **190,** 54–56.

Breckenridge, L. J., and Almers, W. (1987). *Proc. Natl. Acad. Sci. USA* **84,** 1945–1949.

Breckenridge, W. C., Morgan, I. G., Zanetta, J. P., and Vincendon, G. (1973). *Biochim. Biophys. Acta* **320,** 681–686.

Brethes, D., Dayanithi, G., Letellier, L., and Nordmann, J. J. (1987). *Proc. Natl. Acad. Sci. USA* **84,** 1439–1443.

Brocklehurst, K. W., and Hutton, J. C. (1983). *Biochem. J.* **210,** 533–539.

Brocklehurst, K. W., and Hutton, J. C. (1984). *Biochem. J.* **220,** 283–290.

Brocklehurst, K. W., and Pollard, H. B. (1985). *FEBS Lett.* **183,** 107–110.

Brocklehurst, K. W., and Pollard, H. B. (1988a). *Curr. Top. Membr. Transp.* **32,** 203–225.

Brocklehurst, K. W., and Pollard, H. B. (1988b). *FEBS Lett.* **234,** 439–445.

Brocklehurst, K. W., Morita, K., and Pollard, H. B. (1985). *Biochem. J.* **228,** 35–42.

Brooks, J. C., Treml, S., and Brooks, M. (1984). *Life Sci.* **35,** 569–574.

Brown, E. M., Redgrave, J., and Thatcher, J. (1984). *FEBS Lett.* **175,** 72–75.

Burger, K. N. J., Knoll, G., and Verkleij, A. J. (1988). *Biochim. Biophys. Acta* **939,** 89–101.

Burgoyne, R. D. (1984). *Biochim. Biophys. Acta* **779,** 201–216.

Burgoyne, R. D. (1987). *Nature* (*London*) **328**, 112–113.
Burgoyne, R. D., and Cheek, T. R. (1987). *Biosci. Rep.* **7**, 281–288.
Burgoyne, R. D., and Geisow, M. J. (1982). *J. Neurochem.* **39**, 1387–1396.
Burgoyne, R. D., and Geisow, M. J. (1989). *Cell Calcium* **10**, 1–10.
Burgoyne, R. D., Cheek, T. R., and Norman, K. M. (1986). *Nature* (*London*) **319**, 68–70.
Burgoyne, R. D., Morgan, A., and O'Sullivan, A. J. (1988). *FEBS Lett.* **238**, 151–155.
Burnham, D. B., and Williams, J. A. (1982). *J. Biol. Chem.* **257**, 10523–10528.
Burnham, D. B., Sung, C. K., Munowitz, P., and Williams, J. A. (1988). *Biochim. Biophys. Acta* **969**, 33–39.
Burridge, K., and Phillips, J. H. (1975). *Nature* (*London*) **254**, 256–258.
Busa, W. B., and Nuccitelli, R. (1985). *J. Cell Biol.* **100**, 1325–1329.
Busa, W. B., Ferguson, J. E., Joseph, S. K., Williamson, J. R., and Nuccitelli, R. (1985). *J. Cell Biol.* **101**, 677–682.
Cameron, R. S., and Castle, J. D. (1984). *J. Membr. Biol.* **79**, 127–144.
Campbell, A. K. (1983). "Intracellular Calcium Its Universal Role as Regulator." Wiley, Chichester, England.
Campbell, K. P., Leung, A. T., and Sharp, A. H. (1988). *Trends Neurosci.* **11**, 425–430.
Carafoli, E. (1987). *Annu. Rev. Biochem.* **56**, 305–433.
Caratsch, C. G., Schumacher, S., Grassi, F., and Eusebi, F. (1988). *Naunyn-Schmiedeberg's Arch. Pharmacol.* **337**, 9–12.
Casey, P. J., and Gilman, A. G. (1988). *J. Biol. Chem.* **263**, 2577–2580.
Cazalis, M., Dayanithi, G., and Nordmann, J. J. (1987). *J. Physiol.* (*London*) **390**, 71–91.
Chan, K. M., and Junger, K. D. (1983). *J. Biol. Chem.* **258**, 4404–4410.
Chandler, D. E. (1984). *J. Cell Sci.* **72**, 23–36.
Chandler, D. E. (1988). *Curr. Top. Membr. Transp.* **32**, 169–202.
Chandler, D. E., and Heuser, J. E. (1979). *J. Cell Biol.* **83**, 91–108.
Chandler, D. E., and Heuser, J. E. (1980). *J. Cell Biol.* **86**, 666–674.
Chandler, D. E., and Kazilek, C. J. (1987). *Biochim. Biophys. Acta* **931**, 175–179.
Cheek, T. R., and Burgoyne, R. D. (1987). *J. Biol. Chem.* **262**, 11663–11666.
Clapham, D. E., and Neher, E. (1984). *J. Physiol.* (*London*) **353**, 541–564.
Cobbold, P. H., Cheek, T. R., Cuthbertson, K. S. R., and Burgoyne, R. D. (1987). *FEBS Lett.* **211**, 44–48.
Cockcroft, S., and Stutchfield, J. (1988). *Philos. Trans. R. Soc. London, Ser. B* **320**, 247–265.
Cockcroft, S., Howell, T. W., and Gomperts, B. D. (1987). *J. Cell Biol.* **105**, 2745–2750.
Cohen, F. S., Zimmerberg, J., and Finkelstein, A. (1980). *J. Gen. Physiol.* **75**, 251–270.
Cohen, J., and Beisson, J. (1980). *Genetics* **95**, 797–818.
Cohen, J., and Beisson, J. (1988). *In* "Paramecium" (H. D. Görtz, ed.), pp. 363–392. Springer-Verlag, Berlin.
Cohen, P. (1988). *Proc. R. Soc. London, Ser. B.* **234**, 115–144.
Cohn, J. A., Kinder, B., Jamieson, J. D., Delahunt, N. G., and Gorelick, F. S. (1987). *Biochim. Biophys. Acta* **928**, 320–331.
Collins, S. A., Pon, D. J., and Sen, A. K. (1987). *Biochim. Biophys. Acta* **927**, 392–401.
Cooper, J. A. (1987). *J. Cell Biol.* **105**, 1473–1478.
Côté, A., Doucet, J. P., and Trifaró, J. M. (1986). *Neuroscience* **19**, 629–645.
Couteaux, R. (1978). "Recherches Morphologiques et Cytochimiques sur l'Organisation des Tissus Excitables." Robin & Mareuge, Paris.
Crabb, J. H., and Jackson, R. C. (1985). *J. Cell Biol.* **101**, 2263–2273.
Cran, D. G., Moor, R. M., and Irvine, R. F. (1988). *J. Cell Sci.* **91**, 139–144.
Creutz, C. E. (1981). *J. Cell Biol.* **91**, 247–257.
Creutz, C. E., and Pollard, H. B. (1983). *J. Auton. Nerv. Syst.* **7**, 13–18.

Creutz, C. E., Scott, J. H., Pazoles, C. J., and Pollard, H. B. (1982). *J. Cell. Biochem.* **18,** 87–97.

Creutz, C. E., Dowling, L. G., Sando, J. J., Villar-Palasi, C., Whipple, J. H., and Zaks, W. J. (1983). *J. Biol. Chem.* **258,** 14664–14674.

Creutz, C. E., Dowling, L. G., Kyger, E. M., and Franson, R. C. (1985). *J. Biol. Chem.* **260,** 7171–7173.

Creutz, C. E., Zaks, W. J., Hamman, H. C., Crane, S., Martin, W. H., Gould, K. L., Oddie, K. M., and Parsons, S. J. (1987a). *J. Biol. Chem.* **262,** 1860–1868.

Creutz, C. E., Zaks, W. J., Hamman, H. C., and Martin, W. H. (1987b). *In* "Cell Fusion" (A. E. Sowers, ed.), pp. 45–68. Plenum, New York.

Crompton, M. R., Moss, S. E., and Crumpton, M. J. (1988). *Cell* **55,** 1–3.

Crossley, I., Swann, K., Chambers, E., and Whitaker, M. (1988). *Biochem. J.* **252,** 257–262.

Cullis, P. R., DeKruijff, B., Verkleij, A. J., and Hope, M. J. (1986). *Biochem. Soc. Trans.* **14,** 242–245.

Dainaka, J., Ichikawa, A., Koibuchi, Y., Nakagawa, M., and Tomita, K. (1986). *Biochem. Pharmacol.* **35,** 3739–3744.

DaPrada, M., Pletscher, A., and Tranzer, J. P. (1972). *Biochem. J.* **127,** 681–683.

Dartt, D. A., Guerina, V. J., Donowitz, M., Taylor, L., and Sharp, G. W. G. (1982). *Biochem. J.* **202,** 799–802.

Das, S., and Rand, R. P. (1984). *Biochem. Biophys. Res. Commun.* **124,** 491–496.

Datyner, N. B., and Gage, P. W. (1980). *J. Physiol. (London)* **303,** 299–314.

Davis, B., and Lazarus, N. R. (1976). *J. Physiol. (London)* **256,** 709–729.

Dean, P. M. (1974). *In* "Secretory Mechanisms of Exocrine Glands" (N. A. Thorn and O. H. Petersen, eds.), pp. 152–161. Munksgaard, Copenhagen.

DeBlock, J., and DePotter, W. (1987). *FEBS Lett.* **222,** 358.

Decker, G. L., and Lennarz, W. J. (1979). *J. Cell Biol.* **81,** 92–103.

Decker, S. J., and Kinsey, W. H. (1983). *Dev. Biol.* **96,** 37–45.

DeLisle, R. C., and Williams, J. A. (1986). *Annu. Rev. Physiol.* **48,** 225–238.

DeLorenzo, R. J. (1982). *In* "Calcium and Cell Function" (W. Y. Cheung, ed.), pp. 271–309. Academic Press, New York.

DeLorenzo, R. J., Burdette, S., and Holderness, J. (1981). *Science* **213,** 545–549.

DeRiemer, S. A., Strong, J. A., Albert, K. A., Greengard, P., and Kaczmarek, L. K. (1985). *Nature (London)* **313,** 313–316.

DeRobertis, E., and Vaz Ferreira, A. (1957). *Exp. Cell Res.* **12,** 568–574.

Detering, N. K., Decker, G. L., Schmell, E. D., and Lennarz, W. J. (1977). *J. Cell Biol.* **75,** 899–914.

Diaz, R., Mayorga, L., and Stahl, P. (1988). *J. Biol. Chem.* **263,** 6093–6100.

DiVirgilio, F., Lew, D. P., and Pozzan, T. (1984). *Nature (London)* **310,** 691–693.

Dormer, R. L., Brown, G. R., Doughney, C., and McPherson, M. A. (1987). *Biosci. Rep.* **7,** 333–344.

Doughney, C., Brown, G. R., McPherson, M. A., and Dormer, R. L. (1987). *Biochim. Biophys. Acta* **928,** 341–348.

Douglas, W. W. (1968). *Br. J. Pharmacol.* **34,** 451–474.

Douglas, W. W. (1974). *Biochem. Soc. Symp.* **39,** 1–28.

Douglas, W. W., and Poisner, A. M. (1962). *J. Physiol. (London)* **162,** 385–392.

Dowling, L. G., and Creutz, C. E. (1985). *Biochem. Biophys. Res. Commun.* **132,** 382–389.

Drenckhahn, D., Gröschel-Stewart, U., and Unsicker, K. (1977). *Cell Tissue Res.* **183,** 273–279.

Drenckhahn, D., Gröschel-Stewart, U., Kendrick-Jones, J., and Scholey, J. M. (1983). *Eur. J. Cell Biol.* **30,** 100–111.

Drust, D. S., and Creutz, C. E. (1988). *Nature* (*London*) **331**, 88-91.
Düzgünes, N. (1985). *In* "Subcellular Biochemistry" (D. B. Roodyn, ed.), Vol. 11, pp. 195-286. Plenum, New York.
Düzgünes, N., Paiement, J., Freeman, K. B., Lopez, N. G., Wilschut, J., and Papahadjopoulos, D. (1984). *Biochemistry* **23**, 3486-3494.
Düzgünes, N., Hong, K., Baldwin, P. A., Bentz, J., Nir, S., and Papahadjopoulos, D. (1987). *In* "Cell Fusion" (A. E. Sowers, ed.), pp. 241-267. Plenum, New York.
Dunlop, M., and Larkins, R. G. (1988). *Biochem. J.* **253**, 67-72.
Elhai, J., and Scandella, C. J. (1983). *Exp. Cell Res.* **148**, 63-71.
Ellens, H., Bentz, J., and Szoka, F. C. (1986). *Biochemistry* **25**, 4141-4147.
Epel, D., and Vacquier, V. D. (1978). *In* "Membrane Fusion" (G. Poste and G. L. Nicolson, eds.), pp. 1-63. North-Holland Publ., Amsterdam.
Erne, P., Schachter, M., Fabbro, D., Miles, C. M. M., and Sever, P. S. (1987). *Biochem. Biophys. Res. Commun.* **145**, 66-72.
Evered, D., and Collins, G. M. (1982). *Ciba Found. Symp.* **92**.
Ewald, D. A., Williams, A., and Levitan, I. B. (1985). *Nature* (*London*) **315**, 503-506.
Farach, H. A., Mundi, D. I., Strittmatter, W. J., and Lennarz, W. J. (1987). *J. Biol. Chem.* **262**, 5483-5481.
Farber, L. H., Wilson, F. J., and Wolff, D. J. (1987). *J. Neurochem.* **49**, 404-414.
Ferguson, J. E., and Shen, S. S. (1984). *Gamete Res.* **9**, 329-338.
Fernandez, J. M., Neher, E., and Gomperts, B. D. (1984). *Nature* (*London*) **312**, 453-455.
Fernandez, J. M., Lindau, M., and Eckstein, F. (1987). *FEBS Lett.* **216**, 89-93.
Fewtrell, C., and Sherman, E. (1987). *Biochemistry* **26**, 6995-7003.
Finkelstein, A., Zimmerberg, J., and Cohen, F. S. (1986). *Annu. Rev. Physiol.* **48**, 163-174.
Foreman, J. C., and Lichtenstein, L. M. (1980). *Biochim. Biophys. Acta* **629**, 587-603.
Fournier, S., and Trifaró, J. M. (1988a). *J. Neurochem.* **50**, 27-37.
Fournier, S., and Trifaró, J. M. (1988b). *J. Neurochem.* **51**, 1599-1609.
Freedman, S. D., and Jamieson, J. D. (1982). *J. Cell Biol.* **95**, 909-917.
Frei, E., and Zahler, P. (1979). *Biochim. Biophys. Acta* **550**, 450-463.
Frischenschlager, I., Schmidt, W., and Winkler, H. (1983). *J. Neurochem.* **41**, 1480-1483.
Frye, R. A., and Holz, R. W. (1983). *Mol. Pharmacol.* **23**, 547-550.
Frye, R. A., and Holz, R. W. (1984). *J. Neurochem.* **43**, 146-150.
Frye, R. A., and Holz, R. W. (1985). *J. Neurochem.* **44**, 265-273.
Gaardsvoll, H., Obendorf, D., Winkler, H., and Bock, E. (1988). *FEBS Lett.* **242**, 117-120.
Geisow, M. J., and Burgoyne, R. D. (1982). *J. Neurochem.* **38**, 1735-1741.
Geisow, M. J., Walker, J. H., Boustead, C., and Taylor, W. (1987). *Biosci. Rep.* **7**, 289-298.
Gething, M. J., Doms, R. W., York, D., and White, J. (1986). *J. Cell Biol.* **102**, 11-23.
Gill, D. L., Ueda, T., Chueh, S. H., and Noel, M. W. (1986). *Nature* (*London*) **320**, 461-464.
Gill, D. L., Mullaney, J. M., and Ghosh, T. K. (1988). *J. Exp. Biol.* **139**, 105-133.
Gilligan, D. M., and Satir, B. H. (1982). *J. Biol. Chem.* **257**, 13903-13906.
Gilligan, D. M., and Satir, B. H. (1983). *J. Cell Biol.* **97**, 224-234.
Gilman, A. G. (1987). *Annu. Rev. Biochem.* **56**, 615-649.
Glossmann, H., and Striessnig, J. (1988). *ISI Atlas Sci. Pharmacol.* **2**, 202-210.
Goldenring, J. R., Gonzalez, B., McGuire, J. S., and DeLorenzo, R. J. (1983). *J. Biol. Chem.* **258**, 12632-12640.
Gomperts, B. D., Barrowman, M. M., and Cockcroft, S. (1986). *Fed. Proc.* **45**, 2156-2161.
Gomperts, B. D., Cockcroft, S., Howell, T. W., Nüsse, O., and Tatham, P. E. R. (1987). *Biosci. Rep.* **7**, 369-381.
Gorelick, F. S., Wang, J. K. T., Lai, Y., Nairn, A. C., and Greengard, P. (1988). *J. Biol. Chem.* **263**, 17209-17212.

Goto, S., Matsukado, Y., Mihara, Y., Inoue, N., and Miyamoto, E. (1986). *Brain Res.* **397**, 161–172.

Goud, B., Salminen, A., Walworth, N. C., and Novick, P. J. (1988). *Cell* **53**, 753–768.

Grapengiesser, E., Gylfe, E., and Hellman, B. (1989). *Arch. Biochem. Biophys.* **268**, 404–407.

Gray, E. G. (1983). *Proc. R. Soc. London, Ser. B* **218**, 253–258.

Guraya, S. S. (1982). *Int. Rev. Cytol.* **78**, 257–360.

Gustin, M., and Hennessey, T. M. (1988). *Biochim. Biophys. Acta* **940**, 99–104.

Gutierrez, L. M., Ballesta, J. J., Hidalgo, M. J., Gandia, L., García, A. G., and Reig, J. A. (1988). *J. Neurochem.* **51**, 1023–1030.

Haacke, B., and Plattner, H. (1984). *Exp. Cell Res.* **151**, 21–28.

Hall, P. F. (1982). *In* "Cellular Regulation of Secretion and Release" (P. M. Conn, ed.), pp. 195–222. Academic Press, New York.

Harris, B., Cheek, T. R., and Burgoyne, R. D. (1986). *Biochim. Biophys. Acta* **889**, 1–5.

Harrison, D. E., and Ashcroft, S. J. H. (1982). *Biochim. Biophys. Acta* **714**, 313–319.

Hashimoto, Y., King, M. M., and Soderling, T. R. (1988). *Proc. Natl. Acad. Sci. USA* **85**, 7001–7005.

Hauser, H., Levine, B. A., and Williams, R. J. P. (1976). *Trends Biochem. Sci.* **1**, 278–281.

Haycock, J. W., Browning, M. D., and Greengard, P. (1988). *Proc. Natl. Acad. Sci. USA* **85**, 1677–1681.

Hempstead, B. L., Parker, C. W., and Kulczycki, A. (1983). *Proc. Natl. Acad. Sci. USA* **80**, 3050–3053.

Henne, V., and Söling, H. D. (1986). *FEBS Lett.* **202**, 267–273.

Hermann, J., Cayla, X., Dumortier, K., Goris, J., Ozon, R., and Merlevede, W. (1988). *Eur. J. Biochem.* **173**, 17–25.

Hescheler, J., Kameyama, M., and Trautwein, W. (1986). *Eur. J. Physiol.* **407**, 182–189.

Heuser, J. E., Reese, T. S., Dennis, M. J., Jan, Y., Jan, L., and Evans, L. (1979). *J. Cell Biol.* **81**, 275–300.

Höhne, B., Plattner, H., and Hofer, W. (1989). In preparation.

Hollenberg, M. D., Valentine-Braun, K. A., and Northup, J. K. (1988). *Trends Pharmacol. Sci.* **9**, 63–66.

Holz, R. W. (1986). *Annu. Rev. Physiol.* **48**, 175–189.

Holz, R. W., and Senter, R. A. (1986). *J. Neurochem.* **46**, 1835–1842.

Holz, R. W., and Senter, R. A. (1988). *Cell. Mol. Neurobiol.* **8**, 115–128.

Hong, K., Düzgünes, N., Ekerdt, R., and Papahadjopoulos, D. (1982). *Proc. Natl. Acad. Sci. USA* **79**, 4642–4644.

Hong, K., Düzgünes, N., Meers, P. R., and Papahadjopoulos, D. (1987). *In* "Cell Fusion" (A. E. Sowers, ed.), pp. 269–284. Plenum, New York.

Hopkins, C. R., and Duncan, C. J. (1979). "Secretory Mechanisms." Cambridge Univ. Press, London.

Howell, S. L., and Tyhurst, M. (1977). *J. Cell Sci.* **27**, 289–301.

Howell, T. W., and Gomperts, B. D. (1987). *Biochim. Biophys. Acta* **927**, 177–183.

Howell, T. W., Cockcroft, S., and Gomperts, B. D. (1987). *J. Cell Biol.* **105**, 191–197.

Huang, K., Wallner, B. P., Mattaliano, R. J., Tizard, R., Burne, C., Frey, A., Hession, C., McGray, P., Sinclair, L. K., Chow, E. P., Browning, J. L., Ramachandran, K. L., Tang, J., Smart, J. E., and Pepinsky, R. B. (1986). *Cell* **46**, 191–199.

Huang, S. K., and Hui, S. W. (1986). *Biochim. Biophys. Acta* **806**, 539–548.

Hui, S. W., Stewart, T. P., Boni, L. T., and Yeagle, P. L. (1981). *Science* **212**, 921–923.

Hui, S. W., Isac, T., Boni, L. T., and Sen, A. (1985). *J. Membr. Biol.* **84**, 137–146.

Hui, S. W., Nir, S., Stewart, T. P., Boni, L. T., and Huang, S. K. (1988). *Biochim. Biophys. Acta* **941**, 130–140.

Imai, S., and Onozuka, M. (1988). *Comp. Biochem. Physiol. C* **91C**, 535–540.

Ingebritsen, T. S., and Cohen, P. (1983). *Eur. J. Biochem.* **132**, 255–261.

Inomata, M., Kasai, Y., Nakamura, M., and Kawashima, S. (1988). *J. Biol. Chem.* **263**, 19783–19787.

Irvine, R. F. (1988). *ISI Atlas Sci. Biochem.* **1**, 337–342.

Irvine, R. F., and Moore, R. M. (1987). *Biochem. Biophys. Res. Commun.* **146**, 284–290.

Ishizuka, Y., and Nozawa, Y. (1983). *Biochem. Biophys. Res. Commun.* **117**, 710–717.

Iwamatsu, T., Yashimoto, Y., and Hiramoto, Y. (1988). *Dev. Biol.* **129**, 191–197.

Iyengar, R., and Birnbaumer, L. (1987). *ISI Atlas Sci. Pharmacol.* **1**, 213–221.

Jackson, R. C., and Crabb, J. H. (1988). *Curr. Top. Membr. Transp.* **32**, 45–85.

Jaffe, L. A., Turner, P. R., Kline, D., Kado, R. T., and Shilling, F. (1988). *In* "Regulatory Mechanisms in Developmental Processes" (G. Eguchi, T. S., Okada, and L. Saxén, eds.), pp. 15–18. Elsevier, Amsterdam.

Jahn, R., Schiebler, W., Ouimet, C., and Greengard, P. (1985). *Proc. Natl. Acad. Sci. USA* **82**, 4137–4141.

Jain, M. K., and Vaz, W. L. C. (1987). *Biochim. Biophys. Acta* **905**, 1–8.

Janssens, P. M. W. (1988). *Comp. Biochem. Physiol. A* **90A**, 209–223.

Jones, P. M., Stutchfield, J., and Howell, S. L. (1985). *FEBS Lett.* **191**, 102–106.

Jones, P. M., Salmon, D. M. W., and Howell, S. L. (1988). *Biochem. J.* **254**, 397–403.

Kaczmarek, L. K. (1987). *Trends Neurosci.* **10**, 30–34.

Kaibuchi, K., Sano, K., Hoshijima, M., Takai, Y., and Nishizuka, Y. (1982). *Cell Calcium* **3**, 323–335.

Kamel, L. C., Bailey, J., Schoenbaum, L., and Kinsey, W. (1985). *Lipids* **20**, 350–356.

Kaneshiro, E. S. (1987). *J. Lipid Res.* **28**, 1241–1258.

Kanno, T., Cockrane, D. E., and Douglas, W. W. (1973). *Can. J. Physiol. Pharmacol.* **51**, 1001–1004.

Kao, L. S., and Schneider, A. S. (1986). *J. Biol. Chem.* **261**, 4881–4888.

Katsu, T., Tasaka, K., and Fujita, Y. (1983). *FEBS Lett.* **151**, 219–222.

Kawasaki, H., Imajoh, S., Kawashima, S., Hayashi, H., and Suzuki, K. (1986). *J. Biochem.* (*Tokyo*) **99**, 1525–1532.

Kelly, R. B. (1985). *Science* **230**, 25–32.

Kenigsberg, R. L., and Trifaró, J. M. (1985). *Neuroscience* **14**, 335–347.

Kern, H. F., Bieger, W., Völkl, A., Rohr, G., and Adler, G. (1979). *In* "Secretory Mechanisms" (C. R. Hopkins and C. J. Duncan, eds.), pp. 79–99. Cambridge Univ. Press, London.

Kersken, H., Momayezi, M., Braun, C., and Plattner, H. (1986a). *J. Histochem. Cytochem.* **34**, 455–465.

Kersken, H., Vilmart-Seuwen, J., Momayezi, M., and Plattner, H. (1986b). *J. Histochem. Cytochem.* **34**, 443–454.

Kesteven, N. T., and Knight, D. E. (1982). *J. Physiol.* (*London*) **328**, 57P.

Kielian, M., and Helenius, A. (1985). *J. Cell Biol.* **101**, 2284–2291.

Kikkawa, U., and Nishizuka, Y. (1986). *Annu. Rev. Cell Biol.* **2**, 149–178.

Kirshner, N., and Smith, W. J. (1966). *Science* **154**, 422–423.

Kissmehl, R., Momayezi, M., and Plattner, H. (1989). In preparation.

Klee, C. B. (1988). *Biochemistry* **27**, 6645–6653.

Klee, C. B., Crouch, T. H., and Krinks, M. H. (1979). *Proc. Natl. Acad. Sci. USA* **76**, 6270–6273.

Kline, D. (1988). *Dev. Biol.* **126**, 346–361.

Knight, D. E. (1987). *Biosci. Rep.* **7**, 355–367.

Knight, D. E., and Baker, P. F. (1982). *J. Membr. Biol.* **68**, 107–140.

Knight, D. E., and Baker, P. F. (1983). *FEBS Lett.* **160**, 98–100.

Knight, D. E., and Baker, P. F. (1985). *FEBS Lett.* **189**, 345–349.
Knight, D. E., and Scrutton, M. C. (1984). *Biochem. Soc. Trans.* **12**, 969–972.
Knight, D. E., and Scrutton, M. C. (1986). *Biochem. J.* **234**, 497–506.
Knight, D. E., Tonge, D. A., and Baker, P. F. (1985). *Nature (London)* **317**, 719–721.
Knight, D. E., Sugden, D., and Baker, P. F. (1988). *J. Membr. Biol.* **104**, 21–34.
Knoll, G., and Plattner, H. (1989). *In* "Electron Microscope Analysis of Subcellular Dynamics" (H. Plattner, ed.), Chap. 6. CRC Press, Boca Raton, Florida.
Knoll, G., Verkleij, A. J., and Plattner, H. (1987). *In* "Cryotechniques in Biological Electron Microscopy" (R. A. Steinbrecht and K. Zierold, eds.), pp. 258–271. Springer-Verlag, Berlin.
Knoll, G., Lumpert, C., Braun, C., Gras, V., and Plattner, H. (1989). In preparation.
Konings, F., and DePotter, W. (1982). *Biochem. Biophys. Res. Commun.* **104**, 254–258.
Kreis, T. E., Allan, V. J., Matteoni, R., and Ho, W. C. (1988). *Protoplasma* **145**, 153–159.
Kung, C., and Saimi, Y. (1982). *Annu. Rev. Physiol.* **44**, 519–534.
Kurihara, H., and Uchida, K. (1987). *Histochemistry* **87**, 223–227.
Kuznetzov, S. A., and Gelfand, V. I. (1986). *Proc. Natl. Acad. Sci. USA* **83**, 8530–8534.
Lacy, P. E. (1975). *Am. J. Pathol.* **79**, 170–187.
Lacy, P. E., Howell, S. L., Young, D. A., and Fink, C. J. (1968). *Nature (London)* **219**, 1177–1179.
Ladona, M. G., Bader, M. F., and Aunis, D. (1987). *Biochim. Biophys. Acta* **927**, 18–25.
Lagunoff, D., and Chi, E. Y. (1980). *In* "Cell Biology of Inflammation" (G. Weissmann, ed.), pp. 217–265. Elsevier/North-Holland, Amsterdam.
Landis, D. M. D. (1988). *J. Electron Microsc. Tech.* **10**, 129–151.
Langley, O. K., Perrin, D., and Aunis, D. (1986). *J. Histochem. Cytochem.* **34**, 517–525.
Lapetina, E. G., Silió, J., and Ruggiero, M. (1985). *J. Biol. Chem.* **260**, 7078–7083.
Lapidot, M., Nussbaum, O., and Loyter, A. (1987). *J. Biol. Chem.* **262**, 13736–13741.
Launay, J. F., Stock, C., and Grenier, J. F. (1980). *In* "Biology of Normal and Cancerous Exocrine Pancreatic Cells" (A. Ribet, L. Pradayrol, and C. Susini, eds.), pp. 267–272. Elsevier/North-Holland, Amsterdam.
Lefort-Tran, M., Aufderheide, K., Pouphile, M., Rossignol, M., and Beisson, J. (1981). *J. Cell Biol.* **88**, 301–311.
Lelkes, P. I., and Pollard, H. B. (1987). *J. Biol. Chem.* **262**, 15496–15505.
Lelkes, P. I., Lavie, E., Naquira, D., Schneeweiss, F., Schneider, A. S., and Rosenheck, K. (1980). *FEBS Lett.* **115**, 129–133.
Lew, P. D., Monod, A., Waldvogel, F. A., Dewald, B., Baggiolin, M., and Pozzan, T. (1986). *J. Cell Biol.* **102**, 2197–2204.
Light, P. E., Sahaf, Z. Y., and Publicover, S. J. (1988). *Naunyn-Schmiedeberg's Arch. Pharmacol.* **338**, 339–344.
Llinás, R., Steinberg, Z., and Walton, K. (1981). *Biophys. J.* **33**, 323–352.
Llinás, R., McGuinness, T. L., Leonard, C. S., Sugimori, M., and Greengard, P. (1985). *Proc. Natl. Acad. Sci. USA* **82**, 3035–3039.
Lo, W. W. Y., and Hughes, J. (1987). *FEBS Lett.* **224**, 1–3.
Lochrie, M. A., and Simon, M. I. (1988). *Biochemistry* **27**, 4957–4965.
Lohse, M. J., Klotz, K. N., Salzer, M. J., and Schwabe, U. (1988). *Proc. Natl. Acad. Sci. USA* **85**, 8875–8879.
Lopez Vinals, A. E., Farias, R. N., and Morero, R. D. (1987). *Biochem. Biophys. Res. Commun.* **143**, 403–409.
Low, P. S., Lloyd, D. H., Stein, T. M., and Rogers, J. A. (1979). *J. Biol. Chem.* **254**, 4119–4125.
Lowe, A. W., Madeddu, L., and Kelly, R. B. (1988). *J. Cell Biol.* **106**, 51–59.

Lucy, J. A. (1978). *In* "Membrane Fusion" (G. Poste and G. L. Nicolson, eds.), pp. 268–304. Elsevier, Amsterdam.

Lucy, J. A., and Ahkong, Q. F. (1986). *FEBS Lett.* **199,** 1–11.

Luini, A., and Axelrod, J. (1985). *Proc. Natl. Acad. Sci. USA* **82,** 1012–1014.

Luini, A., and DeMatteis, M. A. (1988). *Cell. Mol. Neurobiol.* **8,** 129–138.

Lumpert, C. J., Kersken, H., Gras, U., and Plattner, H. (1987). *Cell Biol. Int. Rep.* **11,** 405–414.

Lumpert, C. J., Kersken, H., and Plattner, H. (1989). In preparation.

McArdle, H., Mullaney, I., Magee, A., Unson, C., and Milligan, G. (1988). *Biochem. Biophys. Res. Commun.* **152,** 243–251.

Macer, D. R. J., and Koch, G. L. E. (1988). *J. Cell Sci.* **91,** 61–70.

McGuinness, T. L., Lai, Y., Ouimet, C. C., and Greengard, P. (1985). *In* "Calcium in Biological Systems" (R. P. Rubin, G. B. Weiss, and J. W. Putney, eds.), pp. 291–305. Plenum, New York.

Machado-DeDomenech, E., and Söling, H. D. (1987). *Biochem. J.* **242,** 749–754.

Machemer, H. (1988). *In* "Paramecium" (H. D. Görtz, ed.), pp. 185–215. Springer-Verlag, Berlin.

McKay, D. B., Aronstam, R. S., and Schneider, A. S. (1985). *Mol. Pharmacol.* **28,** 10–16.

Martin, W. H., and Creutz, C. E. (1987). *J. Biol. Chem.* **262,** 2803–2810.

Matsumoto, Y., Perry, G., Scheibel, L. W., and Aikawa, M. (1987). *Eur. J. Cell Biol.* **45,** 36–43.

Matsuoka, I., Syoto, B., Kurihara, K., and Kubo, S. (1987). *FEBS Lett.* **216,** 295–299.

Matt, H., and Plattner, H. (1983). *Cell Biol. Int. Rep.* **7,** 1025–1031.

Matt, H., Bilinski, M., and Plattner, H. (1978). *J. Cell Sci.* **32,** 67–86.

Matter, K., Dreyer, F., and Aktories, K. (1989). *J. Neurochem.* **52,** 370–376.

Matthew, W. D., Tsavaler, L., and Reichardt, L. F. (1986). *J. Cell Biol.* **91,** 257–269.

Matthews, E. K. (1975). *In* "Calcium Transport in Contraction and Secretion" (E. Carafoli, F. Clementi, W. Drabikowski, and A. Margreth, eds.), pp. 203–210. North-Holland Publ., Amsterdam.

Matthies, H. J. G., Palfrey, H. C., and Miller, R. J. (1988). *FEBS Lett.* **229,** 238–242.

Means, A. R., Tash, J. S., and Chafouleas, J. G. (1982). *Physiol. Rev.* **62,** 1–39.

Meers, P., Bentz, J., Alford, D., Nir, S., Papahadjopoulos, D., and Hong, K. (1988a). *Biochemistry* **27,** 4430–4439.

Meers, P., Hong, K., and Papahadjopoulos, D. (1988b). *Biochemistry* **27,** 6784–6794.

Melançon, P., Glick, B. S., Malhotra, V., Weidman, P. J., Serafini, T., Gleason, M. L., Orci, L., and Rothman, J. E. (1987). *Cell* **51,** 1053–1062.

Meldolesi, J., and Ceccarelli, B. (1988). *Curr. Top. Membr. Transp.* **32,** 139–168.

Meldolesi, J., and Pozzan, T. (1987). *Exp. Cell Res.* **171,** 271–283.

Meldolesi, J., Jamieson, J. D., and Palade, G. E. (1971). *J. Cell Biol.* **49,** 130–149.

Meldolesi, J., DeCamilli, P., and Peluchetti, D. (1974). *In* "Secretory Mechanisms in Exocrine Glands" (N. Thorn and O. H. Petersen, eds.), pp. 137–151. Munksgaard, Copenhagen.

Meldolesi, J., Borgese, N., DeCamilli, P., and Ceccarelli, B. (1978). *In* "Cell Surface Reviews" (G. Poste and G. L. Nicolson, eds.), Vol. 5, pp. 509–627. North-Holland Publ., Amsterdam.

Meldolesi, J., Scheer, H., Madeddu, L., and Wanke, E. (1986). *Trends Pharmacol. Sci.* **7,** 151–155.

Meldolesi, J., Gatti, G., Ambrosini, A., Pozzan, T., and Westhead, E. W. (1988a). *Biochem. J.* **255,** 761–768.

Meldolesi, J., Volpe, P., and Pozzan, T. (1988b). *Trends Neurosci.* **10,** 449–452.

Merritt, J. E., and Rink, T. J. (1987a). *J. Biol. Chem.* **262,** 4958–4960.
Merritt, J. E., and Rink, T. J. (1987b). *J. Biol. Chem.* **262,** 14912–14916.
Merritt, J. E., Taylor, C. W., Putney, J. W., and Rubin, R. P. (1986a). *Biochem. Soc. Trans.* **14,** 605–606.
Merritt, J. E., Taylor, C. W., Rubin, R. P., and Putney, J. W. (1986b). *Biochem. J.* **236,** 337–343.
Metz, S., VanRollins, M., Strife, R., Fujimoto, W., and Robertson, R. P. (1983). *J. Clin. Invest.* **71,** 1191–1205.
Meyer, D., and Burger, M. (1979). *J. Biol. Chem.* **254,** 9854–9859.
Meyer, T., Holowka, D., and Stryer, L. (1988). *Science* **240,** 653–656.
Michaelson, D. M., Barkai, G., and Barenholz, Y. (1983). *Biochem. J.* **211,** 155–162.
Michell, R. H. (1975). *Biochim. Biophys. Acta* **415,** 81–147.
Michell, R. H., and Putney, J. W. (1987). "Inositol Lipids in Cellular Signalling. Current Communications in Molecular Biology." Cold Spring Harbor Lab., Cold Spring Harbor, New York.
Michell, R. H., Kirk, C. J., Jones, L. M., Downes, C. P., and Creba, J. A. (1981). *Philos. Trans. R. Soc. London, Ser. B* **296,** 123–137.
Michener, M. L., Dawson, W. B., and Creutz, C. E. (1986). *J. Biol. Chem.* **261,** 6548–6555.
Miledi, R. (1973). *Proc. R. Soc. London, Ser. B* **183,** 421–425.
Miller, R. J. (1985). *Trends Neurosci.* **8,** 463–465.
Miller, R. J. (1987). *Science* **235,** 46–52.
Miyamoto, S., and Fujime, S. (1988). *FEBS Lett.* **238,** 67–70.
Miyazaki, S. J. (1988). *J. Cell Biol.* **106,** 345–353.
Molski, T. F. P., Tao, W., Becker, E. L., and Sha'afi, R. I. (1988). *Biochem. Biophys. Res. Commun.* **151,** 836–843.
Momayezi, M., Kersken, H., Gras, U., Vilmart-Seuwen, J., and Plattner, H. (1986). *J. Histochem. Cytochem.* **34,** 1621–1638.
Momayezi, M., Girwert, A., Wolf, C., and Plattner, H. (1987a). *Eur. J. Cell Biol.* **44,** 247–257.
Momayezi, M., Lumpert, C. J., Kersken, H., Gras, U., Plattner, H., Krinks, M. H., and Klee, C. B. (1987b). *J. Cell Biol.* **105,** 181–189.
Morel, N., Israel, M., Lesbats, B., Birman, S., and Manaranche, R. (1987). *Ann. N.Y. Acad. Sci.* **493,** 151–154.
Morita, K., Brocklehurst, K. W., Tomares, S. M., and Pollard, H. B. (1985). *Biochem. Biophys. Res. Commun.* **129,** 511–516.
Morris, A. P., Gallacher, D. V., Irvine, R. F., and Petersen, O. H. (1987). *Nature (London)* **330,** 653–655.
Morris, S. J., and Bradley, D. (1984). *Biochemistry* **23,** 4642–4650.
Muff, R., Nemeth, E. F., Haller-Brem, S., and Fischer, J. A. (1988). *Arch. Biochem. Biophys.* **265,** 128–135.
Mundy, D. I., and Strittmatter, W. J. (1985). *Cell* **40,** 645–656.
Murtaugh, T. J., Gilligan, D. M., and Satir, B. H. (1987). *J. Biol. Chem.* **262,** 15734–15739.
Nagy, A., Baker, R. R., Morris, S. J., and Whittaker, V. P. (1976). *Brain Res.* **109,** 285–309.
Nakamura, T., and Ui, M. (1985). *J. Biol. Chem.* **260,** 3584–3593.
Navone, F., Jahn, R., DiGioia, G., Stukenbrok, H., Greengard, P., and DeCamilli, P. (1986). *J. Cell Biol.* **103,** 2511–2527.
Neer, E. J., and Clapham, D. E. (1988). *Nature (London)* **333,** 129–134.
Neher, E. (1974). *Biochim. Biophys. Acta* **373,** 327–336.
Neher, E. (1986). *J. Physiol. (London)* **381,** 71P.
Neher, E. (1988). *J. Physiol. (London)* **395,** 193–214.

Neher, E., and Almers, W. (1986). *EMBO J.* **5**, 51–53.
Neher, E., and Marty, A. (1982). *Proc. Natl. Acad. Sci. USA* **79**, 6712–6716.
Nestler, E. J., and Greengard, P. (1984). "Protein Phosphorylation in the Nervous System." Wiley, New York.
Nicchitta, C. V., Joseph, S. K., and Williamson, J. R. (1987). *Biochem. J.* **248**, 741–747.
Nichols, R. A., Haycock, J. W., Wang, J. K. T., and Greengard, P. (1987). *J. Neurochem.* **48**, 615–621.
Nishioka, D., Porter, D. C., Trimmer, J. S., and Vacquier, V. (1987). *Exp. Cell Res.* **173**, 628–632.
Nishizuka, Y. (1984). *Nature (London)* **308**, 693–697.
Nordmann, J. J., Dayanithi, G., and Lemos, J. R. (1987). *Biosci. Rep.* **7**, 411–426.
Obendorf, D., Schwarzenbrunner, U., Fischer-Colbrie, R., Laslop, A., and Winkler, H. (1988a). *J. Neurochem.* **51**, 1573–1580.
Obendorf, D., Schwarzenbrunner, U., Fischer-Colbrie, R., Laslop, A., and Winkler, H. (1988b). *Neuroscience* **25**, 343–351.
Oberdorf, J. A., Lebeche, D., Head, J. F., and Kaminer, B. (1988). *J. Biol. Chem.* **263**, 6806–6809.
Ohki, S. (1987). *In* "Molecular Mechanisms of Membrane Fusion" (S. Ohki, D. Doyle, T. D. Flanagan, S. W. Hui, and E. Mayhew, eds.), pp. 123–138. Plenum, New York.
Ohki, S., Doyle, D., Flanagan, T. D., Hui, S. W., and Mayhew, E., eds. (1988). "Molecular Mechanisms of Membrane Fusion." Plenum, New York.
Oishi, K., Raynor, R. L., Charp, P. A., and Kuo, J. F. (1988). *J. Biol. Chem.* **263**, 6865–6871.
Okano, Y., Takagi, H., Nakashima, S., Tohmatsu, T., and Nozawa, Y. (1985). *Biochem. Biophys. Res. Commun.* **132**, 110–117.
Olbricht, K., Plattner, H., and Matt, H. (1984). *Exp. Cell Res.* **151**, 14–20.
OpDenKamp, J. A. F. (1979). *Annu. Rev. Biochem.* **48**, 47–71.
Orci, L., and Perrelet, A. (1978). *In* "Membrane Fusion" (G. Poste and G. L. Nicolson, eds.), pp. 629–656. Elsevier/North-Holland, Amsterdam.
Orci, L., Gabbay, K. H., and Malaisse, W. J. (1972). *Science* **175**, 1128–1130.
Orci, L., Amherdt, M., Roth, J., and Perrelet, A. (1979). *Diabetologia* **16**, 135–138.
Orci, L., Glick, B. S., and Rothmann, J. E. (1986). *Cell* **46**, 171–184.
Orgad, S., Dudai, Y., and Cohen, P. (1987). *Eur. J. Biochem.* **164**, 31–38.
Ornberg, R. L., and Reese, T. S. (1981). *J. Cell Biol.* **90**, 40–54.
Palade, G. E. (1959). *In* "Subcellular Particles" (T. Hayashi, ed.), pp. 64–80. Ronald Press, New York.
Palade, G. E. (1975). *Science* **189**, 347–358.
Palade, G. E., and Bruns, R. R. (1968). *J. Cell Biol.* **37**, 633–649.
Pallen, C. J., and Wang, J. H. (1985). *Arch. Biochem. Biophys.* **237**, 281–291.
Pandol, S. J., Schoeffield, M. S., Sachs, G., and Muallen, S. (1985). *J. Biol. Chem.* **260**, 10081–10086.
Pang, D. T., Wang, J. K. T., Valtorta, F., Benfenati, F., and Greengard, P. (1988). *Proc. Natl. Acad. Sci. USA* **85**, 762–766.
Papahadjopoulos, D. (1978). *In* "Membrane Fusion" (G. Poste and G. L. Nicolson, eds.), pp. 765–790. Elsevier/North-Holland, Amsterdam.
Papahadjopoulos, D., Portis, A., Pangborn, W., and Newton, C. (1978). *In* "Transport of Macromolecules in Cellular Systems" (S. C. Silverstein, ed.), Life Sci. Res. Rep. No. 11, pp. 413–430. Dahlem Konf., Berlin.
Papahadjopoulos, D., Meers, P. R., Hong, K., Ernst, J. D., Goldstein, I. M., and Düzgünes, N. (1987). *In* "Molecular Mechanisms of Membrane Fusion" (S. Ohki, D. Doyle, T. D. Flanagan, S. W. Hui, and E. Mayhew, eds.), pp. 1–16. Plenum, New York.

Pape, R., and Plattner, H. (1985). *Eur. J. Cell Biol.* **36**, 38–47.

Pape, R., and Plattner, H. (1989). In preparation.

Pape, R., Haacke-Bell, B., Lüthe, N., and Plattner, H. (1988). *J. Cell Sci.* **90**, 37–49.

Parente, R. A., Nir, S., and Szoka, F. C. (1988). *J. Biol. Chem.* **263**, 4724–4730.

Parsegian, V. A., Rand, R. P., and Gingell, D. (1984). In "Cell Fusion" (D. Evered and J. Whelan, eds.), Ciba Found. Symp., pp. 9–27. Pitman, London.

Partridge, L. D., and Swandulla, D. (1988). *Trends Neurosci.* **11**, 69–72.

Penner, R., and Neher, E. (1988). *J. Exp. Biol.* **139**, 329–345.

Penner, R., Neher, E., and Dreyer, F. (1986). *Nature (London)* **324**, 76–78.

Penner, R., Matthews, G., and Neher, E. (1988). *Nature (London)* **334**, 499–504.

Peplinsky, R. B., Tizard, R., Mattaliano, R. J., Sinclair, L. K., Miller, G. T., Browning, J. L., Chow, E. P., Burne, C., Huang, K. S., Pratt, D., Wachter, L., Hession, C., Frey, A. Z., and Wallner, B. (1988). *J. Biol. Chem.* **263**, 10799–10811.

Perrin, D., and Aunis, D. (1985). *Nature (London)* **315**, 589–592.

Perrin, D., Langley, O. K., and Aunis, D. (1987). *Nature (London)* **326**, 498–501.

Petrucci, T. C., and Morrow, J. S. (1987). *J. Cell Biol.* **105**, 1355–1363.

Pfaffinger, P. J., Martin, J. M., Hunter, D. D., Nathanson, N. M., and Hille, B. (1985). *Nature (London)* **317**, 536–538.

Pfeiffer, A., Gagnon, C., and Heisler, S. (1984). *Biochem. Biophys. Res. Commun.* **122**, 413–419.

PintoDaSilva, P. (1988). In "Molecular Mechanisms of Membrane Fusion" (S. Ohki, D. Doyle, T. D. Flanagan, S. W. Hui, and E. Mayhew, eds.), pp. 521–530. Plenum, New York.

PintoDaSilva, P., and Nogueira, M. L. (1977). *J. Cell Biol.* **73**, 161–181.

Plattner, H. (1974). *Nature (London)* **252**, 722–724.

Plattner, H. (1981). *Cell Biol. Int. Rep.* **5**, 435–459.

Plattner, H. (1987). In "Cell Fusion" (A. E. Sowers, ed.), pp. 69–98. Plenum, New York.

Plattner, H. (1989a). In "Freeze–Fracture Studies on Membranes" (S. W. Hui, ed.), Chap. 8. CRC Press, Boca Raton, Florida. In press.

Plattner, H. (1989b). In "Cellular Membrane Fusion: Fundamental Mechanisms and Application of Membrane Fusion Techniques" (J. Wilschut and D. Hoekstra, eds.). Dekker, New York. In press.

Plattner, H., and Bachmann, L. (1982). *Int. Rev. Cytol.* **79**, 237–304.

Plattner, H., and Knoll, G. (1987). *Scanning Microsc.* **1**, 1199–1216.

Plattner, H., Miller, F., and Bachmann, L. (1973). *J. Cell Sci.* **13**, 687–719.

Plattner, H., Reichel, K., and Matt, H. (1977). *Nature (London)* **267**, 702–704.

Plattner, H., Reichel, K., Matt, H., Beisson, J., Lefort-Tran, M., and Pouphile, M. (1980). *J. Cell Sci.* **46**, 17–40.

Plattner, H., Westphal, C., and Tiggemann, R. (1982). *J. Cell Biol.* **92**, 368–377.

Plattner, H., Pape, R., Haacke, B., Olbricht, K., Westphal, C., and Kersken, H. (1985). *J. Cell Sci.* **77**, 1–17.

Plattner, H., Lumpert, C. J., Gras, U., Vilmart-Seuwen, J., Stecher, B., Höhne, B., Momayezi, M., Pape, R., and Kersken, H. (1988). In "Molecular Mechanisms of Membrane Fusion" (S. Ohki, D. Doyle, T. D. Flanagan, S. W. Hui, and E. Mayhew, eds.), pp. 477–494. Plenum, New York.

Pocotte, S. L., Frye, R. A., Senter, R. A., Terbush, D. R., Lee, S. A., and Holz, R. W. (1985). *Proc. Natl. Acad. Sci. USA* **82**, 930–934.

Poisner, A. M., and Trifaró, J. M. (1967). *Mol. Pharmacol.* **3**, 561–571.

Pollard, H. B., and Rojas, E. (1988). *Proc. Natl. Acad. Sci. USA* **85**, 2974–1978.

Pollard, H. B., Pazoles, C. J., Creutz, C. E., and Zinder, O. (1979). *Int. Rev. Cytol.* **58,** 159–197.

Pollard, H. B., Pazoles, C. J., Creutz, C. E., and Zinder, O. (1980). *In* "Monographs of Neural Science" (M. M. Cohen, ed.), Vol. 7, pp. 106–116. Karger, Basel.

Pollard, H. B., Creutz, C. E., Fowler, V., Scott, J., and Pazoles, C. J. (1981). *Cold Spring Harbor Symp. Quant. Biol.* **46,** 819–834.

Pollard, H. B., Burns, A. L., Stutzin, A., Rojas, E., Lelkes, P. I., and Morita, K. (1987). *In* "Stimulus–Secretion Coupling in Chromaffin Cells" (K. Rosenheck and P. I. Lelkes, eds.), Vol. 2, pp. 1–13. CRC Press, Boca Raton, Florida.

Pollard, H. B., Burns, A. L., and Rojas, E. (1988a). *J. Exp. Biol.* **139,** 267–286.

Pollard, H. B., Rojas, E., Burns, A. L., and Parra, C. (1988b). *In* "Molecular Mechanisms of Membrane Fusion" (S. Ohki, D. Doyle, T. D. Flanagan, S. W. Hui, and E. Mayhew, eds.), pp. 341–355. Plenum, New York.

Portis, A., Newton, C., Pangborn, W., and Papahadjopoulos, D. (1979). *Biochemistry* **18,** 780–790.

Poste, G., and Allison, A. C. (1973). *Biochim. Biophys. Acta* **300,** 421–465.

Poste, G., and Nicolson, G. L., eds. (1978). "Membrane Fusion." North-Holland Publ., Amsterdam.

Pouphile, M., Lefort-Tran, M., Plattner, H., Rossignol, M., and Beisson, J. (1986). *Biol. Cell.* **56,** 151–162.

Powers, R. E., Saluja, A. K., Houlihan, M. J., and Steer, M. L. (1985). *Biochem. Biophys. Res. Commun.* **13,** 284–288.

Pozzan, T., Volpe, P., Zorzato, F., Bravin, M., Krause, K. H., Lew, D. P., Hashimoto, S., Bruno, B., and Meldolesi, J. (1988). *J. Exp. Biol.* **139,** 181–193.

Pralong, W. F., Wollheim, C. B., and Bruzzone, R. (1988). *FEBS Lett.* **242,** 79–84.

Prentki, M., Deeney, J. T., Matschinsky, F. M., and Joseph, S. K. (1986). *FEBS Lett.* **197,** 285–288.

Pribluda, U. S., and Metzger, H. (1987). *J. Biol. Chem.* **262,** 11449–11454.

Publicover, S. J. (1985). *Comp. Biochem. Physiol. A* **82A,** 7–11.

Pumplin, D. W., Reese, T. S., and LLinás, R. (1981). *Proc. Natl. Acad. Sci. USA* **78,** 7210–7213.

Putney, J. W. (1986). *Cell Calcium* **7,** 1–12.

Putney, J. W. (1987). *Trends Pharmacol. Sci.* **8,** 481–486.

Putney, J. W. (1988). *J. Exp. Biol.* **139,** 135–150.

Rand, R. P., and Parsegian, V. A. (1986). *Annu. Rev. Physiol.* **48,** 201–212.

Rand, R. P., Kachar, B., and Reese, T. S. (1985). *Biophys. J.* **47,** 483–489.

Rash, J. E., Walrond, J. P., and Morita, M. (1988). *J. Electron Microsc. Tech.* **10,** 153–185.

Rasmussen, H., and Goodman, D. B. P. (1977). *Physiol. Rev.* **57,** 421–509.

Rindler, M. J., Ivanov, I. E., and Sabatini, D. D. (1987). *J. Cell Biol.* **104,** 231–241.

Rindlisbacher, B., Reist, M., and Zahler, P. (1987). *Biochim. Biophys. Acta* **905,** 349–357.

Rink, T. J., and Knight, D. E. (1988). *J. Exp. Biol.* **139,** 1–30.

Rink, T. J., Sanchez, A., and Hallam, T. (1983). *Nature (London)* **305,** 317–319.

Roberts, M. L., and Butcher, F. R. (1983). *Biochem. J.* **210,** 353–359.

Roe, J. L., Farach, H. A., Strittmatter, W. J., and Lennarz, W. J. (1988). *J. Cell Biol.* **107,** 539–544.

Rogalski, A. A., Bergmann, J. E., and Singer, S. J. (1984). *J. Cell Biol.* **99,** 1101–1109.

Rojas, E., and Pollard, H. B. (1987). *FEBS Lett.* **217,** 25–31.

Roos, D. S., Robinson, J. M., and Davidson, R. L. (1983). *J. Cell Biol.* **97,** 909–917.

Rosenheck, K., and Plattner, H. (1986). *Biochim. Biophys. Acta* **856,** 373–382.

Rosenthal, W., Hescheler, J., Trautwein, W., and Schultz, G. (1988). *FASEB J.* **2,** 2784–2790.

Rothman, J. E., and Lenard, J. (1977). *Science* **195,** 743–753.

Rotrosen, D., Gallin, J. I., Spiegel, A. M., and Malech, H. L. (1988). *J. Biol. Chem.* **263,** 10958–10964.

Roufogalis, B. D. (1982). *In* "Calcium and Cell Function" (W. Y. Cheung, ed.), pp. 129–159. Academic Press, New York.

Rubin, R. P., Weiss, G. B., and Putney, G. W., eds. (1985). "Calcium in Biological Systems." Plenum, New York.

Ruggiero, M., Zimmerman, T. P., and Lapetina, E. G. (1985). *Biochem. Biophys. Res. Commun.* **131,** 620–627.

Sage, S. O., and Rink, T. J. (1987). *J. Biol. Chem.* **262,** 16364–16369.

Saimi, Y., Martinac, B., Gustin, M. C., Culbertson, M. R., Adler, J., and Kung, C. (1988). *Trends Biochem. Sci.* **13,** 304–309.

Salas, P. J. I., Misek, D. E., Vega-Salas, D. E., Gundersen, D., Cereijido, M., and Rodriguez-Boulan, E. (1986). *J. Cell Biol.* **102,** 1853–1867.

Salzberg, B. M., and Obaid, A. L. (1988). *J. Exp. Biol.* **139,** 195–231.

Sarafian, T., Aunis, D., and Bader, M. F. (1987). *J. Biol. Chem.* **262,** 16671–16676.

Sasaki, H. (1984). *Dev. Biol.* **101,** 125–135.

Schacht, J. (1976). *J. Neurochem.* **27,** 1119–1124.

Schäfer, R., Christian, A. L., and Schulz, I. (1988). *Biochem. Biophys. Res. Commun.* **155,** 1051–1059.

Schäfer, T., Karli, U. O., Schweizer, F. E., and Burger, M. M. (1987). *Biosci. Rep.* **7,** 269–279.

Schatzman, R. C., Turner, R. S., and Kuo, J. F. (1984). *In* "Calcium and Cell Function" (W. Y. Cheung, ed.), Vol. 5, pp. 33–66. Academic Press, New York.

Schilling, K., and Gratzl, M. (1988). *FEBS Lett.* **233,** 22–24.

Schlegel, R. (1987). *In* "Cell Fusion" (A. E. Sowers, ed.), pp. 33–43. Plenum, New York.

Schmidt, W., Patzak, A., Lingg, G., Winkler, H., and Plattner, H. (1983). *Eur. J. Cell Biol.* **32,** 31–37.

Schnefel, S., Banfic, H., Eckhardt, L., Schultz, G., and Schulz, I. (1988). *FEBS Lett.* **230,** 125–130.

Schneider, A. S., Cline, H. T., Rosenheck, K., and Sonenberg, M. (1981). *J. Neurochem.* **37,** 567–575.

Schön, E. A., and Decker, G. L. (1981). *J. Ultrastruct. Res.* **76,** 191–201.

Schroer, T. A., Schnapp, B. J., Reese, T. S., and Sheetz, M. P. (1988). *J. Cell Biol.* **107,** 1785–1792.

Schulman, H. (1984). *Trends Pharmacol. Sci.* **5,** 188–192.

Schworer, C. M., Colbran, R. J., Keefer, J. R., and Soderling, T. R. (1988). *J. Biol. Chem.* **263,** 13486–13489.

Scott, J. H., Creutz, C. E., Pollard, H. B., and Ornberg, R. (1985). *FEBS Lett.* **180,** 17–23.

Scott, R. H., and Dolphin, A. C. (1987). *Nature (London)* **330,** 760–762.

Setoguti, T., Inoue, Y., and Kato, K. (1981). *Cell Tissue Res.* **219,** 457–467.

Shapira, R., Silberberg, S. D., Ginsburg, S., and Rahamimoff, R. (1987). *Nature (London)* **325,** 58–60.

Shenolikar, S. (1988). *FASEB J.* **2,** 2753–2764.

Shimomura, H., Terada, A., Hashimoto, Y., and Soderling, T. R. (1988). *Biochem. Biophys. Res. Commun.* **150,** 1309–1314.

Shotton, D. M. (1978). *Nature (London)* **272,** 16–17.

Siegel, D. P. (1984). *Biophys. J.* **45,** 399–420.

Siegel, D. P. (1986). *Biophys. J.* **49**, 1171–1183.

Siegel, D. P. (1987). *In* "Cell Fusion" (A. E. Sowers, ed.), pp. 181–207. Plenum, New York.

Sieghart, W., Theoharides, T. C., Alper, S. L., Douglas, W. W., and Greengard, P. (1978). *Nature (London)* **275**, 329–331.

Simon, S. M., and Llinás, R. R. (1986). *Biophys. J.* **48**, 485–498.

Simons, K., and VanMeer, G. (1988). *Biochemistry* **27**, 6197–6202.

Smith, S. J., and Augustine, G. J. (1988). *Trends Neurosci.* **11**, 458–464.

Smith, V. L., and Dedman, J. R. (1986). *J. Biol. Chem.* **261**, 15815–15818.

Smolen, J. E., Stoehr, S. J., and Boxer, L. A. (1986). *Biochim. Biophys. Acta* **886**, 1–17.

Soifer, D., ed. (1986). "Dynamic Aspects of Microtubule Biology." *Ann. N.Y. Acad. Sci.* **466**.

Sokolov, Y. V., Chanturia, A. N., and Lishko, V. K. (1987). *Biochim. Biophys. Acta* **900**, 295–299.

Somlyo, A. P., Bond, M., and Somlyo, A. V. (1985). *Nature (London)* **314**, 622–624.

Sontag, J. M., Aunis, D., and Bader, M. F. (1988). *Eur. J. Cell Biol.* **46**, 316–326.

Sowers, A. E., ed. (1987). "Cell Fusion." Plenum, New York.

Spear, P. G. (1987). *In* "Cell Fusion" (A. E. Sowers, ed.), pp. 3–32. Plenum, New York.

Spearman, T. N., Hurley, K. P., Olivas, R., Ulrich, R. G., and Butcher, F. R. (1984). *J. Cell Biol.* **99**, 1354–1363.

Specian, R. D., and Neutra, M. R. (1980). *J. Cell Biol.* **85**, 626–640.

Stecher, B., Höhne, B., Gras, U., Momayezi, M., Glas-Albrecht, R., and Plattner, H. (1987). *FEBS Lett.* **223**, 25–32.

Steinhardt, R. A., and Alderton, J. M. (1981). *J. Cell Biol.* **91**, 180a.

Steinhardt, R. A., and Alderton, J. M. (1982). *Nature (London)* **295**, 154–155.

Sternweis, P. C., and Robishaw, J. D. (1984). *J. Biol. Chem.* **259**, 13806–18813.

Stossel, T. P., Bretscher, M. S., Ceccarelli, B., Dales, S., Helenius, A., Heuser, J. E., Hubbard, A. L., Kartenbeck, J., Kinne, R., Papahadjopoulos, D., Pearse, B., Plattner, H., Pollard, T. D., Reuter, W., Satir, B. H., Schliwa, M., Schneider, Y. J., Silverstein, S. C., and Weber, K. (1978). *In* "Transport of Macromolecules in Cellular Systems" (S. C. Silverstein, ed.), Life Sci. Res. Rep. No. 11, pp. 503–516. Dahlem Konf., Berlin.

Stossel, T. P., Hartwig, J. H., Yin, H. L., Zaner, K. S., and Stendahl, O. I. (1981). *Cold Spring Harbor Symp. Quant. Biol.* **46**, 569–578.

Streb, H., Irvine, R. F., Berridge, M. J., and Schulz, I. (1983). *Nature (London)* **306**, 67–69.

Strittmatter, W. J. (1988). *Cell. Mol. Neurobiol.* **8**, 19–25.

Strittmatter, W. J., Couch, C. B., and Mundy, D. I. (1987). *In* "Cell Fusion" (A. E. Sowers, ed.), pp. 99–121. Plenum, New York.

Stryer, L., and Broune, H. R. (1986). *Annu. Rev. Cell Biol.* **2**, 391–419.

Stutchfield, J., and Cockcroft, S. (1988). *Biochem. J.* **250**, 375–382.

Stutchfield, J., Jones, P. M., and Howell, S. L. (1986). *Biochem. Biophys. Res. Commun.* **136**, 1001–1006.

Stutzin, A., Cabantchik, Z. I., Lelkes, P. I., and Pollard, H. B. (1987). *Biochim. Biophys. Acta* **905**, 205–212.

Summers, T. A., and Creutz, C. E. (1985). *J. Biol. Chem.* **260**, 2437–2443.

Swann, K., and Whitaker, M. (1986a). *J. Cell Biol.* **103**, 2333–2342.

Swann, K., and Whitaker, M. (1986b). *J. Physiol. (London)* **377**, 50P.

Swilem, A. M. F., Yagisawa, H., and Hawthorne, J. N. (1986). *J. Physiol. (Paris)* **81**, 246–251.

Tamaoki, T., Momoto, N., Takahashi, I., Kato, Y., Morimoto, M., and Tomita, F. (1986). *Biochem. Biophys. Res. Commun.* **135**, 397–402.

Tartakoff, A. M. (1987). "The Secretory and Endocytic Paths." Wiley, New York.

Tash, J. S., Krinks, M., Patel, J., Means, R. L., Klee, C. B., and Means, A. R. (1988). *J. Cell Biol.* **106**, 1625–1633.

Tatham, P. E. R., and Gomperts, B. D. (1989). *Biosci. Rep.* **9**, 99–109.

Taylor, C. W., Merritt, J. E., Rubin, R. P., and Putney, J. W. (1986). *Biochem. Soc. Trans.* **14**, 604–605.

Terbush, D. R., Bittner, M. A., and Holz, R. W. (1988). *J. Biol. Chem.* **263**, 18873–18879.

Thayer, S. A., Sturek, M., and Miller, R. J. (1988). *Eur. J. Physiol.* **412**, 216–223.

Theoharides, T. C., Sieghart, W., Greengard, P., and Douglas, W. W. (1980). *Science* **207**, 80–82.

Thomas, L., Hartung, K., Langosch, D., Rehm, H., Bamberg, E., Franke, W. W., and Betz, H. (1988). *Science* **242**, 1050–1053.

Thuret-Carnahan, J., Bossu, J. L., Feltz, A., Langley, K., and Aunis, D. (1985). *J. Cell Biol.* **100**, 1863–1974.

Torri-Tarelli, F., Grohavaz, F., Fesce, R., and Ceccarelli, B. (1985). *J. Cell Biol.* **101**, 1386–1399.

Toutant, M., Aunis, D., Bockaert, J., Homburger, V., and Rouot, B. (1987a). *FEBS Lett.* **215**, 339–344.

Toutant, M., Aunis, D., Bockaert, J., Homburger, V., and Rouot, B. (1987b). *FEBS Lett.* **222**, 51–55.

Treiman, M., Weber, W., and Gratzl, M. (1983). *J. Neurochem.* **40**, 661–669.

Trifaró, J. M., and Kenigsberg, R. L. (1987). *In* "Stimulus–Secretion Coupling in Chromaffin Cells" (K. Rosenheck and P. I. Lelkes, eds.), Vol. 1, pp. 125–153. CRC Press, Boca Raton, Florida.

Tsien, R. Y. (1988). *Trends Neurosci.* **11**, 419–424.

Turner, P. R., Jaffe, L. A., and Fein, A. (1986). *J. Cell Biol.* **102**, 70–76.

Turner, P. R., Jaffe, L. A., and Primakoff, P. (1987). *Dev. Biol.* **120**, 577–583.

Vallar, L., Biden, T. J., and Wollheim, C. B. (1987). *J. Biol. Chem.* **262**, 5049–5056.

Valtorta, F., Villa, A., Jahn, R., DeCamilli, P., Greengard, P., and Ceccarelli, B. (1988). *Neuroscience* **24**, 593–603.

VanDenBosch, H. (1980). *Biochim. Biophys. Acta* **604**, 191–246.

VanMeer, G., and Simons, K. (1988). *J. Cell. Biochem.* **36**, 51–58.

Vegesna, R. V. K., Wu, H. L., Mong, S., and Crooke, S. T. (1988). *Mol. Pharmacol.* **33**, 537–542.

Verkleij, A. J. (1984). *Biochim. Biophys. Acta* **779**, 43–63.

Verkleij, A. J. (1986). *In* "Phospholipid Research and the Nervous System. Biochemical and Molecular Pharmacology" (L. A., Horrocks, L. Freysz, and G. Toffano, eds.), pp. 207–216. Liviana Press, Padua.

Verkleij, A. J., Mombers, C., Gerritsen, W. J., Leunissen-Bijvelt, J., and Cullis, P. R. (1979). *Biochim. Biophys. Acta* **555**, 358–361.

Vicentini, L. M., Ambrosini, A., DiVirgilio, F., Pozzan, T., and Meldolesi, J. (1985). *J. Cell Biol.* **100**, 1330–1333.

Vilmart, J., and Plattner, H. (1983). *J. Histochem. Cytochem.* **31**, 626–632.

Vilmart-Seuwen, J., Kersken, H., Stürzl, R., and Plattner, H. (1986). *J. Cell Biol.* **103**, 1279–1288.

Virtanen, I., and Vartio, T. (1986). *Eur. J. Cell Biol.* **42**, 281–287.

Viveros, O. H. (1975). *In* "Handbook of Physiology" (H. Blaschko, G. Sayers, and A. D. Smith, eds.), Sect. 7, Vol. 6, pp. 389–426. Am. Physiol. Soc., Washington, D.C.

Wakade, A. R., Malhotra, R. K., and Wakade, T. D. (1986). *Nature* (*London*) **321**, 698–700.

Walker, J. H., and Agoston, D. V. (1987). *Biochem. J.* **247**, 249–258.

Wells, E., and Mann, J. (1983). *Biochem. Pharmacol.* **32**, 837–842.

Westhead, E. W. (1987). *Ann. N.Y. Acad. Sci.* **493**, 92–100.
Westphal, C., and Plattner, H. (1981). *Biol. Cell.* **42**, 125–140.
Westrum, L. E., Gray, E. G., Burgoyne, R. D., and Barron, J. (1983). *Cell Tissue Res.* **231**, 93–102.
Whalley, T., and Whitaker, M. (1988). *Biosci. Rep.* **8**, 335–343.
Whitaker, M. (1984). *Nature (London)* **312**, 636–638.
Whitaker, M. (1985). *FEBS Lett.* **189**, 137–140.
Whitaker, M. (1987). *Biosci. Rep.,* **7**, 383–397.
Whitaker, M., and Aitchison, M. (1985). *FEBS Lett.* **182**, 119–124.
Whitaker, M. J., and Baker, P. F. (1983). *Proc. R. Soc. London, Ser. B* **218**, 397–413.
Whitaker, M., and Zimmerberg, J. (1987). *J. Physiol. (London)* **389**, 527–539.
White, J., Kielian, M., and Helenius, A. (1983). *Q. Rev. Biophys.* **16**, 151–195.
White, J. R., Ishizaka, T., Ishizaka, K., and Shaafi, R. I. (1984). *Proc. Natl. Acad. Sci. USA* **81**, 3978–3982.
White, M. F., Takayama, S., and Kahn, C. R. (1985). *J. Biol. Chem.* **260**, 9470–9478.
Wiedenmann, B., and Franke, W. W. (1985). *Cell* **41**, 1017–1028.
Wiedenmann, B., Franke, W. W., Kuhn, C., Moll, R., and Gould, V. E. (1986). *Proc. Natl. Acad. Sci. USA* **83**, 3500–3504.
Wildenauer, D. B., and Zeeb-Wälde, B. C. (1983). *Biochem. Biophys. Res. Commun.* **116**, 469–477.
Winkler, H. (1976). *Neuroscience* **1**, 65–80.
Winkler, H. (1988). *In* "Handbook of Experimental Pharmacology" (U. Trendelenburg and N. Weiner, eds.), Vol. 90/I, pp. 43–118. Springer-Verlag, Berlin.
Wise, B. C., and Costa, E. (1985). *J. Neurochem.* **45**, 227–234.
Wise, B. C., and Kuo, J. F. (1983). *Biochem. Pharmacol.* **32**, 1259–1265.
Wooten, M. W., and Wrenn, R. W. (1984). *FEBS Lett.* **171**, 183–186.
Yagisawa, H., Swilem, A. M., and Hawthorne, J. N. (1986). *Biochem. Soc. Trans.* **14**, 613–614.
Zieseniss, E., and Plattner, H. (1985). *J. Cell Biol.* **101**, 2028–2035.
Zimmerberg, J. (1987). *Biosci. Rep.* **7**, 251–268.
Zimmerberg, J. (1988). *In* "Molecular Mechanisms of Membrane Fusion" (S. Ohki, D. Doyle, T. D. Flanagan, S. W. Hui, and E. Mayhew, eds.), pp. 181–195. Plenum, New York.
Zimmerberg, J., and Liu, J. (1988). *J. Membr. Biol.* **101**, 199–207.
Zimmerberg, J., Cohen, F. S., and Finkelstein, A. (1980). *Science* **210**, 906–908.
Zimmerberg, J., Sardet, C., and Epel, D. (1985). *J. Cell Biol.* **101**, 2398–2410.
Zimmerberg, J., Curran, M., Cohen, F. S., and Brodwick, M. (1987). *Proc. Natl. Acad. Sci. USA* **84**, 1985–1989.
Zingsheim, H. P., and Plattner, H. (1976). *In* "Methods in Membrane Biology" (E. D. Korn, ed.), Vol. 7, pp. 1–146. Plenum, New York.
Zucker, R. S., and Haydon, P. G. (1988). *Nature (London)* **335**, 360–362.
Zwiller, J., Ogasawara, E. M., Nakamoto, S. S., and Boynton, A. L. (1988). *Biochem. Biophys. Res. Commun.* **155**, 767–772.

NOTE ADDED IN PROOF. Polar epithelia vesicle delivery to the apical, but not to the basolateral cell membrane portion requires guidance by microtubules (Achler *et al.*, 1989; Eilers *et al.*, 1989). As stated in Sections V and VI, specific recognition sites independent of the arrangement of microtubules are, therefore, requisite (at least for constitutive exocytosis) and microtubules may be long-range targeting pathways only in regulated exocytosis. This has been shown, in fact, with VS-virus DNA transfected AtT-20 cells (Rivas and Moore, 1989).

Transfection studies suggest that the different G proteins of the type described in Section VIII,B are coupled to different receptors and second messenger systems even within one cell (Ashkenasi et al., 1989).

G-proteins of the type described in Sections VIII,A and VIII,C are now assumed to be involved in connecting different Ca^{2+} pools: for coupling of a subplasmalemmal pool 1 with the cell membrane and of an internal pool 3 (possibly calciosomes) and of pool 1 with pool 2 (rough ER), respectively (Gill et al., 1989; Ghosh et al., 1989). Only pool 2 would be IP_3 sensitive (in contrast to Meldolesi et al., 1988b, op. cit.; Volpe et al., 1988). These G proteins might be related to G proteins that, in yeast, allow for vesicle interaction in the Golgi area (YPT1p gene product: Segev et al., 1988) or between secretory vesicles and the cell membrane (SEC4 gene product: Goud et al., 1988, op. cit.). [Another possible equivalent is the ras-type gene product (Section VIII,B)]. GTP hydrolysis is required in these cases. Any relationship to G_E (end of Section VIII,C) remains to be established, particularly with regard to opposite effects of γ-S-GTP. G proteins of low M_r (18,000–29,000), i.e., different from other G proteins mentioned in Section VIII,B, are attached at chromaffin granules (Burgoyne and Morgan, 1989; Doucet et al., 1989). They might play a role not only in constitutive (as assumed for the SEC4 protein), but possibly also in regulated secretion.

Other putative docking (and possibly fusion) proteins have also been identified in chromaffin cells; a 51-kDa protein (absent from fibroblasts) is anchored in the cell membrane (Schweizer et al., 1989) and a CaM-BP of 65 kDa is present on the vesicle and the cell membrane. In the latter case, Trifaró et al. (1989) ascertained its identity with p65 (Matthew et al., 1986, op. cit.) and they stressed its functional equivalence to PP65 from Paramecium (Momayezi et al., 1987b, op. cit.).

Distinct IP_3 receptors on the surface of Ca^{2+} storage compartments bind only the physiological stereoisomer (Nahorski et al., 1989). Immunocytochemically, IP_3 receptors were found on the ER membranes (Ross et al., 1989). This would correspond to pool 2 of Gill et al. (1989) and Ghosh et al. (1989).

A new aspect is the possible relevance of an only locally increased PI turnover which might not result in a large overall increase of products. This possibility is suggested by data obtained via photoflash activation of caged IP_3 (Parker, 1988), whereby small threshold values of IP_3 suffice to release Ca^{2+} from internal pools in oocytes. Effects of this kind might explain some of the discrepancies discussed in Section IX.

Similarly, model experiments with liposomes assign a high fusogenic capacity to DAG, already in a rather small percentage of the total lipid (Siegel et al., 1989a). Contact and fusion between liposomes, as analyzed by fast freezing, involves formation of $\approx$15–20 nm large pedestal-like structures (Siegel et al., 1989b) quite similar to those mentioned in Sections II and XVI.

Our concept that ATP primes fusogenic sites and that protein dephosphorylation might induce membrane fusion (Sections VII and X) has obtained additional support from work with egg cells (Crossley et al., 1989) and chromaffin cells (Holz et al., 1989); in the latter case replacement of ATP by GTP results in the same secretory activity (Schäfer et al., 1987). In neutrophils ATP depletion results in a depressed time course of secretion, but its extent remains unchanged (Nüsse and Lindau, 1988). These data are in line with those presented in Section VII.

Some role of PKC (Sections VIII and X) in secretion regulation remains undebatable. Support comes from leakage and reconstitution experiments with permeabilized pituitary cells (Nahor et al., 1989). PKC is now also assumed to facilitate access of vesicles to the cell membrane by disassembly of subplasmalemmal F-actin (Burgoyne et al., 1989). Results obtained with alleged PKC activators require caution, since these also affect other cell functions (Hockberger *et al.*, 1989). Any role of PKC in membrane fusion regulation remains questionable.

For viral systems Stegmann *et al.* (1989) give a survey on the quite different conditions required for membrane fusion with different fusogenic proteins. Hence, their diversification in different exocytotic systems (as speculated in Section XII) is to be expected.

Using recent methodological developments, Cheek (1989) and O'Sullivan *et al.* (1989) stress the importance of local pCa_i detection, since pCa_i increase can be locally restricted to certain subplasmalemmal areas or it may comprise the whole cytosolic compartment, depending on the type of receptor activated in chromaffin cells. Along these lines the occasional absence of exocytosis despite a pCa_i transient (Neher and Almers, 1986, *op. cit.*; Section XIII) could be due to a pCa_i increase remote from exocytosis sites. Since Smith and Augustine (1988, *op. cit.*) derived the irrelevance of CaM for the end plate activation simply from Ca^{2+} diffusion calculations, it is interesting to see that predictions from similar calculations do not fit the experimental data (Parnas *et al.*, 1989). In this work, therefore, similar reservations are formulated as in Section XIII. In fact, fura-2 revealed only a ≈ 10-fold increase of pCa_i (i.e., much less than calculated) at active zones of nerve terminals than elsewhere along the cell membrane (Smith *et al.*, 1988, as cited in Cheek, 1988, *op. cit.*).

Since GTP-mediated intracellular fusion processes (ER-Golgi) require at least resting pCa_i levels (Beckers and Balch, 1989), one might expect similar regulation processes for nonregulated exocytosis (as argued in Section XIII), since both are of the PF-PF fusion type.

In the rough ER of different cell types a high affinity CaBP has been further identified (Fliegel *et al.*, 1989), which is similar to, but immunologically different from a calsequestrin-like protein (CSLP) detected in calciosomes of liver and exocrine pancreas (Hashimoto *et al.*, 1988). With sea urchin eggs, CSLP has been localized, however, to a smooth, branched ER network (Henson *et al.*, 1989).

The irrelevance of cytosolic pH in the range between 6.6 and 7.2 for exocytosis (Section XVII) has been stressed by Schäfer *et al.* (1987, *op. cit.*) using surface capacitance recordings. In contrast to this, alkalinizing agents applied to whole cells, also chromaffin cells, reduced secretory contents release within the same pH range (Kuijpers *et al.*, 1989), but possibly by impairing the extrusion of secretory contents rather than by inhibiting membrane fusion. Again it appears important to check membrane fusion by EM or electrophysiology, as discussed in Section XIV.

Achler *et al.* (1989) *J. Cell Biol.* **109,** 179.
Ashkenasi *et al.* (1989). *Cell* **56,** 487.
Beckers and Balch. (1989). *J. Cell Biol.* **108,** 1245.
Burgoyne and Morgan. (1989). *FEBS Lett.* **245,** 122.
Burgoyne *et al.* (1989). *Cell Signaling* **1,** 323.
Cheek. (1989). *J. Cell Sci.* **93,** 211.
Crossley *et al.* (1989). *J. Physiol.* **415,** 90.
Doucet *et al.* (1989). *FEBS Lett.* **247,** 127.
Eilers *et al.* (1989). *J. Cell Biol.* **108,** 13.
Fliegel *et al.* (1989). *Biochem. Biophys. Acta* **982,** 1.
Ghosh *et al.* (1989). *Nature (London)* **340,** 236.
Gill *et al.* (1989). *Cell Calcium* **10,** 363.
Hashimoto *et al.* (1988). *J. Cell Biol.* **107,** 2523.
Henson *et al.* (1989). *J. Cell Biol.* **109,** 149.
Hockberger. (1989). *Nature (London)* **338,** 340.
Holz *et al.* (1989). *J. Biol. Chem.* **264,** 5412.

Kuijpers *et al.* (1989). *J. Biol. Chem.* **264,** 693.

Nahor *et al.* (1989). *Proc. Hatl. Acad. Science U.S.A.* **86,** 4501.

Nahorski *et al.* (1989). *Trends Pharmacol. Sci.* **4,** 139.

Nüsse and Lindau. (1988). *J. Cell Biol.* **107,** 2117.

O'Sullivan *et al.* (1989). *EMBO J.* **8,** 401.

Parker. (1988). *J. Physiol.* **407,** 95.

Parnas *et al.* (1989). *Biophys. J.* **55,** 859.

Rivas and Moore. (1989). *J. Cell Biol.* **109,** 51.

Ross *et al.* (1989). *Nature (London)* **339,** 468.

Schäfer *et al.* (1987). *J. Neurochem.* **49,** 1697.

Schweizer *et al.* (1989). *Nature (London)* **339,** 709.

Segev *et al.* (1988). *Cell* **52,** 915.

Siegel *et al.* (1989a). *Biochemistry* **28,** 3703.

Siegal *et al.* (1989b). *Biophys. J.* **56,** 161.

Stegmann *et al.* (1989). *Annu. Rev. Biophys. Chem.* **18,** 187.

Trifaró *et al.* (1989). *Neurosciences* **29,** 1.

Volpe *et al.* (1988). *Proc. Natl. Acad. Sci. U.S.A.* **85,** 1091.

Index

287

nuclear assembly and disassembly, 87–89

SAR mapping at functional elements in *Saccharomyces cerevisiae,* 78–84
ARS elements, 80–81
centromere, 81–82
and nucleosomal organization, 83
scaffold binding, contribution of activity of ARS, CEN, or silencer, 83–84
silent mating-type loci, 82
specific protein components of matrices and scaffolds, 65–71
c-myc, 69
nuc2 protein, 70
proteins common to metaphase and interphase scaffolds
p62, 67
topoisomerase II, 66–67
repressor-activator binding protein-1 (RAP-1), 70–71
RNA polymerase, 67–69
v-myc, 69
specific scaffold-associated DNA regions, 71–78
presence of topoisomerase II consenses and A/T-rich motifs in SAR, 73–74
proximity of SAR to promoters and enhancers, 75–76
SAR, close proximity of to tissue-specific enhancers, 76
SAR, interphase, binding of by metaphase scaffold, 782
SAR, mapping of to boundaries of active domains, 77–78
saturable and reversible binding to scaffolds, 74–75
Scaffold-attached fragments (SAR), identified in *S. cerevisiae,* summary of, 79, *see also* SAR
Scaffold binding, essentiality of, 83–84
Sectioned eggs, biochemical analysis, 164–165
Secretion, chemosmotic hypothesis, 218, 259
Secretory organelles, docking of and assembly of components relevant for exocytosis, 213–218

Seminiferous epithelium, mammal cells of
actin filaments, 3–4
organization of, 1–3
Sertoli cells, mammalian, actin filaments in, 23–52
ectoplasmic specializations, 30–52
definition, 30–34
function, 41–51
intercellular attachment, 41–46
positioning of spermatid in semiferous epithelium, 46–48
unifying hypothesis, 48–51
in nonmammalian vertebrates, 51–52
structure, 35–41
tubulobulbar complexes, 23–30
Silencer activity, and scaffold binding, 83–84
Skin breakdown of tail, 105–108
Skin transformation, studies of larval organs, 127–139
body skin structure, 128
epidermal cells metamorphic changes, 128–132
keratin genes, expression, 136–139
type A antigens, 132–136
Small intestine transformation, studies, 139–143
epithelium-mesenchymal interactions, 141–143
transformation, 140–141
Spermatogenic cells, mammalian, actin filaments in, 4–23
intercellular bridges, 5–9
spermatozoa, 15–23
actin function in, 21–23
biochemical studies results, 15–17
immunofluorescence studies, 17–20
morphological data of, 20–21
subacrosomal space, 9–15
Spermatozoa, function of actin in, 21–23, *see also* Post-spermatogenic cells
Staphylococcus a-toxin, 207
Styela eggs, localized mRNA, 160–162, 169
cytoskeleton, 170–173
Subacrosomal space of differentiating spermatid, 9–15
pattern of filament distribution, 13–14
presence of F-actin in, 14–15